Spring Boot 企业级应用开发实战

柳伟卫◎著

北京大学出版社
PEKING UNIVERSITY PRESS

内容提要

本书围绕如何整合以目前最新的 Spring Boot 2 版本为核心的技术栈，来实现一个完整的企业级博客系统而展开。读者可以通过学习构建这个博客系统的整个过程，来达到设计和实现一个企业级 Java EE 应用开发的目的。该博客系统是一个类似于 WordPress 的专注于博客功能的博客平台，支持多用户访问和使用。该博客系统所涉及的相关技术有 Spring Boot、Spring、Spring MVC、Spring Security 、Spring Data、Hibernate、Gradle、Bootstrap、jQuery、HTML5、JavaScript、CSS、Thymeleaf、MySQL、H2、Elasticsearch、MongoDB 等，技术点较为丰富，内容富有前瞻性。

本书面向实战，除了给出基本的原理外，会辅以大量的案例和源码，利于读者理论联系实践。全书对于技术讲解的安排，是按照渐进式的教学方式来进行的。按照学习的难度，大致可以分为实战入门阶段、实战进阶阶段、实战高级阶段三个部分，内容包括 Spring Boot 概述、Spring 框架核心概念、Spring MVC 及常用 MediaType、集成 Thymeleaf、数据持久化、全文搜索、架构设计与分层、集成 Bootstrap、博客系统的需求分析与设计、集成 Spring Security、博客系统的整体框架实现、用户管理实现、角色管理实现、权限管理实现、文件服务器实现、博客管理实现、评论管理实现、点赞管理实现、分类管理实现、标签管理实现、首页搜索实现等。

本书主要面向的用户是 Java 开发者，以及对 Spring Boot 及企业级开发感兴趣并有一定了解的读者。

图书在版编目(CIP)数据

Spring Boot 企业级应用开发实战 / 柳伟卫著 — 北京:北京大学出版社，2018.3
ISBN 978-7-301-29230-3

Ⅰ.①S… Ⅱ.①柳… Ⅲ.①JAVA语言—程序设计 Ⅳ.①TP312.8

中国版本图书馆CIP数据核字(2018)第028100号

书　　名　Spring Boot 企业级应用开发实战
SPRING BOOT QIYE JI YINGYONG KAIFA SHIZHAN
著作责任者　柳伟卫　著
责任编辑　尹　毅
标准书号　ISBN 978-7-301-29230-3
出版发行　北京大学出版社
地　　址　北京市海淀区成府路205 号　100871
网　　址　http://www.pup.cn　　新浪微博:@ 北京大学出版社
电子信箱　pup7@ pup.cn
电　　话　邮购部 62752015　发行部 62750672　编辑部 62570390
印 刷 者　北京大学印刷厂
经 销 者　新华书店
787毫米×1092毫米　16开本　28.25印张　656千字
2018年3月第1版　2018年4月第2次印刷
印　　数　3001–6000册
定　　价　98.00 元

谨将此书献给我的小女儿伊然，愿她永远天真烂漫！

写作背景

对于 Spring Boot 知识的整理归纳，最早是在笔者的第一本书《分布式系统常用技术及案例分析》的微服务章节中，作为微服务的技术实现方式来展开的。由于篇幅限制，当时讲解的案例深度和广度也比较有限。其后，笔者又在 GitHub 上，以开源方式撰写了《Spring Boot 教程》系列课程①，为网友们提供了更加丰富的使用案例。在 2017 年年初，笔者应邀给慕课网做了一个关于 Spring Boot 实战的系列视频课程②。视频课程上线后受到了广大的 Spring Boot 技术爱好者的关注，课程的内容也引发了热烈的反响。很多该课程的学员，通过学习该课程，不但技术能力提高了，而且在如何采用新技术来实现企业级应用上有了更深刻的理解，最重要的是提升了自己在市场上的价值。

鉴于 Spring Boot 技术人才在社会上的需求依然很旺盛，而市面上有关 Spring Boot 学习资料，大多停留在“Hello World”级别的案例，缺乏使用 Spring Boot 来构建完整企业级应用实战的能力。故笔者将以往系列课程中的技术做了总结和归纳，采用目前最新的 Spring Boot 2 技术来重新编写了整个教学案例，整理成书，希望能够弥补 Spring Boot 在实战方面的空白，使广大 Spring Boot 爱好者都能受益。

源代码

本书提供源代码下载，下载地址为 https://github.com/waylau/spring-boot-enterprise-application-development。

本书所涉及的技术及相关版本

技术版本是非常重要的，特别是对于实战内容而言。因为不同的版本之间，是存在兼容性问题的，而且不同的版本，软件所对应的功能也是不同的。本书所列出的技术，版本上相对比较新，都是经过笔者自己大量实际测试的。这样，读者在自行搭建博客系统时，可以参考本书所列出的版本，从而可以避免很多因为版本兼容性所产生的问题。建议读者将相关开发环境设置成与本书所采用的一致，或者不低于本书所列的配置。详细的版本配置，可以参阅本书的“附录 E”内容。

本书示例采用 Eclipse 编写，但示例源码与具体的 IDE 无关，读者可以自行选择适合自己的 IDE，如 IntelliJ IDEA、NetBeans 等。

---

① 有关该教程介绍，可参见 https://github.com/waylau/spring-boot-tutorial。

② 有关该课程介绍，可参见 http://coding.imooc.com/class/125.html。

勘误和交流

本书如有勘误，会在 https://github.com/waylau/spring-boot-enterprise-application-development 上进行发布。由于笔者能力有限，时间仓促，错漏之处在所难免，欢迎读者批评指正。

读者也可以直接通过以下方式联系。

博客：https://waylau.com

邮箱：waylau521@gmail.com

微博：http://weibo.com/waylau521

开源：https://github.com/waylau

致谢

感谢北京大学出版社工作人员为本书在出版过程中所做出的努力。

感谢我的父母、妻子和两个女儿。由于撰写本书，牺牲了很多陪伴家人的时间。感谢家人对我工作的理解和支持。

最后，感谢 Spring Boot 团队为 Java 社区提供了这么优秀的框架。由衷地希望 Spring Boot 框架发展得越来越好！

柳伟卫

目录

Contents

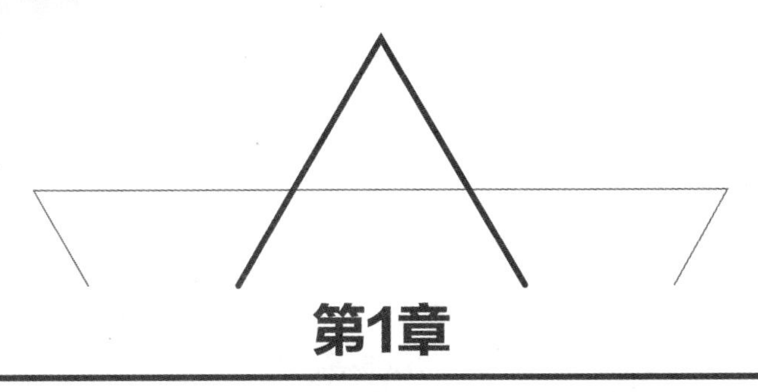

第1章 Spring Boot 概述

1.1 传统企业级应用开发之痛与革新

作为一门“长寿”的编程语言，Java 语言经历了 20 多年的发展，已成为开发者首选的利器。在最新的 TIOBE 编程语言排行榜中， Java 位居榜首。回顾历史，Java 语言的排行也一直是名列三甲。图 1-1 展示的是 2017 年 8 月 TIOBE 编程语言排行榜情况。

Aug 2017	Aug 2016	Change	Programming Language	Ratings	Change
1	1		Java	12.961%	-6.05%
2	2		C	6.477%	-4.83%
3	3		C++	5.550%	-0.25%
4	4		C#	4.195%	-0.71%
5	5		Python	3.692%	-0.71%
6	8	↑	Visual Basic .NET	2.569%	+0.05%
7	6	↓	PHP	2.293%	-0.88%
8	7	↓	JavaScript	2.098%	-0.61%
9	9		Perl	1.995%	-0.52%
10	12	↑	Ruby	1.965%	-0.31%
11	14	↑	Swift	1.825%	-0.16%
12	11	↓	Delphi/Object Pascal	1.825%	-0.45%
13	13		Visual Basic	1.809%	-0.24%
14	10	⇊	Assembly language	1.805%	-0.56%
15	17	↑	R	1.766%	+0.16%
16	20	⇈	Go	1.645%	+0.37%
17	18	↑	MATLAB	1.619%	+0.08%
18	15	↓	Objective-C	1.505%	-0.38%
19	22	↑	Scratch	1.481%	+0.43%
20	26	⇈	Dart	1.273%	+0.30%

图1-1　TIOBE 编程语言排行榜

然而，作为当今企业级应用的首选编程语言，Java 的发展也并非一帆风顺。

1.1.1 Java 大事件

最初，SUN 公司准备用一种新的语言来设计用于智能家电类（如机顶盒）的程序开发。“Java 之父”James Gosling 创造出了这种全新的语言，并被他命名为“Oak”（橡树），以他的办公室外的树来命名。然而，由于当时机顶盒的项目并没有成功拿下，于是 Oak 被阴差阳错地应用到了万维网。作为原型，SUN 公司的工程师编写了一个小型万维网浏览器 WebRunner（后来改名为 Hot-

Java），该浏览器可以直接用来运行 Java 小程序（即 Java Applet）。同年，Oak 改名为 Java。由于 Java Applet 程序可以产生一般网页所不能实现的效果，从而引起业界对 Java 的热捧，于是当时很多操作系统都预装了 Java 虚拟机。

1997 年 4 月 2 日，JavaOne 会议召开，参与者逾 1 万人，创下当时全球同类会议规模最大之纪录。

1998 年 12 月 8 日，Java 2 企业平台 J2EE 发布，正式进军企业级应用开发领域。

1999 年 6 月，随着 Java 的快速发展，SUN 公司将 Java 分为 3 个版本，即标准版（J2SE）、企业版（J2EE）和微型版（J2ME）。从版本的划分可以看出当时 Java 语言的野心，企图统治桌面应用、服务器端应用及移动端应用。

2004 年 9 月 30 日 J2SE 1.5 发布，成为 Java 语言发展史上的又一里程碑。为了表示该版本的重要性，J2SE 1.5 更名为 Java SE 5.0。

2005 年 6 月，JavaOne 大会召开，SUN 公司发布了 Java SE 6。此时，Java 的各种版本已经更名，以取消其中的数字“2”，J2EE 更名为 Java EE，J2SE 更名为 Java SE，J2ME 更名为 Java ME。

2009 年 4 月 20 日，Oracle 公司以 74 亿美元收购了 SUN 公司，从此 Java 归属于 Oracle 公司。

2011 年 7 月 28 日，Oracle 公司发布 Java 7 的正式版。该版本新增了诸如 try-with-resources 语句、增强 switch-case 语句支持字符串类型等特性。

2011 年 6 月中旬，Oracle 公司正式发布了 Java EE 7。该版本的目标在于提高开发人员的生产力，满足苛刻的企业需求 。

2014 年 3 月 19 日，Oracle 公司发布 Java 8 的正式版。该版本中的 Lambdas 表达式、Streams 流式计算框架等广受开发者关注。

由于 Java 9 中计划开发的模板化项目（或称为 Jigsaw）存在较大的技术难度，JCP 执行委员会内部成员也无法达成共识，所以造成了该版本的发布一再延迟。Java 9 及 Java EE 8 终于在 2017 年 9 月份发布，并将 Java EE 8 移交给了开源组织 Eclipse 基金会。

图 1-2 所示为 Java EE 8 整体架构图。

Java EE Web Profile
EJB 3.2
JPA 2.2
Bean Validation 2.0
JTA 1.2
Servlet 4.0
JSP 2.3
EL 3.0
JSTL 1.2
JSF 2.3
WebSocket 1.1
JAX-RS 2.1
JSON-P 1.1
JSON-B 1.0
CDI 2.0
DI 1.0
Annotations 1.3
Managed Beans 1.0
Interceptors 1.2
Security 1.0
JASPIC 1.1
JAX-WS 2.2
Web Services 1.4
JAX-RPC 1.1
SAAJ 1.3
JAX-R 1.0
JAXB 2.2
JAX-P 1.3
StAX 1.0
JACC 1.5
JAAS 1.0
JNDI 1.2
JMX 1.2
Deployment 1.2
Management 1.1
JDBC 4.2
JTS 1.0
Concurrency 1.0
Batch 1.0
JMS 2.0
Connector 1.7
Mail 1.6
Activation 1.1

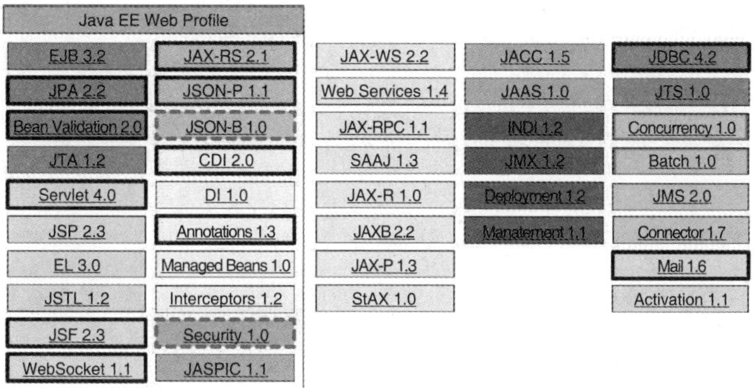

图1-2　Java EE 8整体架构图

1.1.2 Java 企业级应用现状

作为 Java 企业级应用开发的规范——Java EE，从诞生之初就饱受争议。特别是 EJB（Enterprise Java Beans）作为 Java 企业级应用开发的核心，由于其设计的复杂性，使之在 J2EE 架构中的表现一直不是很好。EJB 大概是 J2EE 架构中唯一没有兑现其能够简化开发并提高生产力承诺的组件。

正当 Java 开发者无法忍受 EJB 的臃肿不堪时，Spring 应运而生。Spring 框架打破了传统 EJB 开发模式中以 Bean 为重心的强耦合、强侵入性的弊端，采用依赖注入和 AOP（Aspect Oriented Programming，面向切面编程）等技术，来解耦对象间的依赖关系，无须继承复杂 Bean，只需要 POJOs（Plain Old Java Objects，简单的 Java 对象），就能快速实现企业级应用的开发。为此，"Spring 之父" Rod Johnson 还特意撰写了《Expert one-on-one J2EE Development without EJB》一书，来向 EJB 宣战，从而业界掀起了以 Spring 为核心的轻量级应用开发的狂潮。

Spring 框架最初的 Bean 管理是通过 XML 文件来描述的。然后随着业务的增加，应用里面存在了大量的 XML 配置，这些配置除包括 Spring 框架自身的 Bean 配置外，还包括了其他框架的集成配置等，到最后 XML 文件变得臃肿不堪，难以阅读和管理。同时，XML 文件内容本身不像 Java 文件那样能够在编译期事先做类型校验，所以也就很难排查 XML 文件中的错误配置。

以下展示的是在 Spring 应用中非常常见的 XML 配置管理 Bean 的方式。

```
<beans xmlns="http://www.springframework.org/schema/beans"
    xmlns:xsi="http://www.w3.org/2001/XMLSchema-instance"
    xmlns:p="http://www.springframework.org/schema/p"
    xsi:schemaLocation="http://www.springframework.org/schema/beans
        http://www.springframework.org/schema/beans/spring-beans.xsd">

    <bean name="john-classic" class="com.example.Person">
        <property name="name" value="Way Lau"/>
        <property name="spouse" ref="jane"/>
    </bean>

    <bean name="john-modern"
        class="com.example.Person"
        p:name="Way Lau"
        p:spouse-ref="jane"/>

    <bean name="jane" class="com.example.Person">
        <property name="name" value="Jane Doe"/>
    </bean>
</beans>
```

1.1.3 革新

针对上面传统企业级应用开发过程中的痛点，Java 及 Spring 框架也都提出了变革。例如，Java 5 引入的注解技术，就能更好描述 Java 程序；EJB 3 也向 Hibernate 等框架吸收了大量的精华（甚至请 Hibernate 的作者来设计 EJB 3），从而大大改善其实体 Bean 的重耦合的现状；从 Spring 3 开始，也引入了 Java 配置的方式来管理 Bean，从而大量减少了 XML 的使用，甚至是零配置。

1.1.4 约定大于配置

实现程序的零配置，其核心思想就是“约定大于配置”（Convention over Configuration）。约定大于配置是一个简单的概念，即系统、类库、框架应该假定合理的默认值，而无须提供不必要的配置。在大部分情况下，用户会发现使用框架提供的默认值会让自己所涉及的项目运行得更快。

零配置并不是完全没有配置，而是通过约定来减少配置，特别是减少 XML 文件的数量。

实现约定大于配置主要从以下几个方面入手。

1. 约定代码结构或命名规范来减少配置数量

如果模型中有个名为 Sale 的类，那么数据库中对应的表就会默认命名为“sale”。只有在偏离这一约定时，如将该表命名为“products_sold”，才需编写有关这个名称的配置。

例如，EJB3 持久化，将一个特殊的 Bean 持久化，设计者所需要做的只是将这个类标注为 @Entity。框架将会假定表名和列名是基于类名和属性名。系统也提供了一些钩子，当有需要时设计者可以重新编写这些名称。下面是一个表与实体映射的例子。

```
@Entity // 实体
public class User implements UserDetails {

    private static final long serialVersionUID = 1L;

    @Id // 主键
    @GeneratedValue(strategy=GenerationType.IDENTITY) // 自增策略
    private Long id; // 实体一个唯一标识

    @Column(nullable = false, length = 20) // 映射为字段，值不能为空
    private String name;

    @Column(nullable = false, length = 50, unique = true)
    private String email;

    // 以下省略 getter/setter 方法
}
```

例如，Maven 项目约定，在没有自定义的情况下，源代码假定是在 /src/main/java 目录，资源文件假定是在 /src/main/resources 目录。测试代码假定是在 /src/test 目录。项目假定会产生一个 JAR

文件。Maven 假定想要把编译好的字节码放到 /target/classes，并且在 /target 创建一个可分发的 JAR 文件。Maven 对约定优于配置的应用不仅仅是简单的目录位置，Maven 的核心插件使用了一组通用的约定，以用来编译源代码，打包可分发的构件，生成 Web 站点，以及许多其他的过程。Maven 的力量来自它的“武断”，它有一个定义好的生命周期和一组知道如何构建和装配软件的通用插件。如果遵循这些约定，开发者只需将自己开发的源代码放到正确的目录，而后 Maven 将会帮你处理剩下的事情。

例如，Yeoman 创建项目，只需一行代码：

```
yeoman init angular
```

就会创建整个详细结构，包括渲染路由的框架、单元测试等。

例如，HTML 5 Boilerplate（http://html5boilerplate.com) 项目，提供了制作 APP 的默认模板及文件路径规范，无论是网站或者富 UI 的 APP，都可以采用这个模板作为起步。HTML5 Boilerplate 的模板核心部分不过 30 行，但是每一行都可谓千锤百炼，可以用最小的消耗解决一些前端的顽固问题。以下是 HTML5 Boilerplate 初始化项目的目录结构。

```
C:.
|   .gitattributes
|   .gitignore
|   .htaccess
|   404.html
|   apple-touch-icon-114x114-precomposed.png
|   apple-touch-icon-144x144-precomposed.png
|   apple-touch-icon-57x57-precomposed.png
|   apple-touch-icon-72x72-precomposed.png
|   apple-touch-icon-precomposed.png
|   apple-touch-icon.png
|   CHANGELOG.md
|   CONTRIBUTING.md
|   crossdomain.xml
|   favicon.ico
|   humans.txt
|   index.html
|   LICENSE.md
|   README.md
|   robots.txt
|
├─css
|       main.css
|       normalize.css
|
├─doc
|       crossdomain.md
|       css.md
|       extend.md
```

```
│       faq.md
│       html.md
│       js.md
│       misc.md
│       TOC.md
│       usage.md
│
├─img
│       .gitignore
│
└─js
    │   main.js
    │   plugins.js
    │
    └─vendor
            jquery-1.9.1.min.js
            modernizr-2.6.2.min.js
```

2. 采用更简洁的配置方式来替代 XML

很多配置方式都比 XML 简洁。

例如，hibernate.properties 的配置方式如下。

```
hibernate.connection.driver_class = org.postgresql.Driver
hibernate.connection.url = jdbc:postgresql://localhost/mydatabase
hibernate.connection.username = myuser
hibernate.connection.password = secret
hibernate.c3p0.min_size=5
hibernate.c3p0.max_size=20
hibernate.c3p0.timeout=1800
hibernate.c3p0.max_statements=50
hibernate.dialect = org.hibernate.dialect.PostgreSQL82Dialect
```

例如，Apache Shiro 的 ini 配置方式如下。

```
[users]
root = secret, admin
guest = guest, guest
presidentskroob = 12345, president
darkhelmet = ludicrousspeed, darklord, schwartz
lonestarr = vespa, goodguy, schwartz

[roles]
admin = *
schwartz = lightsaber:*
goodguy = winnebago:drive:eagle5
```

又如，Hibernate 通过以下 Java 代码来进行配置。

```
Configuration cfg = new Configuration()
```

```
    .addClass(org.hibernate.auction.Item.class)
    .addClass(org.hibernate.auction.Bid.class)
    .setProperty("hibernate.dialect", "org.hibernate.dialect.MySQLIn
noDBDialect")
    .setProperty("hibernate.connection.datasource", "java:comp/env/
jdbc/test")
    .setProperty("hibernate.order_updates", "true");
```

3. 用 Gradle 替代 Maven

大部分 Java 应用都采用 Maven 来进行软件的项目管理，但目前其实已经有了更好的选择，如 Gradle。观察下面 Maven 的配置。

```
<parent>
    <groupId>org.springframework.boot</groupId>
    <artifactId>spring-boot-starter-parent</artifactId>
    <version>2.0.0.M2</version>
</parent>
<dependencies>
    <dependency>
        <groupId>org.springframework.boot</groupId>
        <artifactId>spring-boot-starter-web</artifactId>
    </dependency>
</dependencies>
```

如果是采用 Gradle 来配置，仅仅只需一行。

```
compile(“org.springframework.boot:spring-boot-starter-web:2.0.0.M2”)
```

从这个小例子就能看出，XML 对于 Gradle 的配置脚本而言，是多么低效和冗余！

注意：本书所有的示例，都是采用 Gradle 来进行项目的管理。如有需要，读者也可以将项目源码自行转化为 Maven 等管理方式。

4. 通过注解约定其含义来减少配置数量

Spring 会自动搜索某些路径下的 Java 类，并将这些 Java 类注册为 Bean 实例，这样就省去了将所有 Bean 都配置在 XML 中的麻烦。

```
<?xml version="1.0" encoding="UTF-8"?>
<beans xmlns="http://www.springframework.org/schema/beans"
    xmlns:xsi="http://www.w3.org/2001/XMLSchema-instance"
    xmlns:context="http://www.springframework.org/schema/context"
    xsi:schemaLocation="http://www.springframework.org/schema/beans
        http://www.springframework.org/schema/beans/spring-beans.xsd
        http://www.springframework.org/schema/context
        http://www.springframework.org/schema/context/spring-context.
xsd">

    <context:component-scan base-package="com.waylau.rest"/>

```

```
</beans>
```

注意：如果配置了 <context:component-scan/>，那么 <context:annotation-config/> 标签就可以不用在 XML 中配置了，因为前者包含了后者。另外，<context:component-scan/> 还提供了两个子标签 <context:include-filter> 、<context:exclude-filter>，用来控制扫描文件的颗粒度，如下所示。

```
<beans>
    <context:component-scan base-package="com.waylau.rest">
    <context:include-filter type="regex" expression=".*Stub.*Reposito-
ry"/>
    <context:exclude-filter type="annotation"expression="org.springframe-
work.stereotype.Repository"/>
    </context:component-scan>
</beans>
```

如果采用 Java 代码来配置，其实现如下。

```
public static void main(String[] args) {
    AnnotationConfigApplicationContext ctx = new AnnotationConfigApplica
tionContext();
    ctx.scan("com.waylau.rest");
    ctx.refresh();
    MyService myService = ctx.getBean(MyService.class);
}
```

Spring 会适当地将显示指定路径下的类全部注册成 Spring Bean。Spring 通过使用特殊的注解来标注 Bean 类。

- @Component：标注一个普通的 Spring Bean 类。
- @Controller：标注一个控制器组件类。
- @Service：标注一个服务组件类。
- @Repository：标注一个仓库组件类。

甚至 SQL 语句也可以注解。

```
@Entity
@Table(name="USER")
@SQLInsert( sql="INSERT INTO USER(size, name, nickname, id) VALUES(?,up-
per(?),?,?)")
@SQLUpdate( sql="UPDATE USER SET size = ?, name = upper(?), nickname =
? WHERE id = ?")
@SQLDelete( sql="DELETE USER WHERE id = ?")
@SQLDeleteAll( sql="DELETE USER")
@Loader(namedQuery = "user")
@NamedNativeQuery(name="user", query="select id, size, name, lower(
nickname ) as nickname from USER where xml:id= ?", resultClass = USER.
class)
public class USER {
```

```
    @Id
    private Long id;
    private Long size;
    private String name;
    private String nickname;
    //...
}
```

以下是 Jersey 2.x 通过注解方式来实现 REST 和 MVC 模式。

```
@POST
@Produces({"text/html"})
@Consumes(MediaType.APPLICATION_FORM_URLENCODED)
@Template(name = "/short-link") @ErrorTemplate(name = "/error-form")
@Valid
public ShortenedLink createLink(@NotEmpty @FormParam("link") final String
link) {
    // ...
}
```

以下是 Shiro 没有加注解的情况。

```
Subject currentUser =
    SecurityUtils.getSubject();
if (currentUser.hasRole("administrator")) {
    // ....
} else {
    // ....
}
```

用了注解，整个代码都简洁很多。

```
@RequiresRoles( "administrator" )
public void openAccount( Account acct ) {
    // ...
}
```

当年风靡一时的 SSH 框架之一的 Struts 2.x 也提供了注解的支持。

```
package com.waylau.actions;

import com.opensymphony.xwork2.ActionSupport;
import org.apache.struts2.convention.annotation.Action;
import org.apache.struts2.convention.annotation.Actions;
import org.apache.struts2.convention.annotation.Result;
import org.apache.struts2.convention.annotation.Results;

@Results({
    @Result(name="failure", location="fail.jsp")
})
public class HelloWorld extends ActionSupport {
```

```
    @Action(value="/different/url",
    results={@Result(name="success", location="http://struts.apache.
org", type="redirect")}
    )
    public String execute() {
    return SUCCESS;
    }

    @Action("/another/url")

    public String doSomething() {
    return SUCCESS;
    }
}
```

5. 定制开箱即用的 Starter

在这方面，Spring Boot 做到了极致。Spring Boot 提供了各种开箱即用的 Starter，旨在最大化减少应用的配置。例如，spring-boot-starter-web 就提供了全栈式 Web 开发的支持，其默认配置就已经包括 Tomcat、Spring MVC、Hibernate 等常用的 Web 开发框架的集成。用户需要做的仅仅是将 spring-boot-starter-web 依赖包纳入进来即可。

```
dependencies {
    compile("org.springframework.boot:spring-boot-starter-web:2.0.0.M2")

}
```

1.2 Spring Boot 2 总览

Spring Boot 可以说是近几年来 Spring 乃至整个 Java 社区比较有影响力的项目之一，也被视为 Java EE 开发的颠覆者，下一代的企业级应用开发的首选框架。Spring Boot 是伴随着 Spring 4 而诞生的，在继承了 Spring 一切优点的基础上，其最大的特色就是简化了 Spring 应用的集成、配置、开发，提供开箱即用的极速体验。所以 Spring Boot 一经推出，就引起了巨大的反响，受到了业界极大的关注，并在 2016 年 10 月 11 日获得了 JAX Innovation Awards 2016 大奖[①]。同时，Spring Boot 以其快速的开发方式、极简的启动配置深受广大 Java 爱好者的好评，在 GitHub 上的点赞量已经超过了 14 000 次。

下面就来讨论 Spring Boot 产生的背景。

① 有关该奖项的报道，可参见 https://jaxenter.com/winners-jax-innovation-awards-2016-jax-london-129588.html。

1.2.1 解决传统 Spring 开发过程中的痛点

正如上一节所介绍的，多年以来，传统企业级应用开发中存在很多问题，其中，Spring 平台饱受非议的一点就是大量的 XML 配置及复杂的依赖管理。随着 Spring 3.0 的发布，Spring IO 团队逐渐开始摆脱 XML 配置文件，并且在开发过程中大量使用“约定大于配置”的思想（大部分情况下就是 Java Config 的方式）来摆脱 Spring 框架中各类纷繁复杂的配置。

Spring Boot 正是在这样的一个背景下被抽象出来的开发框架，它本身并不提供 Spring 框架的核心特性及扩展功能，只是用于快速、敏捷地开发新一代基于 Spring 框架的应用程序。也就是说，它并不是用来替代 Spring 的解决方案，而是和 Spring 框架紧密结合，用于提升 Spring 开发者体验的工具。同时，Spring Boot 集成了大量常用的第三方库的配置，Spring Boot 应用为这些第三方库提供了几乎可以零配置的开箱即用的能力。这样，大部分 Spring Boot 应用都只需要非常少量的配置代码，从而使开发者能够更加专注于业务逻辑，而无须进行诸如框架的整合等这些只有高级开发者或者架构师才能胜任的工作。

从最根本上来讲，Spring Boot 就是一些依赖库的集合，它能够被任意项目的构建系统所使用。在追求开发体验的提升方面，Spring Boot，甚至可以说整个 Spring 生态系统都使用到了 Groovy 编程语言。Spring Boot 所提供的众多便捷功能，都是借助于 Groovy 强大的 MetaObject 协议、可插拔的 AST 转换过程及内置了解决方案引擎所实现的依赖。在其核心的编译模型之中，Spring Boot 使用 Groovy 来构建工程文件，所以它可以使用通用的导入和样板方法（如类的 main 方法）对类所生成的字节码进行装饰（Decorate）。这样使用 Spring Boot 编写的应用就能非常简洁，却依然可以提供众多的功能。

1.2.2 Spring Boot 的目标

Spring Boot 简化了基于 Spring 的应用开发，通过少量的代码就能创建一个独立的、产品级别的 Spring 应用。 Spring Boot 为 Spring 平台及第三方库提供开箱即用的设置，这样开发者就可以有条不紊地来进行应用的开发。多数 Spring Boot 应用只需要很少的 Spring 配置。

开发者可以使用 Spring Boot 创建 Java 应用，并使用 java -jar 启动它或者也可以采用传统的 WAR 部署方式。同时 Spring Boot 也提供了一个运行“spring 脚本”的命令行工具。

Spring Boot 主要的目标如下。

（1）为所有 Spring 开发提供一个更快更广泛的入门体验。

（2）开箱即用，不合适时也可以快速抛弃。

（3）提供一系列大型项目常用的非功能性特征，如嵌入式服务器、安全性、度量、运行状况检查、外部化配置等。

（4）零配置。无冗余代码生成和 XML 强制配置，遵循“约定大于配置”。

Spring Boot 内嵌如表 1-1 所示的容器以支持开箱即用。

表1-1 Spring Boot内嵌容器

名称	Servlet版本	Java版本
Tomcat 8.5	3.1	Java 8+
Tomcat 8	3.1	Java 7+
Tomcat 7	3.0	Java 6+
Jetty 9.4	3.1	Java 8+
Jetty 9.3	3.1	Java 8+
Jetty 9.2	3.1	Java 7+
Jetty 8	3.0	Java 6+
Undertow 1.3	3.1	Java 7+

开发者也可以将 Spring Boot 应用部署到任何兼容 Servlet 3.0+ 的容器。需要注意的是，Spring Boot 2 要求不低于 Java 8 版本。

简言之，Spring Boot 抛弃了传统 Java EE 项目烦琐的配置、学习过程，让开发过程变得更容易。

1.2.3 Spring Boot 不是Spring的替代者

Spring Boot 并不是要成为 Spring 平台里面众多“基础层”（Foundation）项目的替代者。Spring Boot 的目标不是为已解决的问题域提供新的解决方案，而是为平台带来另一种开发体验，从而简化对这些已有技术的使用。对于已经熟悉 Spring 生态系统的开发人员来说，Spring Boot 是一个很理想的选择，而对于采用 Spring 技术的新人来说，Spring Boot 提供了一种更简洁的方式来使用这些技术。图 1-3 所示为 Spring Boot 与其他框架的关系。

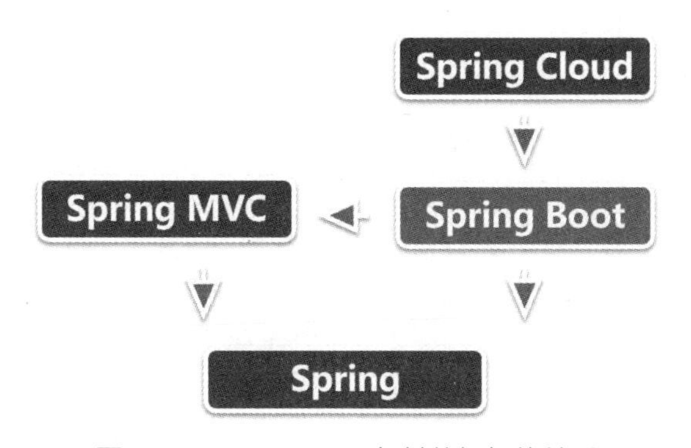

图1-3 Spring Boot 与其他框架的关系

1. Spring Boot 与 Spring 框架的关系

Spring 框架是通过 IoC 机制来管理 Bean 的。Spring Boot 依赖 Spring 框架来管理对象的依赖。Spring Boot 并不是 Spring 的精简版本，而是为使用 Spring 做好各种产品级准备。

2. Spring Boot 与 Spring MVC 框架的关系

Spring MVC 实现了 Web 项目中的 MVC 模式。如果 Spring Boot 是一个 Web 项目，就可以选择采用 Spring MVC 来实现 MVC 模式。当然也可以选择其他类似的框架来实现。

3. Spring Boot 与 Spring Cloud 框架的关系

Spring Cloud 框架可以实现一整套分布式系统的解决方案（当然其中也包括微服务架构的方案），包括服务注册、服务发现、监控等，而 Spring Boot 只是作为开发单一服务的框架的基础。

1.2.4 Spring Boot 2 新特性

目前，Spring Boot 团队已经紧锣密鼓地开发 Spring Boot 2 版本，截至目前，Spring Boot 最新版本为 2.0.0，本书的所有示例源码都是基于 Spring Boot 2 版本来编写的。

Spring Boot 2 相比于 Spring Boot 1 增加了如下新特性。

（1）对 Gradle 插件进行了重写。

（2）基于 Java 8 和 Spring Framework 5。

（3）支持响应式的编程方式。

（4）对 Spring Data、Spring Security、Spring Integration、Spring AMQP、Spring Session、Spring Batch 等都做了更新。

1.2.5 Gradle 插件

Spring Boot 的 Gradle 插件用于支持在 Gradle 中方便构建 Spring Boot 应用。它允许开发人员将应用打包成可执行的 jar 或 war 文件，运行 Spring Boot 应用程序，以及管理 Spring Boot 应用中的依赖关系。 Spring Boot 2 需要 Gradle 的版本不低于 3.4。

那么如何来安装 Gradle 插件?

安装 Gradle 插件，需要添加以下内容。

```
buildscript {
    repositories {
        maven { url 'https://repo.spring.io/libs-milestone' }
    }

    dependencies {
        classpath 'org.springframework.boot:spring-boot-gradle-plugin:
2.0.0.M2'
    }
```

```
}

apply plugin: 'org.springframework.boot'
```

独立地添加应用插件对项目的改动几乎很少。同时，插件会检测何时应用某些其他插件，并会相应地进行响应。例如，当应用 java 插件时，将自动配置用于构建可执行 jar 的任务。

一个典型的 Spring Boot 项目将至少应用 groovy 插件、java 插件或 org.jetbrains.kotlin.jvm 插件和 io.spring.dependency-management 插件。例如，

```
apply plugin: 'java'
apply plugin: 'io.spring.dependency-management'
```

使用 Gradle 插件来运行 Spring Boot 应用，只需简单地执行以下命令。

```
$ ./gradlew bootRun
```

1.2.6 基于最新的 Java 8 和 Spring Framework 5

Spring Boot 2 基于最新的 Java 8 和 Spring Framework 5，这意味着 Spring Boot 2 拥有构建现代应用的能力。

最新发布的 Java 8 中的 Streams API、Lambda 表达式等，都极大地改善了开发体验，并让编写并发程序更加容易。

核心的 Spring Framework 5.0 已经利用 Java 8 所引入的新特性进行了修订。这些内容包括以下特点。

（1）基于 Java 8 的反射增强，Spring Framework 5.0 中的方法参数可以更加高效地进行访问。

（2）接口提供基于 Java 8 的默认方法构建的选择性声明。

（3）支持候选组件索引作为类路径扫描的替代方案。

（4）当然，最为重要的是，此次 Spring Framework 5.0 推出了新的响应式堆栈 WEB 框架。这个堆栈是完全的响应式且非阻塞，适合于事件循环风格的处理，可以进行少量线程的扩展。

总之，最新 Spring Boot 2 让开发企业级应用更加简单，可以更加方便地构建响应式编程模型。

1.2.7 Spring Boot 周边技术栈的更新

相应地，Spring Boot 2 会集成最新的技术栈，包括 Spring Data、Spring Security、Spring Integration、Spring AMQP、Spring Session、Spring Batch 等都做了更新，其他的第三方依赖也会尝试使用最新的版本，如本课程中所使用的 Thymeleaf、Spring Data Elasticsearch、Spring Data Mongodb 等。

使用 Spring Boot 2 使开发人员有机会接触最新的技术框架。

1.3 快速开启第一个 Spring Boot 项目

本节将带领各位读者一起开启第一个 Spring Boot 项目。正如 Spring Boot 所承诺的那样，使用 Spring Boot 可以最大化减少项目的配置，真正做到开箱即用。现在就来给大家演示一下，如何创建第一个 Spring Boot 项目。

1.3.1 配置环境

本例子采用的开发环境为 JDK 8、Gradle 4.0。其中，JDK 的安装，可以参阅笔者所著的开源书《Java 编程要点》（https://github.com/waylau/essential-java）；Gradle 的安装可以参阅《Gradle 用户指南》（https://github.com/waylau/gradle-user-guide）。

检查 JDK 版本情况，确保不低于 Java 8 版本。

```
$ java -version
java version "1.8.0_112"
Java(TM) SE Runtime Environment (build 1.8.0_112-b15)
Java HotSpot(TM) 64-Bit Server VM (build 25.112-b15, mixed mode)
```

检查 Gradle 版本情况。

```
$ gradle -v

------------------------------------------------------------
Gradle 4.0
------------------------------------------------------------

Build time:   2017-06-14 15:11:08 UTC
Revision:     316546a5fcb4e2dfe1d6aa0b73a4e09e8cecb5a5

Groovy:       2.4.11
Ant:          Apache Ant(TM) version 1.9.6 compiled on June 29 2015
JVM:          1.8.0_112 (Oracle Corporation 25.112-b15)
OS:           Windows 10 10.0 amd64
```

1.3.2 通过 Spring Initializr 初始化一个 Spring Boot 原型

Spring Initializr 是用于初始化 Spring Boot 项目的可视化平台。虽然通过 Maven 或者 Gradle 来添加 Spring Boot 提供的 Starter 使用起来非常简单，但是由于组件和关联部分众多，有这样一个可视化的配置构建管理平台对于用户来说非常友好。下面将演示如何通过 Spring Initializr 初始化一个 Spring Boot 项目原型。

访问 Spring 提供的官方 Spring Initializr 网站，当然用户也可以搭建自己的 Spring Initializr 平台。

按照页面提示，输入相应的项目元数据（Project Metadata）资料，并选择依赖。由于要初始化一个 Web 项目，因此在依赖搜索框中输入关键字“web”，并且选择“Web:Full-stack web development with Tomcat and Spring MVC”选项。顾名思义，该项目将会采用 Spring MVC 作为 MVC 的框架，并且集成了 Tomcat 作为内嵌的 Web 容器。图 1-4 所示为 Spring Initializr 的管理界面。

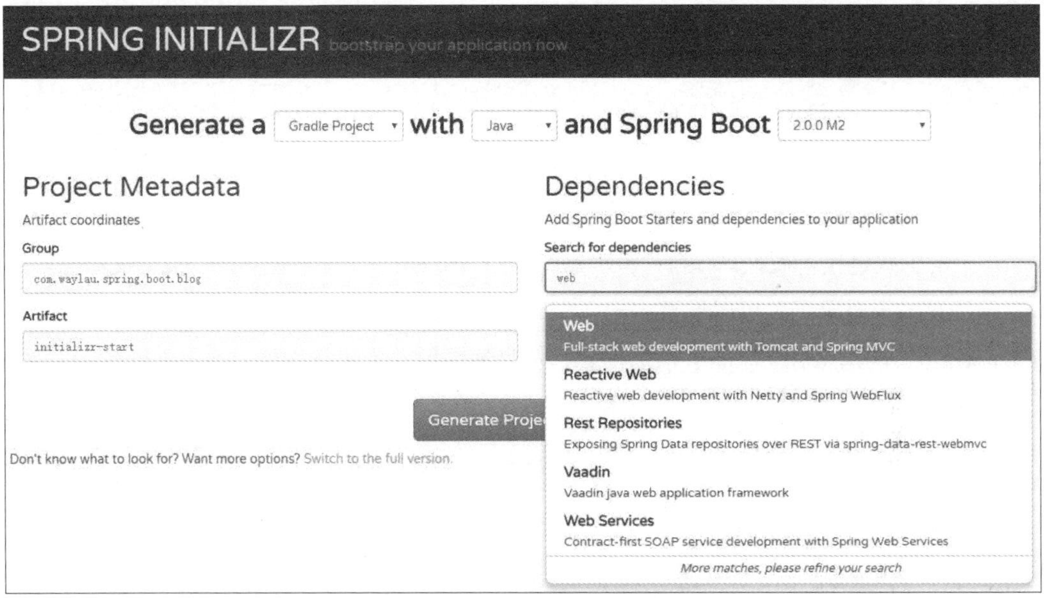

图1-4 Spring Initializr 管理界面

这里采用 Gradle 作为项目管理工具，Spring Boot 版本选型为 2.0.0.M2，Group 的信息填为“com.waylau.spring.boot.blog”，Artifact 填为“initializr-start”。最后，单击“Generate Project”按钮，此时可以下载以项目“initializr-start”命名的 zip 包。该压缩包包含了这个原型项目的所有源码及配置。

1.3.3 用 Gradle 编译项目

在项目根目录 initializr-start 下，执行 gradle build 命令来对项目进行构建，构建过程如下。

```
$ gradle build
Starting a Gradle Daemon, 1 busy Daemon could not be reused, use --status
for details
Download https://repo.spring.io/milestone/org/springframework/boot/
spring-boot-starter/2.0.0.M2/spring-boot-starter-2.0.0.M2.pom
Download https://repo1.maven.org/maven2/org/yaml/snakeyaml/1.18/
snakeyaml-1.18.pom
Download https://repo.spring.io/milestone/org/springframework/boot/
spring-boot/2.0.0.M2/spring-boot-2.0.0.M2.pom

...
```

```
> Task :test
2017-06-30 00:49:37.813  INFO 11572 --- [       Thread-5]
o.s.w.c.s.Ge nericWebApplicationContext    : Closing org.springframework.
web.context.support.GenericWebApplicationContext@6a26f94c: startup date
[Fri Jun 30 00:49:36 CST 2017]; root of
 context hierarchy

BUILD SUCCESSFUL in 2m 2s
5 actionable tasks: 5 executed
```

在编译开始阶段，Gradle 会解析项目配置文件，而后去找相关的依赖，并下载到本地。速度快慢取决于用户本地的网络。控制台会打印整个下载、编译的过程，当然，在这里为了节省篇幅，省去了大部分的下载过程。最后，看到“BUILD SUCCESSFUL”字样，说明已经编译成功了。

返回项目的根目录下，可以发现多出了一个 build 目录，在该目录 build/libs 下可以看到一个 initializr-start-0.0.1-SNAPSHOT.jar，该文件就是项目编译后的可执行文件。通过下面的命令来运行该文件。

```
java -jar build/libs/initializr-start-0.0.1-SNAPSHOT.jar
```

成功运行后，可以在控制台看到如下输出。

```
$ java -jar build/libs/initializr-start-0.0.1-SNAPSHOT.jar

...

2017-06-30 00:55:27.874  INFO 11468 --- [           main]
o.s.b.w.embedded.   tomcat.TomcatWebServer  : Tomcat started on port(s):
8080 (http)2017-06-30 00:55:27.874  INFO 11468 --- [           main]
c.w.s.b.b.i.InitializrStartApplication    : Started InitializrStart
Application in 4.27 seconds (JVM running for 5.934)
```

从输出内容可以看到，该项目使用的是 Tomcat 容器，使用的端口号是 8080。

在控制台输入“Ctrl + C”，可以关闭该程序。

1.3.4 探索项目

在启动项目后，在浏览器中输入“ http://localhost:8080/”，可以得到如下信息。

```
Whitelabel Error Page

This application has no explicit mapping for /error, so you are seeing
this as a fallback.
Fri Jun 30 00:56:44 CST 2017
There was an unexpected error (type=Not Found, status=404).
No message available
```

由于在项目中还没有任何对请求的处理程序，因此 Spring Boot 会出现上述默认的错误提示信息。

观察一下 initializr-start 项目的目录结构。

```
initializr-start
│  .gitignore
│  build.gradle
│  gradlew
│  gradlew.bat
│
├─.gradle
│  ├─4.0
│  │  ├─fileChanges
│  │  │      last-build.bin
│  │  │
│  │  ├─fileContent
│  │  │      annotation-processors.bin
│  │  │      fileContent.lock
│  │  │
│  │  ├─fileHashes
│  │  │      fileHashes.bin
│  │  │      fileHashes.lock
│  │  │      resourceHashesCache.bin
│  │  │
│  │  └─taskHistory
│  │          fileSnapshots.bin
│  │          taskHistory.bin
│  │          taskHistory.lock
│  │
│  └─buildOutputCleanup
│          built.bin
│          cache.properties
│          cache.properties.lock
│
├─build
│  ├─classes
│  │  └─java
│  │      ├─main
│  │      │  └─com
│  │      │      └─waylau
│  │      │          └─spring
│  │      │              └─boot
│  │      │                  └─blog
│  │      │                      └─initializrstart
│  │      │                              InitializrStartApplication.
class
│  │      │
│  │      └─test
│  │          └─com
```

```
│   │                   └─waylau
│   │                       └─spring
│   │                           └─boot
│   │                               └─blog
│   │                                   └─initializrstart
│   │                                           InitializrStartApplication
Tests.class
│   │
│   ├─libs
│   │       initializr-start-0.0.1-SNAPSHOT.jar
│   │
│   ├─reports
│   │   └─tests
│   │       └─test
│   │           │   index.html
│   │           │
│   │           ├─classes
│   │           │       com.waylau.spring.boot.blog.initializrstart.
InitializrStartApplicationTests.html
│   │           │
│   │           ├─css
│   │           │       base-style.css
│   │           │       style.css
│   │           │
│   │           ├─js
│   │           │       report.js
│   │           │
│   │           └─packages
│   │                   com.waylau.spring.boot.blog.initializrstart.html
│   │
│   ├─resources
│   │   └─main
│   │       │   application.properties
│   │       │
│   │       ├─static
│   │       └─templates
│   ├─test-results
│   │   └─test
│   │       │   TEST-com.waylau.spring.boot.blog.initializrstart.Initial
izrStartApplicationTests.xml
│   │       │
│   │       └─binary
│   │               output.bin
│   │               output.bin.idx
│   │               results.bin
│   │
│   └─tmp
│       ├─bootJar
│       │       MANIFEST.MF
```

```
|       |
|       ├─compileJava
|       └─compileTestJava
├─gradle
|   └─wrapper
|          gradle-wrapper.jar
|          gradle-wrapper.properties
|
└─src
    ├─main
    |   ├─java
    |   |   └─com
    |   |       └─waylau
    |   |           └─spring
    |   |               └─boot
    |   |                   └─blog
    |   |                       └─initializrstart
    |   |                               InitializrStartApplication.java
    |   |
    |   └─resources
    |       |   application.properties
    |       |
    |       ├─static
    |       └─templates
    └─test
        └─java
            └─com
                └─waylau
                    └─spring
                        └─boot
                            └─blog
                                └─initializrstart
                                        InitializrStartApplicationTests.
java
```

1. build.gradle 文件

在项目的根目录下可以看到 build.gradle 文件，这个是项目的构建脚本。Gradle 是以 Groovy 语言为基础，面向 Java 应用为主，基于 DSL（领域特定语言）语法的自动化构建工具。Gradle 这个工具集成了构建、测试、发布及常用的其他功能，如软件打包、生成注释文档等。跟以往 Maven 等构架工具不同，配置文件不需要烦琐的 XML，而是简洁的 Groovy 语言脚本。

对于本项目的 build.gradle 文件中配置的含义，笔者已经添加了详细注释。

```
buildscript { // buildscript代码块中脚本优先执行

    // ext用于定义动态属性
    ext {
        springBootVersion = '2.0.0.M2'
```

```
    }

    // 使用了Maven的中央仓库（也可以指定其他仓库）
    repositories {
        mavenCentral()
        maven { url "https://repo.spring.io/snapshot" }
        maven { url "https://repo.spring.io/milestone" }
    }

    // 依赖关系
    dependencies {

        // classpath 声明说明了在执行其余脚本时，ClassLoader可以使用这些依赖项
        classpath("org.springframework.boot:spring-boot-gradle-plugin:
${springBootVersion}")
    }
}

// 使用插件
apply plugin: 'java'
apply plugin: 'eclipse'
apply plugin: 'org.springframework.boot'
apply plugin: 'io.spring.dependency-management'

// 指定了生成的编译文件的版本，默认是打成了jar包
version = '0.0.1-SNAPSHOT'

// 指定编译.java 文件的JDK版本
sourceCompatibility = 1.8

//使用了Maven 的中央仓库（也可以指定其他仓库）
repositories {
    mavenCentral()
    maven { url "https://repo.spring.io/snapshot" }
    maven { url "https://repo.spring.io/milestone" }
}

// 依赖关系
dependencies {

    // 该依赖用于编译阶段
    compile('org.springframework.boot:spring-boot-starter-web')

    // 该依赖用于测试阶段
    testCompile('org.springframework.boot:spring-boot-starter-test')
}
```

2. gradlew 和 gradlew.bat文件

gradlew 和 gradlew.bat 这两个文件是 Gradle Wrapper 用于构建项目的脚本。使用 Gradle Wrapper

的好处在于，可以使项目组成员不必预先在本地安装好 Gradle 工具，在用 Gradle Wrapper 构建项目时，Gradle Wrapper 首先会去检查本地是否存在 Gradle，如果不存在，会根据配置上的 Gradle 的版本和安装包的位置来自动获取安装包并构建项目。使用 Gradle Wrapper 的另外一个好处在于，所有的项目组成员能够统一项目所使用的 Gradle 版本，从而规避了由于环境不一致导致编译失败的问题。对于 Gradle Wrapper 的使用，在类似 UNIX 的平台上（如 Linux 和 Mac OS），直接运行 gradlew 脚本，就会自动完成 Gradle 环境的搭建。而在 Windows 环境下，则执行 gradlew.bat 文件。

3. build 和 .gradle 目录

build 和 .gradle 目录都是在 Gradle 对项目进行构建后生成的目录、文件。

4. Gradle Wrapper

Gradle Wrapper 免去了用户在使用 Gradle 进行项目构建时需要安装 Gradle 的烦琐步骤。 每个 Gradle Wrapper 都绑定到一个特定版本的 Gradle，所以当用户第一次在给定 Gradle 版本下运行上面的命令之一时，它将下载相应的 Gradle 发布包，并使用它来执行构建。默认 Gradle Wrapper 的发布包指向官网的 Web 服务地址，相关配置记录在了 gradle-wrapper.properties 文件中。用户可以查看一下 Sring Boot 提供的 Gradle Wrapper 配置，参数“distributionUrl”就是用于指定发布包的位置。

```
distributionBase=GRADLE_USER_HOME
distributionPath=wrapper/dists
zipStoreBase=GRADLE_USER_HOME
zipStorePath=wrapper/dists
distributionUrl=https\://services.gradle.org/distributions/gradle-
3.5.1-bin.zip
```

从上述配置可以看出，当前 Spring Boot 采用的是 Gradle 3.5.1 版本。用户也可以自行来修改版本和发布包存放的位置。如下面的例子指定了发布包的位置在本地的文件系统中。

```
distributionUrl=file\:/D:/software/webdev/java/gradle-4.0-all.zip
```

5. src目录

如果用户用过 Maven，那么肯定对 src 目录不陌生。Gradle 约定了该目录下的 main 目录下是程序的源码，test 下是测试用的代码。

1.3.5 如何提升 Gradle 的构建速度

由于 Gradle 工具是舶来品，因此对于国人来说，很多时候会觉得编译速度非常慢。这里面很大一部分原因是网络的限制，毕竟 Gradle 及 Maven 的中央仓库都架设在国外，国内要访问，速度上肯定会有一些限制。下面介绍几个配置技巧来提升 Gradle 的构建速度。

1. Gradle Wrapper 指定本地

正如之前提到的，Gradle Wrapper 是为了便于统一版本。如果项目组成员都明确了 Gradle

Wrapper，尽可能事先将 Gradle 放置到本地，而后修改 Gradle Wrapper 配置，将参数“distributionUrl”指向本地文件。例如，将 Gradle 放置到 D 盘的某个目录。

```
#distributionUrl=https\://services.gradle.org/distributions/gradle-3.5.1
-bin.zipdistributionUrl=file\:/D:/software/webdev/java/gradle-4.0-all.zip
```

2. 使用国内 Maven 镜像仓库

Gradle 可以使用 Maven 镜像仓库。使用国内的 Maven 镜像仓库可以极大提升依赖包的下载速度。下面演示了使用自定义镜像的方法。

```
repositories {
    //mavenCentral()
    maven { url "https://repo.spring.io/snapshot" }
    maven { url "https://repo.spring.io/milestone" }
    maven { url "http://maven.aliyun.com/nexus/content/groups/public/"
}
}
```

上面注释掉了下载缓慢的中央仓库，改用自定义的镜像仓库。

1.3.6 示例源码

本节示例源码在 initializr-start 目录下。

1.4 如何进行 Spring Boot 项目的开发及测试

1.3 节介绍了如何使用 Spring Initializr 来初始化一个 Spring Boot 项目原型 initializr-start。在该项目原型的基础上来编写一个简单的 Web 项目。当访问项目时，界面会打印出“Hello World”字样。

1.4.1 编写项目构建信息

复制样例程序“initializr-start”到新的 hello-world 目录下，当然相关的编译文件（比如 build、.gradle 等目录下的文件）就不需要复制过来了。最终，新项目的根目录下会有 gradle、src 目录及 build.gradle、gradlew.bat、gradlew 文件。在该项目上做一点小变更，就能生成一个新项目的构建信息。

打开 build.gradle 文件，做相应的修改变更。默认 Spring Boot 的版本都是 0.0.1-SNAPSHOT，这里改为 1.0.0 意味着是一个成熟的可用的项目。

```
version = '1.0.0'
```

先尝试执行 gradle build 来对“hello-world”项目进行构建。如果构建成功，则说明构建信息编写正确。

```
$ gradle build
:compileJava
:processResources
:classes
:bootJar
:jar SKIPPED
:assemble
:compileTestJava
:processTestResources NO-SOURCE
:testClasses
:test
2017-07-01 20:37:20.319  INFO 13816 --- [       Thread-5]
o.s.w.c.s.GenericWebApplicationContext   : Closing org.springframework.
web.context.
support.GenericWebApplicationContext@7d21825:
startup date [Sat Jul 01 20:37:18 CST 2017]; root of
context hierarchy
:check
:build

BUILD SUCCESSFUL

Total time: 14.422 secs
```

1.4.2 编写程序代码

现在开始编写代码，首先进入 hello-world 项目的 src 目录下，应该能够看到 com.waylau.spring.boot.blog.initializrstart 包及 InitializrStartApplication.java 文件。为了规范，将该包名改为“com.waylau.spring.boot.blog”，将 InitializrStartApplication.java 更名为“Application.java”。

1. 观察 Application.java

打开 Application.java 文件，观察以下代码。

```
package com.waylau.spring.boot.blog;

import org.springframework.boot.SpringApplication;
import org.springframework.boot.autoconfigure.SpringBootApplication;

@SpringBootApplication
public class Application {

    public static void main(String[] args) {
        SpringApplication.run(Application.class, args);
    }
}
```

首先看到的是 @SpringBootApplication 注解。对于经常使用 Spring 的用户而言，很多开发者总是使用 @Configuration、@EnableAutoConfiguration 和 @ComponentScan 注解 main 类。由于这些注解被如此频繁地一块使用，Spring Boot 提供了一个方便的 @SpringBootApplication 选择。该 @SpringBootApplication 注解等同于使用 @Configuration、 @EnableAutoConfiguration 和 @ComponentScan 的默认属性，即：

@SpringBootApplication =（默认属性的）@Configuration + @EnableAutoConfiguration + @ComponentScan

它们的含义分别如下。

- @Configuration：经常与 @Bean 组合使用，使用这两个注解就可以创建一个简单的 Spring 配置类，可以用来替代相应的 XML 配置文件。@Configuration 的注解类标识这个类可以使用 Spring IoC 容器作为 Bean 定义的来源。@Bean 注解告诉 Spring，一个带有 @Bean 的注解方法将返回一个对象，该对象应该被注册为在 Spring 应用程序上下文中的 Bean。
- @EnableAutoConfiguration：能够自动配置 Spring 的上下文，试图猜测和配置用户想要的 Bean 类，通常会自动根据用户的类路径和 Bean 定义自动配置。
- @ComponentScan：会自动扫描指定包下的全部标有 @Component 的类，并注册成 Bean，当然也包括 @Component 下的子注解 @Service、@Repository、@Controller。这些 Bean 一般是结合 @Autowired 构造函数来注入。

2. main 方法

main 方法是一个标准的 Java 方法，它遵循 Java 对于一个应用程序入口点的约定。main 方法通过调用 run，将业务委托给了 Spring Boot 的 SpringApplication 类。SpringApplication 将引导用户的应用，启动 Spring，相应地启动被自动配置的 Tomcat Web 服务器。需要将 Application.class 作为参数传递给 run 方法，以此告诉 SpringApplication 谁是主要的 Spring 组件，并传递 args 数组以暴露所有的命令行参数。

3. 编写控制器 HelloController

创建 com.waylau.spring.boot.blog.controller 包，用于放置控制器类。

HelloController.java 的代码非常简单。当请求到 /hello 路径时，将会响应“Hello World!”字样的字符串。代码如下。

```
@RestController
public class HelloController {

    @RequestMapping("/hello")
    public String hello() {
        return "Hello World! Welcome to visit waylau.com!";
    }
}
```

@RestController 等价于 @Controller 与 @ResponseBody 的组合，主要用于返回在 RESTful 应用常用的 JSON 格式数据，即：

@RestController = @Controller + @ResponseBody

其中各项含义如下。

- @ResponseBody：该注解用于将 Controller 的方法返回对象，通过适当的 HttpMessageConverter 转换为指定格式后，写入 Response 对象的 body 数据区。
- @RequestMapping：是一个用来处理请求地址映射的注解，可用于类或方法上。用于类上，它表示类中的所有响应请求的方法都是以该地址作为父路径。根据方法的不同，还可以用 GetMapping、PostMapping、PutMapping、DeleteMapping、PatchMapping 代替。
- @RestController：暗示用户，这是一个支持 REST 的控制器。

1.4.3 编写测试用例

进入 test 目录下，默认产生了测试用例的包 com.waylau.spring.boot.blog.initializrstart 及测试类 InitializrStartApplicationTests.java。将测试用例包更名为“com.waylau.spring.boot.blog”，测试用例类更名为“ApplicationTests.java”文件。

1. 编写 HelloControllerTest.java 测试类

相对于源程序一一对应，在测试用例包下创建 com.waylau.spring.boot.blog.controller 包，用于放置控制器的测试类。

测试类 HelloControllerTest.java 代码如下。

```
@RunWith(SpringRunner.class)
@SpringBootTest
@AutoConfigureMockMvc
public class HelloControllerTest {

    @Autowired
    private MockMvc mockMvc;

    @Test
    public void testHello() throws Exception {
        mockMvc.perform(MockMvcRequestBuilders.get("/hello").accept(Me-
diaType.APPLICATION_JSON))
                .andExpect(status().isOk())
                .andExpect(content().string(equalTo("Hello World! Wel-
come to visit waylau.com!")));
    }
}
```

2. 运行测试类

用 JUnit 运行该测试，绿色表示该代码测试通过。

1.4.4 配置 Gradle Wrapper

Gradle 项目可以使用 Gradle 的安装包进行构建，也可以使用 Gradle Wrapper 来进行构建。使用 Gradle Wrapper 的好处是可以使项目的构建工具版本得到统一。

修改 Wrapper 属性文件（位于 gradle/wrapper/gradle-wrapper.properties）中的 distributionUrl 属性，将其改为指定的 Gradle 版本，这里采用了 Gradle 4 版本。

```
distributionUrl=https\://services.gradle.org/distributions/gradle-4.0-
bin.zip
```

或者也可以指向本地的文件。

```
distributionUrl=file\:/D:/software/webdev/java/gradle-4.0-all.zip
```

这样，Gradle Wrapper 会自动安装 Gradle 的版本。

不同平台，执行不同的命令脚本。

- gradlew（UNIX Shell 脚本）。
- gradlew.bat（Windows 批处理文件）。

1.4.5 运行程序

1. 使用 Gradle Wrapper

执行 gradlew 来对"hello-world"程序进行构建。

```
$ gradlew build
Downloading https://services.gradle.org/distributions/gradle-4.0-bin.
zip
......................................................................
.............................................................
Unzipping C:\Users\Administrator\.gradle\wrapper\dists\gradle-4.0-
all\1gxqawjd4d149ywxp1eb3gi3n\gradle-4.0-all.zip to C:\Users\Administrator\.
gradle\wrapper\dists\gradle-4.0-all\1gxqawjd4d149ywxp1eb3gi3n
Starting a Gradle Daemon, 1 incompatible and 1 stopped Daemons could
not be reused, use --status for details

> Task :test
2017-07-01 22:14:05.566  INFO 3172 --- [       Thread-8] o.s.w.c.s.
GenericWebApplicationContext    : Closing org.springframework.       web.
context.support.GenericWebApplicationContext@68daf86e: startup date [Sat
Jul 01 22:14:04 CST 2017]; root of context hierarchy
2017-07-01 22:14:05.566  INFO 3172 --- [       Thread-5] o.s.w.c.s.
GenericWebApplicationContext    : Closing org.springframework.web.context.
support.GenericWebApplicationContext@15b48aa7: startup date [Sat Jul 01
22:14:02 CST 2017]; root of context hierarchy
```

```
BUILD SUCCESSFUL in 36s
5 actionable tasks: 5 executed
```

如果是首次使用，首先会下载 Gradle 发布包。可以在 $USER_HOME/.gradle/wrapper/dists 下的用户主目录中找到它们。

2. 运行程序

执行 java -jar build/libs/hello-world-1.0.0.jar 来运行程序。

```
$ java -jar build/libs/hello-world-1.0.0.jar

  .   ____          _            __ _ _
 /\\ / ___'_ __ _ _(_)_ __  __ _ \ \ \ \
( ( )\___ | '_ | '_| | '_ \/ _` | \ \ \ \
 \\/  ___)| |_)| | | | | || (_| |  ) ) ) )
  '  |____| .__|_| |_|_| |_\__, | / / / /
 =========|_|==============|___/=/_/_/_/
 :: Spring Boot ::        (v2.0.0.M2)

2017-07-01 22:17:12.172  INFO 13956 --- [           main] com.waylau.
spring.boot.blog.Application  : Starting Application on AGOC3-705091335
with PID 13956 (D:\workspaceGithub\spring-boot-enterprise-applica-
tion-development\samples\hello-world\build\libs\hello-world-1.0.0.jar
started by Administrator in D:\workspaceGithub\spring-boot-enterprise-
application-development\samples\hello-world)
2017-07-01 22:17:12.177  INFO 13956 --- [           main] com.waylau.
spring.boot.blog.Application  : No active profile set, falling back to
default profiles: default
2017-07-01 22:17:12.251  INFO 13956 --- [           main] ConfigServ-
letWebServerApplicationContext : Refreshing org.springframework.boot.
web.servlet.context.AnnotationConfigServletWebServerApplicationContex-
t@18ef96: startup date [Sat Jul 01 22:17:12 CST 2017]; root of context
hierarchy
2017-07-01 22:17:13.666  INFO 13956 --- [           main] o.s.b.w.
embedded.tomcat.TomcatWebServer  : Tomcat initialized with port(s): 8080
(http)
2017-07-01 22:17:13.684  INFO 13956 --- [           main] o.apache.
catalina.core.StandardService   : Starting service [Tomcat]
2017-07-01 22:17:13.686  INFO 13956 --- [           main] org.apache.
catalina.core.StandardEngine  : Starting Servlet Engine: Apache Tomcat/
8.5.15
2017-07-01 22:17:13.796  INFO 13956 --- [ost-startStop-1]
o.a.c.c.C.[Tomcat].[localhost].[/]       : Initializing Spring embedded
WebApplicationContext
2017-07-01 22:17:13.796  INFO 13956 --- [ost-startStop-1] o.s.web.
context.ContextLoader            : Root WebApplicationContext: initial-
ization completed in 1548 ms
2017-07-01 22:17:13.947  INFO 13956 --- [ost-startStop-1] o.s.b.w.servlet.
```

```
ServletRegistrationBean  : Mapping servlet: 'dispatcherServlet' to  [/]

2017-07-01 22:17:13.954  INFO 13956 --- [ost-startStop-1]
o.s.b.w.servlet.FilterRegistrationBean   : Mapping filter: 'characterEn-
codingFilter' to: [/*]
2017-07-01 22:17:13.955  INFO 13956 --- [ost-startStop-1]
o.s.b.w.servlet.FilterRegistrationBean   : Mapping filter: 'hiddenHttp-
MethodFilter' to: [/*]
2017-07-01 22:17:13.955  INFO 13956 --- [ost-startStop-1]
o.s.b.w.servlet.FilterRegistrationBean   : Mapping filter: 'httpPutForm-
ContentFilter' to: [/*]
2017-07-01 22:17:13.956  INFO 13956 --- [ost-startStop-1] o.s.b.w.
servlet.FilterRegistrationBean   : Mapping filter: 'requestContextFilter'
to: [/*]
2017-07-01 22:17:14.284  INFO 13956 --- [           main]
s.w.s.m.m.a.RequestMappingHandlerAdapter : Looking for @ControllerAd-
vice: org.springframework.boot.web.servlet.context.AnnotationConfigServ-
letWebServerApplicationContext@18ef96: startup date [Sat Jul 01 22:17:12
CST 2017]; root of context hierarchy
2017-07-01 22:17:14.358  INFO 13956 --- [           main]
s.w.s.m.m.a.RequestMappingHandlerMapping : Mapped "{[/hello]}" onto pub-
lic java.lang.String com.waylau.spring.boot.blog.controller.HelloCon-
troller.hello()
2017-07-01 22:17:14.366  INFO 13956 --- [           main]
s.w.s.m.m.a.RequestMappingHandlerMapping : Mapped "{[/error]}" onto pub-
lic org.springframework.http.ResponseEntity<java.util.Map<java.lang.
String, java.lang.Object>> org.springframework.boot.autoconfigure.web.
servlet.error.BasicErrorController.error(javax.servlet.http.HttpServle-
tRequest)
2017-07-01 22:17:14.368  INFO 13956 --- [           main]
s.w.s.m.m.a.RequestMappingHandlerMapping : Mapped "{[/error],produces=
[text/html]}" onto public org.springframework.web.servlet.ModelAndView
org.springframework.boot.autoconfigure.web.servlet.error.BasicErrorCon-
troller.errorHtml(javax.servlet.http.
HttpServletRequest,javax.servlet.http.HttpServletResponse)
2017-07-01 22:17:14.416  INFO 13956 --- [           main]
o.s.w.s.handler.SimpleUrlHandlerMapping  : Mapped URL path [/webjars/**]
onto handler of type [class org.springframework.web.servlet.resource.
ResourceHttpRequestHandler]
2017-07-01 22:17:14.416  INFO 13956 --- [           main]
o.s.w.s.handler.SimpleUrlHandlerMapping  : Mapped URL path [/**] onto
handler of type [class org.springframework.web.servlet.resource.Re
sourceHttpRequestHandler]
2017-07-01 22:17:14.459  INFO 13956 --- [           main]
o.s.w.s.handler.SimpleUrlHandlerMapping  : Mapped URL path [/**/favicon.
ico] onto handler of type [class org.springframework.web.servlet.re-
source.ResourceHttpRequestHandler]
2017-07-01 22:17:14.604  INFO 13956 --- [           main]
o.s.j.e.a.AnnotationMBeanExporter        : Registering beans for JMX
```

```
exposure on startup
2017-07-01 22:17:14.684  INFO 13956 --- [           main]
o.s.b.w.em- bedded.tomcat.TomcatWebServer  : Tomcat started on port(s):
8080 (http)2017-07-01 22:17:14.692  INFO 13956 --- [           main]
com.waylau. spring.boot.blog.Application  : Started Application in 2.914
seconds (JVM running for 3.346)
```

3. 访问程序

在浏览器中访问 http://localhost:8080/hello，可以看到页面出现“Hello World! Welcome to visit waylau.com!”字样，如图 1-5 所示。

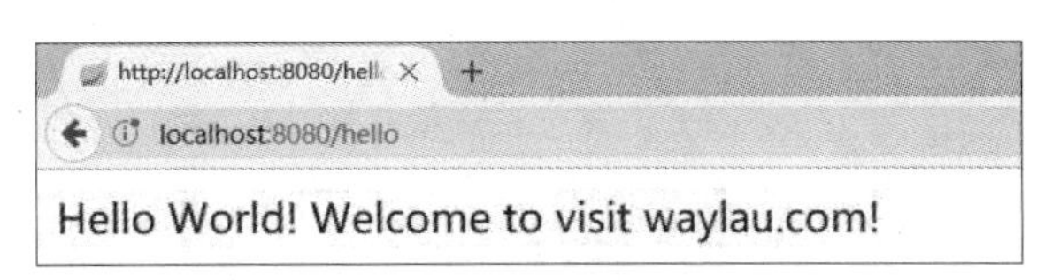

图1-5　浏览器访问界面

由此可知，编写一个 Spring Boot 程序就是这么简单！

1.4.6 其他运行程序的方式

有多种运行 Spring Boot 程序的方式，除了上面介绍的使用 java -jar 命令外，还有其他几种方式。

1. 以“Java Application”运行

hello-world 程序就是一个平常的 Java 程序，所以可以直接在 IDE 中右击项目，以“Java Application”方式来运行程序。这种方式在开发时，非常方便调试程序。

2. 使用 Spring Boot Gradle Plugin 插件运行

Spring Boot 已经内嵌了 Spring Boot Gradle Plugin 插件，所以可以使用 Spring Boot Gradle Plugin 插件来运行程序。在命令行执行方式如下。

```
$ gradle bootRun
```

或者：

```
$ gradlew bootRun
```

1.4.7 如何将项目导入 IDE

由于每位开发者对于 IDE 都会有不同的选择，本书不会对这部分内容进行详细讲解。在本书的后面“附录 A：开发环境的搭建”部分，介绍了将项目导入 Eclipse 来进行开发的方式。

1.4.8 示例源码

本节示例源码在 hello-world 目录下。

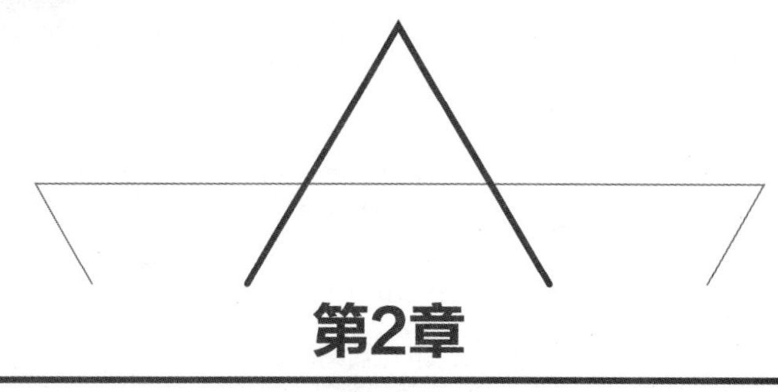

第2章

Spring 框架核心概念

2.1 Spring 框架总览

Spring 框架是整个 Spring 技术栈的核心。Spring 框架实现了对 bean 的依赖管理及 AOP 的编程方式，这些都极大地提升了 Java 企业级应用开发过程中的编程效率，降低了代码之间的耦合。Spring 框架是很好的一站式构建企业级应用的轻量级的解决方案。

Spring Boot 是基于 Spring 框架技术来构建的，所以 Spring Boot 又会使用很多 Spring 框架中的技术。由于市面上讲解 Spring 技术的书籍已经很多了，笔者不再对 Spring 做过多的介绍，只侧重讲解 Spring Boot 中经常会被应用的 Spring 核心技术。如果读者想了解 Spring 框架的全貌，可以参阅笔者所著的开源书《Spring Framework 4.x 参考文档》（https://github.com/waylau/spring-framework-4-reference）。

本书示例所使用的 Spring 框架版本为 5.0.0.RC2。

2.1.1 模块化的 Spring 框架

Spring 框架是模块化的，允许用户自由选择需要使用的部分。例如，用户可以在任何框架上使用 IoC 容器，但也可以只使用 Hibernate 集成代码或 JDBC 抽象层。Spring 框架支持声明式事务管理，通过 RMI 或 Web 服务远程访问用户的逻辑，并支持多种选择来持久化用户的数据，它提供了一个全功能的 Spring MVC 及 Spring WebFlux 框架。同时，它也支持 AOP 透明地集成到用户的软件中。

Spring 框架由二十多个模块组成。图 2-1 所示为 Spring 框架模块主要的组成部分。

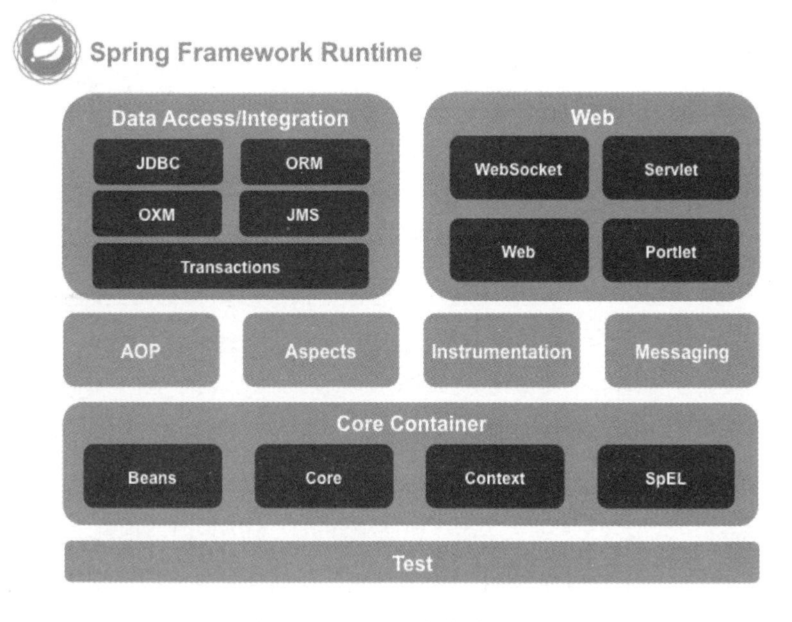

图2-1 Spring 框架模块的组成

2.1.2 使用 Spring 的好处

Spring 框架是一个轻量级的 Java 平台，能够提供完善的基础设施用来支持开发 Java 应用程序。Spring 负责基础设施功能，这样用户就可以专注于应用逻辑的开发。

Spring 可以使用户从 POJO 来构建应用程序，并且将企业服务非侵入性地应用到 POJO，此功能适用于 Java SE 编程模型和完全或部分的 Java EE 模型。

作为一个 Java 应用程序的开发者，可以从 Spring 平台获得以下好处。

（1）使本地 Java 方法可以执行数据库事务，而无须自己去处理事务 API。

（2）使本地 Java 方法可以执行远程过程，而无须自己去处理远程 API。

（3）使本地 Java 方法成为 HTTP 端点，而无须自己处理 Servlet API。

（4）使本地 Java 方法可以拥有管理操作，而无须自己去处理 JMX API。

（5）使本地 Java 方法可以执行消息处理，而无须自己去处理 JMS API。

2.1.3 依赖注入和控制反转

很多人都会被问及“依赖注入”与“控制反转”之间到底有哪些联系和区别。在 Java 应用程序中，不管是受限的嵌入式应用程序，还是多层架构的服务端企业级应用程序，它们通常由来自应用适当的对象进行组合合作。也就是说，对象在应用程序中通过彼此依赖来实现功能。

尽管 Java 平台提供了丰富的应用程序开发功能，但它缺乏组织基本构建块成为一个完整系统的方法。那么，组织系统这个任务最后只能留给架构师和开发人员。开发者可以使用各种设计模式（如 Factory、Abstract Factory、Builder、Decorator 和 Service Locator）来组合各种类和对象实例构成应用程序。虽然这些模式给出了什么样的模式，能解决什么问题，但使用模式一个最大的障碍就是，除非开发者也有非常丰富的经验，否则仍然无法在应用程序中正确地使用它，这就给 Java 开发者带来了一定的技术门槛，特别是那些普通的开发人员。

而 Spring 框架的 IoC（Inversion of Control ，控制反转）组件就能够通过提供正规化的方法来组合不同的组件，使之成为一个完整的可用的应用。Spring 框架将规范化的设计模式作为一级的对象，这样方便开发者将之集成到自己的应用程序，这也是很多组织和机构选择使用 Spring 框架来开发健壮的、可维护的应用程序的原因。开发人员无须手动处理对象的依赖关系，而是交给了 Spring 容器去管理，这极大地提升了开发体验。

那么“依赖注入”与“控制反转”又是什么关系呢？

“依赖注入”（Dependency Injection）是 Martin Fowler 在 2004 年提出的关于“控制反转”的解释[①] 。Martin Fowler 认为“控制反转”一词让人产生疑惑，无法直白地理解“到底哪方面的控制被反转了”。所以，Martin Fowler 建议采用 “依赖注入”一词来代替“控制反转”。

---

① 有关 Martin Fowler 的博客原文，可参见 https://martinfowler.com/articles/injection.html。

“依赖注入”和“控制反转”其实就是一个事物的两种不同的说法而已，本质上是一回事。“依赖注入”是一个程序设计模式和架构模型，有些时候也称为“控制反转”，尽管在技术上来讲，“依赖注入”是一个“控制反转”的特殊实现。“依赖注入”是指一个对象应用另外一个对象来提供一个特殊的能力。例如，把一个数据库连接以参数的形式传到一个对象的结构方法里，而不是在那个对象内部自行创建一个连接。“依赖注入”和“控制反转”的基本思想就是把类的依赖从类内部转化到外部以减少依赖。利用“控制反转”，对象在被创建的时候，会由一个调控系统统一进行对象实例的管理，将该对象所依赖的对象的引用通过调控系统传递给它。也可以说，依赖被注入到对象中。所以，“控制反转”是关于一个对象如何获取它所依赖的对象的引用的过程，而这个过程体现为“谁来传递依赖的引用”的这个职责的反转。

2.1.4 Spring 框架常用模块

Spring 框架基本涵盖了企业级应用开发的各个方面。

1. 核心容器

核心容器（Core Container）由 spring-core、spring-beans、spring-context、spring-context-support 和 spring-expression (Spring Expression Language) 模块组成。

其中，spring-core 和 spring-beans 提供框架的基础部分，包括 IoC 和 Dependency Injection 功能。BeanFactory 是一个复杂的工厂模式的实现。无须编程就能实现单例，并允许开发者将配置和特定的依赖从实际程序逻辑中解耦。

Context（spring-context）模块建立于 Core 和 Beans 模块提供的功能的基础之上，它是一种在框架类型下实现对象存储操作的手段，有点像 JNDI 注册。Context 继承了 Beans 模块的特性，并且增加了对国际化的支持（如用在资源包中）、事件广播、资源加载和创建上下文（如 一个 Servlet 容器）。Context 模块也支持如 EJB、JMX 和基础远程访问这样的 Java EE 特性。ApplicationContext 接口是 Context 模块的主要表现形式。spring-context-support 提供了对常见第三方库的支持，以便集成进 Spring 应用上下文，如缓存（EhCache、JCache）、调度（CommonJ、Quartz）等。

spring-expression 模块提供了一个强大的表达式语言，用来在运行时查询和操作对象图，这是作为 JSP 2.1 规范所指定的统一表达式语言的一种扩展。这种语言支持对属性值、属性参数、方法调用、数组内容存储、收集器和索引、逻辑和算数操作及命名变量，并且通过名称从 Spring 的控制反转容器中取回对象。表达式语言模块还支持列表投影和选择及通用列表聚合。

2. AOP 及 Instrumentation

spring-aop 模块提供 AOP（Aspect Oriented Programming，面向切面编程）的实现，从而能够实现如方法拦截器和切入点完全分离代码。使用源码级别元数据的功能，也可以在代码中加入行为信息，在某种程度上类似于 .NET 属性。

单独的 spring-aspects 模块提供了集成使用 AspectJ。

spring-instrument 模块提供了类 instrumentation 的支持和在某些应用程序服务器使用类加载器实现。spring-instrument-tomcat 用于 Tomcat Instrumentation 代理。

3. 消息

自 Spring 框架 4 版本开始提供 spring-messaging 模块，主要包含从 Spring Integration 项目中抽象出来的，如 Message、MessageChannel、 MessageHandler 及其他用来提供基于消息的基础服务，该模块还包括一组消息映射方法的注解，类似于基于编程模型中的 Spring MVC 的注解。

4. 数据访问 / 集成

数据访问 / 集成（Data Access/Integration） 层由 spring-jdbc（JDBC）、spring-orm（ORM）、spring-oxm（OXM）、spring-jms（JMS）和 spring-tx（Transaction）模块组成。

spring-jdbc 模块提供了一个 JDBC 抽象层，这样开发人员就能避免进行一些烦琐的 JDBC 编码和解析数据库供应商特定的错误代码。

spring-orm 模块为流行的对象关系映射 API 提供集成层，包括 JPA 和 Hibernate。使用 spring-orm 模块，可以将这些 O/R 映射框架与 Spring 提供的所有其他功能结合使用，如前面提到的简单的声明式事务管理功能。

spring-oxm 模块提供了一个支持 Object/XML 映射实现的抽象层，如 JAXB、Castor、JiBX 和 XStream。

spring-jms 模块包含用于生成和使用消息的功能。从 Spring Framework 4.1 起，它提供了与 spring-messaging 的集成。

spring-tx 模块支持用于实现特殊接口和所有 POJO（普通 Java 对象）的类的编程和声明式事务管理。

5. Web

Web 层由 spring-web、spring-webmvc、spring-websocket 和 spring-webmvc-portlet 模块组成 。

spring-web 模块提供了基本的面向 Web 开发的集成功能，如文件上传及用于初始化 IoC 容器的 Servlet 监听和 Web 开发应用程序上下文，它也包含 HTTP 客户端及 Web 相关的 Spring 远程访问的支持。

spring-webmvc 模块（也被称为 Web Servlet 模块）包含 Spring 的 MVC 功能和 REST 服务的功能。

6. Test

spring-test 模块支持通过组合 JUnit 或 TestNG 来实现单元测试和集成测试等功能，它提供了 Spring ApplicationContexts 的持续加载，并能缓存这些上下文，它还提供可用于孤立测试代码的模拟对象（Mock Objects）。

2.2 依赖注入与控制反转

正如前面所介绍的那样，依赖注入（DI）与控制反转（IoC）可以视为同一事物的不同表述。Spring 通过 IoC 容器来管理所有 Java 对象（也称为 bean）及其相互间的依赖关系。

在软件开发过程中，系统的各个对象之间，各个模块之间，软件系统与硬件系统之间或多或少都会存在耦合关系，如果一个系统的耦合度过高，就会造成难以维护的问题。但是完全没有耦合的代码是不能工作的，代码需要相互协作、相互依赖来完成功能。而 IoC 的技术恰好就解决了这类问题，各个对象之间不需要直接关联，而是在需要用到对方的时候由 IoC 容器来管理对象之间的依赖关系，对于开发人员来说只需要维护相对独立的各个对象代码即可。

2.2.1 IoC 容器和 bean

IoC 是一个过程，即对象定义其依赖关系，而其他与之配合的对象只能通过构造函数参数、工厂方法的参数，或者在从工厂方法构造或返回后在对象实例上设置的属性来定义其依赖关系。然后，IoC 容器在创建 bean 时会注入这些依赖项。这个过程在职责上是反转的，就是把原先代码里需要实现的对象创建、依赖的代码，反转给容器来帮忙实现和管理，所以称为“控制反转”。

IoC 的技术原理如下。

（1）反射：在运行状态中，根据提供的类的路径或者类名，通过反射来动态地获取该类的所有属性和方法。

（2）工厂模式：把 IoC 容器当作一个工厂，在配置文件或者注解中给出定义，然后利用反射技术，根据给出的类名生成相应的对象。对象生成的代码及对象之间的依赖关系在配置文件中定义，这样就实现了解耦。

org.springframework.beans 和 org.springframework.context 包是 Spring IoC 容器的基础。BeanFactory 接口提供了能够管理任何类型的对象的高级配置机制。ApplicationContext 是 BeanFactory 的子接口，它更容易与 Spring 的 AOP 功能集成，进行消息资源处理（用于国际化）、事件发布，以及作为应用层特定的上下文（例如，用于 Web 应用程序的 WebApplicationContext）。

简而言之，BeanFactory 提供了基本的配置功能，而 ApplicationContext 在此基础之前增加了更多的企业特定功能。

在 Spring 应用中， bean 是由 Spring IoC 容器来进行实例化、组装并受其管理的对象。bean 和它们之间的依赖关系反映在容器使用的配置元数据中。

2.2.2 配置元数据

配置元数据（Configuration Metadata）描述了 Spring 容器在应用程序中是如何来实例化、配置和组装对象的。

最初，Spring 是用 XML 文件格式来记录配置元数据，从而很好地实现了 IoC 容器本身与实际写入此配置元数据的格式完全分离。

当然，基于 XML 的元数据不是唯一允许的配置元数据形式。目前，比较流行的配置元数据的方式是采用注释或者是基于 Java 的配置。

（1）基于注解的配置：Spring 2.5 引入了支持基于注解的配置元数据。

（2）基于 Java 的配置：从 Spring 3.0 开始，Spring JavaConfig 项目提供了许多功能，并成为 Spring 框架核心的一部分。因此，可以使用 Java 而不是 XML 文件来定义应用程序类外部的 bean。这类注解，比较常用的有 @Configuration、@Bean、@Import 和 @DependsOn 等。

Spring 配置至少需要一个或者多个由容器管理的 bean。基于 XML 的配置元数据，需要用 <beans/> 元素内的 <bean/> 元素来配置这些 bean；而在基于 Java 的配置方式中，通常在使用了 @Configuration 注解的类中使用 @Bean 注解的方法。

以下示例显示了基于 XML 的配置元数据的基本结构。

```
<?xml version="1.0" encoding="UTF-8"?>
<beans xmlns="http://www.springframework.org/schema/beans"
    xmlns:xsi="http://www.w3.org/2001/XMLSchema-instance"
    xsi:schemaLocation="http://www.springframework.org/schema/beans
        http://www.springframework.org/schema/beans/spring-beans.xsd">

    <bean id="..." class="...">
        <!-- 放置这个bean的协作者和配置 -->
    </bean>

    <bean id="..." class="...">
        <!-- 放置这个bean的协作者和配置 -->
    </bean>

    <!-- 省略了更多的bean的配置-->
</beans>
```

在上面的 XML 文件中，id 属性是用于标识单个 bean 定义的字符串。 class 属性定义 bean 的类型，并使用完全限定的类名。id 属性的值是指协作对象。

以下示例显示了基于注解的配置元数据的基本结构。

```
@Configuration
public class AppConfig {

    @Bean
    public MyService myService() {
        return new MyServiceImpl();
    }
}
```

2.2.3 实例化容器

Spring IoC 容器需要在应用启动时进行实例化。在实例化过程中，IoC 容器会从各种外部资源（如本地文件系统、Java 类路径等）加载配置元数据，提供给 ApplicationContext 构造函数。

下面是一个从类路径中加载基于 XML 的配置元数据的例子。

```
ApplicationContext context =
    new ClassPathXmlApplicationContext(new String[] {"services.xml",
"daos.xml"});
```

当系统规模比较大时，通常会让 bean 定义分到多个 XML 文件。这样，每个单独的 XML 配置文件通常就能够表示系统结构中的逻辑层或模块。就如上面的例子所演示的那样。

当某个构造函数需要多个资源位置时，可以使用一个或多个 <import/> 来从另一个文件加载 bean 的定义。例如，

```
<beans>
    <import resource="services.xml"/>
    <import resource="resources/messageSource.xml"/>
    <import resource="/resources/themeSource.xml"/>

    <bean id="bean1" class="..."/>
    <bean id="bean2" class="..."/>
</beans>
```

2.2.4 两种注入方式

在 Spring 框架中，主要有以下两种注入方式。

1. 基于构造函数

基于构造函数的 DI 是通过调用具有多个参数的构造函数的容器来完成的，每个参数表示依赖关系，这个与调用具有特定参数的静态工厂方法来构造 bean 几乎是等效的。以下示例演示了一个只能使用构造函数注入的依赖注入的类，该类是一个 POJO，并不依赖于容器特定的接口、基类或注解。

```
public class SimpleMovieLister {

    // SimpleMovieLister依赖于MovieFinder
    private MovieFinder movieFinder;

    // Spring容器可以通过构造器来注入MovieFinder
    public SimpleMovieLister(MovieFinder movieFinder) {
        this.movieFinder = movieFinder;
    }
```

```
    // 省略使用注入的MovieFinder的具体业务逻辑
}
```

2. 基于 setter 方法

基于 setter 方法的 DI 是通过在调用无参数构造函数或无参数静态工厂方法来实例化 bean 之后，通过容器调用 bean 的 setter 方法完成的。

以下示例演示了一个只能使用 setter 来将依赖进行注入的类。该类是一个 POJO，并不依赖于容器特定的接口、基类或注解。

```
public class SimpleMovieLister {

    // SimpleMovieLister依赖于MovieFinder
    private MovieFinder movieFinder;

    // Spring容器可以通过setter方法来注入MovieFinder
    public void setMovieFinder(MovieFinder movieFinder) {
        this.movieFinder = movieFinder;
    }

    // 省略使用注入的MovieFinder的具体业务逻辑
}
```

2.2.5 bean 范围

默认情况下，所有 Spring bean 都是单例的，意思是在整个 Spring 应用中，bean 的实例只有一个。可以在 bean 中添加 scope 属性来修改这个默认值。scope 属性可用的值如表 2-1 所示。

表2-1 scope属性可用的值

范围	描述
singleton	定义 bean 的范围为每个 Spring 容器一个实例（默认值）
prototype	允许 bean 可以被多次实例化（使用一次就创建一个实例）
request	定义 bean 的范围是 HTTP 请求。每个 HTTP 请求都有自己的实例。只有在使用有 Web 能力的 Spring 上下文时才有效
session	定义 bean 的范围是 HTTP 会话。只有在使用有 Web 能力的 Spring ApplicationContext 时才有效
application	定义了每个 ServletContext 一个实例
websocket	定义了每个 WebSocket 一个实例。只有在使用有 Web 能力的 Spring ApplicationContext 时才有效

下面详细讨论 singleton bean 与 prototype bean 在用法上的差异。

1. singleton bean

对于 singleton bean 来说，IoC 容器只管理一个 singleton bean 的一个共享实例，所有对 id 或 id 匹配该 bean 定义的 bean 的请求都会导致 Spring 容器返回一个特定的 bean 实例。

换句话说，当定义一个 bean 定义并将其定义为 singleton 时，Spring IoC 容器将仅创建一个由该 bean 定义的对象实例。该单个实例存储在缓存中，对该 bean 所有后续请求和引用都将返回缓存中的对象实例。

在 Spring IoC 容器中，singleton bean 是默认的创建 bean 的方式，可以更好地重用对象，节省了重复创建对象的开销。

图 2-2 所示为 singleton bean 使用示意图。

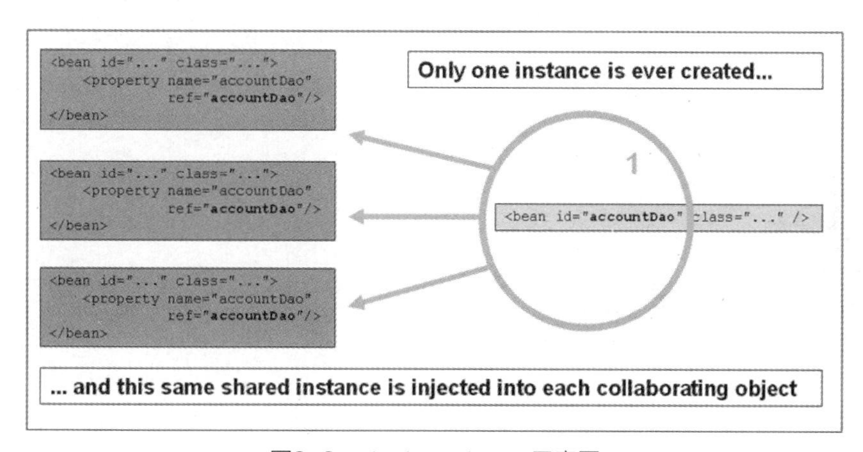

图2-2 singleton bean 示意图

2. prototype bean

对于 prototype bean 来说，IoC 容器导致在每次对该特定 bean 进行请求时创建一个新的 bean 实例。

图 2-3 所示为 prototype bean 使用示意图。

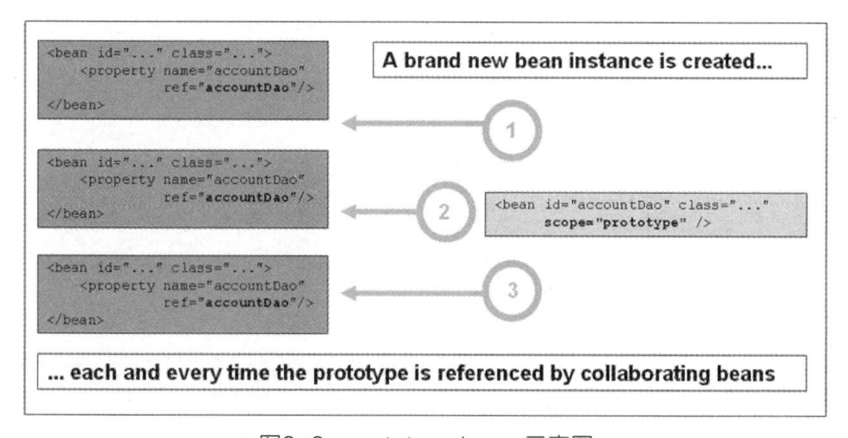

图2-3 prototype bean 示意图

从某种意义上来说，Spring 容器在 prototype bean 上的作用等同于 Java 的 new 操作符。所有过去的生命周期管理都必须由客户端处理。（有关 Spring 容器中 bean 的生命周期的详细信息，请参阅 Lifecycle 回调。）

2.2.6 注意 singleton bean 引用 prototype bean 时的陷阱

大家知道，Spring bean 默认的 scope 是 singleton（单例），但有些场景（如多线程）需要每次调用都生成一个实例，此时 scope 就应该设为 prototype，如下面的例子所示。

```
@Component
@Scope("prototype")
public class DadTask implements Runnable {

    static Logger logger = Logger.getLogger(DadTask.class);

    @Autowired
    DadDao dadDao;

    private String name;

    public DadTask setDadTask(String name) {
        this.name = name;
        return this;
    }

    @Override
    public void run() {
        logger.info("DadTask:"+this + ";DadDao:"+dadDao + ";"+dadDao.
sayHello(name) );
    }
}
```

但是，如果 singleton bean 依赖 prototype bean，通过依赖注入方式，prototype bean 在 singleton bean 实例化时会创建一次（只一次），考虑下面的例子。

```
@Service
public class UserService {

    @Autowired
    private DadTask dadTask;

    public void startTask() {
        ScheduledThreadPoolExecutor  scheduledThreadPoolExecutor   = new
ScheduledThreadPoolExecutor(2);

        scheduledThreadPoolExecutor.scheduleAtFixedRate(dadTask.setDad-
```

```
Task("Lily"), 1000, 2000, TimeUnit.MILLISECONDS);
        scheduledThreadPoolExecutor.scheduleAtFixedRate(dadTask.setDad-
Task("Lucy"), 1000, 2000, TimeUnit.MILLISECONDS);
    }
}
```

调度“Lily”和“Lucy”两个线程，实际上它只初始化一个实例（这样线程就不安全了）。

解决问题的思路如下。

如果 singleton bean 想每次都去创建一个新的 prototype bean 的实例，需要通过方法注入的方式。可以通过实现 ApplicationContextAware 接口来获取到 ApplicationContext 实例，继而通过 getBean 方法来获取到 prototype bean 的实例。程序需要修改如下。

```
@Service
public class UserService implements ApplicationContextAware {

    @Autowired
    private DadTask dadTask;

    private ApplicationContext applicationContext;

    public void startTask() {
        ScheduledThreadPoolExecutor  scheduledThreadPoolExecutor  = new
ScheduledThreadPoolExecutor(2);

        // 每次都拿到DadTask的实例
        dadTask = applicationContext.getBean("dadTask", DadTask.class);
        scheduledThreadPoolExecutor.scheduleAtFixedRate(dadTask.setDad-
Task("Lily"), 1000, 2000, TimeUnit.MILLISECONDS);
        dadTask = applicationContext.getBean("dadTask", DadTask.class);
        scheduledThreadPoolExecutor.scheduleAtFixedRate(dadTask.setDad-
Task("Lucy"), 1000, 2000, TimeUnit.MILLISECONDS);
    }

    @Override
    public void setApplicationContext (ApplicationContext application
-Context) throws BeansException {
        this.applicationContext = applicationContext;
    }
}
```

上述例子，摘自笔者撰写的博客《Spring singleton bean 与 prototype bean 的依赖》[①]，有兴趣的读者可自行参阅。

① 有关该博客原文，可参见 https://waylau.com/spring-singleton-beans-with-prototype-bean-dependencies/。

2.2.7 JSR-330 规范注解

由于 Spring 框架的流行，Java 技术委员会也开始着手完善相关的规范。其中 JSR-330 就是 Java 依赖注入标准规范。自 Spring 3.0 依赖，Spring 支持 JSR-330 规范，Spring 自身很多注解都可以用 JSR-330 规范来代替。

1. @Inject

@javax.inject.Inject 可以代替 @Autowired。考虑下面的例子。

```
import javax.inject.Inject;

public class SimpleMovieLister {

    private MovieFinder movieFinder;

    @Inject
    public void setMovieFinder(MovieFinder movieFinder) {
        this.movieFinder = movieFinder;
    }

    public void listMovies() {
        this.movieFinder.findMovies(...);
        //...
    }
}
```

与 @Autowired 一样，可以在字段级别、方法级别和构造函数参数级别使用 @Inject。

2. @Named 和 @ManagedBean

@javax.inject.Named 或者 javax.annotation.ManagedBean 可以代替 @Component。

```
import javax.inject.Inject;
import javax.inject.Named;

@Named("movieListener") // 等同于 @ManagedBean("movieListener")
public class SimpleMovieLister {

    private MovieFinder movieFinder;

    @Inject
    public void setMovieFinder(MovieFinder movieFinder) {
        this.movieFinder = movieFinder;
    }

    // ...
}
```

注意：javax.annotation.ManagedBean 注解属于 JSR-250 规范。

当使用 @Named 或 @ManagedBean 时，可以以与使用 Spring 注解完全相同的方式使用组件扫描。

```
@Configuration
@ComponentScan(basePackages = "com.waylau")
public class AppConfig  {
    // ...
}
```

与 Spring 原生的注解相比，JSR-330 在用法上还是稍显逊色。例如，与 @Autowired 相比，@Inject 没有“required”属性，不过可以用 Java 8 的 Optional 来代替；与 @Component 相比，JSR-330 的注解并不提供可组合的模型，只是一种识别命名组件的方法。

2.2.8 Spring Boot 中的 bean 及依赖注入

Spring Boot 通常使用基于 Java 的配置，建议主配置是单个 @Configuration 类。通常，定义 main 方法的类作为主要的 @Configuration 类。

Spring Boot 应用了很多 Spring 框架中的自动配置功能。自动配置会尝试根据添加的 jar 依赖关系自动配置 Spring 应用程序。例如，如果 HSQLDB 或者是 H2 在类路径上，并且没有手动配置任何数据库连接 bean，那么 Spring Boot 会自动配置为内存数据库。

要启用自动配置功能，需要通过将 @EnableAutoConfiguration 或 @SpringBootApplication 注解添加到一个 @Configuration 类中。

1. 自动配置

在 Spring Boot 应用中，可以自由使用任何标准的 Spring 框架技术来定义 bean 及其注入的依赖关系。为了简化程序的开发，通常使用 @ComponentScan 来找到 bean，并结合 @Autowired 构造函数来将 bean 进行自动装配注入。这些 bean 涵盖了所有应用程序组件，如 @Component、@Service、@Repository、@Controller 等。下面是一个实际的例子。

```
@Service
public class DatabaseAccountService implements AccountService {

    private final RiskAssessor riskAssessor;

    @Autowired
    public DatabaseAccountService(RiskAssessor riskAssessor) {
        this.riskAssessor = riskAssessor;
    }

    // ...

}
```

如果一个 bean 只有一个构造函数，则可以省略 @Autowired。

```
@Service
public class DatabaseAccountService implements AccountService {

    private final RiskAssessor riskAssessor;

    public DatabaseAccountService(RiskAssessor riskAssessor) {
        this.riskAssessor = riskAssessor;
    }

    // ...

}
```

2. 使用 @SpringBootApplication 注解

正如在第 1 章所演示的那样，由于 Spring Boot 开发人员总是频繁使用 @Configuration、@EnableAutoConfiguration 和 @ComponentScan 来注解它们的主类，并且这些注解经常被一起使用，Spring Boot 提供了一种方便的 @SpringBootApplication 注解来替代。

@SpringBootApplication 注解相当于使用 @Configuration、@EnableAutoConfiguration 和 @ComponentScan 及其默认属性。

```
import org.springframework.boot.SpringApplication;
import org.springframework.boot.autoconfigure.SpringBootApplication;

@SpringBootApplication // 等同于 @Configuration @EnableAutoConfiguration @
ComponentScan
public class Application {

    public static void main(String[] args) {
        SpringApplication.run(Application.class, args);
    }

}
```

2.3 AOP 编程

AOP（Aspect Oriented Programming，面向切面编程）通过提供另一种思考程序结构的方式来补充 OOP（Object Oriented Programming，面向对象编程）。OOP 模块化的关键单元是类，而在 AOP 中，模块化的单元是切面（Aspect）。切面可以实现诸如跨多个类型和对象之间的事务管理、日志等方面的模块化。

Spring 的关键组件之一是 AOP 框架。虽然 Spring IoC 容器不依赖于 AOP，但在 Spring 应用中，经常会使用 AOP 来简化编程。在 Spring 框架中使用 AOP 主要有以下优势。

（1）提供声明式企业服务，特别是作为 EJB 声明性服务的替代品。最重要的是这种服务是声明式事务管理。

（2）允许用户实现自定义切面，在某些不适合用 OOP 编程的场景中，采用 AOP 来补充。

（3）可以对业务逻辑的各个部分进行隔离，从而使业务逻辑各部分之间的耦合度降低，提高程序的可重用性，同时提高了开发的效率。

2.3.1 AOP 核心概念

AOP 概念并非 Spring AOP 特有的，这些概念同样适用于其他 AOP 框架，如 AspectJ。

（1）Aspect（切面）：将关注点进行模块化。某些关注点可能会横跨多个对象，例如，事务管理就是 Java 企业级应用中一个关于横切关注点的很好的例子。在 Spring AOP 中，切面可以使用常规类（基于模式的方法）或使用 @Aspect 注解的常规类来实现切面。

（2）Join Point (连接点）：在程序执行过程中某个特定的点，如某方法调用的时候或者处理异常的时候。在 Spring AOP 中，一个连接点总是代表一个方法的执行。

（3）Advice（通知）：在切面的某个特定的连接点上执行的动作。通知有各种类型，其中包括 “around” “before” 和 “after” 等通知。许多 AOP 框架，包括 Spring，都是以拦截器来实现通知模型的，并维护一个以连接点为中心的拦截器链。

（4）Pointcut（切入点）：匹配连接点的断言。通知和一个切入点表达式关联，并在满足这个切入点的连接点上运行（如当执行某个特定名称的方法时）。切入点表达式如何和连接点匹配是 AOP 的核心。Spring 默认使用 AspectJ 切入点语法。

（5）Introduction（引入）：声明额外的方法或者某个类型的字段。Spring 允许引入新的接口（以及一个对应的实现）到任何被通知的对象。例如，可以使用一个引入来使 bean 实现 IsModified 接口，以便简化缓存机制。在 AspectJ 社区，Introduction 也称 Inter-type Declaration（内部类型声明）。

（6）Target Object（目标对象）：被一个或者多个切面所通知的对象。也有人把它称为 Advised（被通知）对象。既然 Spring AOP 是通过运行时代理实现的，这个对象永远是一个 Proxied（被代理）对象。

（7）AOP Proxy（AOP 代理）：AOP 框架创建的对象，用来实现 Aspect Contract（切面契约）包括通知方法执行等功能。在 Spring 中，AOP 代理可以是 JDK 动态代理或者 CGLIB 代理。

（8）Weaving（织入）：把切面连接到其他的应用程序类型或者对象上，并创建一个 Advised（被通知）的对象。这些可以在编译时（如使用 AspectJ 编译器）、类加载时和运行时完成。Spring 和其他纯 Java AOP 框架一样，在运行时完成织入。其中有关 Advice（通知）的类型主要有以下几种。

（1）Before Advice（前置通知）：在某连接点之前执行的通知，但这个通知不能阻止连接点前的执行（除非它抛出一个异常）。

（2）After Returning Advice（返回后通知）：在某连接点正常完成后执行的通知，例如，一个方法没有抛出任何异常，正常返回。

（3）After Throwing Advice（抛出异常后通知）：在方法抛出异常退出时执行的通知。

（4）After (finally) Advice（最后通知）：当某连接点退出的时候执行的通知（不论是正常返回还是异常退出）。

（5）Around Advice（环绕通知）：包围一个连接点的通知，如方法调用。这是最强大的一种通知类型。环绕通知可以在方法调用前后完成自定义的行为，它也会选择是否继续执行连接点或者直接返回它们自己的返回值或抛出异常来结束执行。

Around Advice（环绕通知）是最常用的一种通知类型。与 AspectJ 一样，Spring 提供所有类型的通知，推荐使用尽量简单的通知类型来实现需要的功能。例如，如果只是需要用一个方法的返回值来更新缓存，虽然使用环绕通知也能完成同样的事情，但是最好使用 After Returning 通知而不是环绕通知。用最合适的通知类型可以使编程模型变得简单，并且能够避免很多潜在的错误。例如，不需要调用 JoinPoint（用于 Around Advice）的 proceed() 方法，就不会有调用的问题。

在 Spring 2.0 中，所有的通知参数都是静态类型，因此可以使用合适的类型（如一个方法执行后的返回值类型）作为通知的参数，而不是使用一个对象数组。

切入点和连接点匹配的概念是 AOP 的关键，这使得 AOP 不同于其他仅仅提供拦截功能的旧技术。切入点使得通知可独立于 OO（Object Oriented，面向对象）层次。例如，一个提供声明式事务管理的 Around Advice（环绕通知）可以被应用到一组横跨多个对象中的方法上（如服务层的所有业务操作）。

2.3.2 Spring AOP 功能和目标

Spring AOP 用纯 Java 实现，它不需要专门的编译过程。Spring AOP 不需要控制类装载器层次，因此它适用于 Servlet 容器或应用服务器。

Spring 目前仅支持使用方法调用作为连接点。虽然可以在不影响 Spring AOP 核心 API 的情况下加入对成员变量拦截器的支持，但 Spring 并没有实现成员变量拦截器。如果需要通知对成员变量的访问和更新连接点，可以考虑其他语言，如 AspectJ。

Spring 实现 AOP 的方法跟其他的框架不同。Spring 并不是要尝试提供最完整的 AOP 实现（尽管 Spring AOP 有这个能力），相反地，它其实侧重于提供一种 AOP 实现和 Spring IoC 容器的整合，用于帮助解决在企业级开发中的常见问题。

因此，Spring AOP 通常都和 Spring IoC 容器一起使用。Aspect 使用普通的 bean 定义语法，与其他 AOP 实现相比这是一个显著的区别。有些事使用 Spring AOP 是无法轻松或者高效完成的，如通知一个细粒度的对象。这种时候，使用 AspectJ 是最好的选择。对于大多数在企业级 Java 应用中遇到的问题，Spring AOP 都能提供一个非常好的解决方案。

Spring AOP 从来没有打算通过提供一种全面的 AOP 解决方案来取代 AspectJ。它们之间的关系应该是互补而不是竞争。Spring 可以无缝地整合 Spring AOP、IoC 和 AspectJ，使所有的 AOP 应用完全融入基于 Spring 的应用体系，这样的集成不会影响 Spring AOP API 或者 AOP Alliance API。Spring AOP 保留了向下兼容性，这体现了 Spring 框架的核心原则——非侵袭性，即 Spring 框架并不强迫在业务或者领域模型中引入框架特定的类和接口。

2.3.3 AOP 代理

Spring AOP 默认使用标准的 JDK 动态代理来作为 AOP 的代理，这样任何接口（或者接口的 set 方法）都可以被代理。

Spring AOP 也支持使用 CGLIB 代理，对于需要代理类而不是代理接口的时候，CGLIB 代理是很有必要的。如果一个业务对象并没有实现一个接口，默认就会使用 CGLIB。此外，面向接口编程也是一个最佳实践，业务对象通常都会实现一个或多个接口。此外，还可以强制地使用 CGLIB，在那些（希望是罕见的）需要通知一个未在接口中声明的方法的情况下，或者需要传递一个代理对象作为一种具体类型的方法的情况下。

2.3.4 使用 @AspectJ

@AspectJ 是用于切面的常规 Java 类注解。Spring 采用了与 AspectJ 相同的注解风格，使用 AspectJ 提供的库进行切入点的解析和匹配。

1. 启用 @AspectJ

可以通过 XML 或者 Java 配置来启用 @AspectJ 支持。在任何情况下，还需要确保 AspectJ 的 aspectjweaver.jar 库在应用程序的类路径中（版本 1.6.8 或以后）。这个库在 AspectJ 发布的 lib 目录中或通过 Maven 的中央存库得到。

下面演示了使用 @Configuration 和 @EnableAspectJAutoProxy 注解来启用 @AspectJ 的例子。

```
@Configuration
@EnableAspectJAutoProxy
public class AppConfig {

}
```

基于 XML 的配置，可以使用 aop:aspectj-autoproxy 元素。

```
<aop:aspectj-autoproxy/>
```

2. 声明 Aspect（切面）

在启用 @AspectJ 支持的情况下，在应用上下文中定义的任意带有一个 @Aspect 注解的切面的

bean 都将被 Spring 自动识别并用于配置 Spring AOP。以下例子展示了一个切面所需要的最小定义。

```
<bean id="myAspect" class="org.xyz.NotVeryUsefulAspect">
    <!-- 配置aspect的属性 -->
</bean>
```

上面这个 bean 指向一个使用了 @Aspect 注解的 bean 类。下面是 NotVeryUsefulAspect 类定义，使用了 org.aspectj.lang.annotation.Aspect 注解。

```
package org.xyz;
import org.aspectj.lang.annotation.Aspect;

@Aspect
public class NotVeryUsefulAspect {

}
```

3. 声明 Pointcut（切入点）

Spring AOP 只支持 Spring bean 方法执行连接点，所以可以把切入点看作是匹配 Spring bean 上的方法执行。一个切入点声明有两个部分，一个包含名称和任意参数的签名，另一个是切入点表达式，该表达式决定了哪个方法的执行。在 @AspectJ 中，一个切入点实际就是一个普通的方法定义提供的一个签名。切入点表达式使用 @Pointcut 注解来表示，需要注意的是，这个方法的返回类型必须是 void。

下面的例子定义了一个切入点“anyOldTransfer”，这个切入点匹配了任意名为 transfer 的方法执行。

```
@Pointcut("execution(* transfer(..))")// Pointcut表达式
private void anyOldTransfer() {}// Pointcut签名
```

切入点表达式也就是 @Pointcut 注解的值，是正规的 AspectJ 5 切入点表达式。如果想要更多了解 AspectJ 的切入点语言，请参见《The AspectJ™ Programming Guide》①。

2.3.5 示例

下面用一个例子来演示 Spring AOP 的用法。

出于某些原因，业务服务的执行有时可能会失败，但希望的是操作能被重试，而不是直接抛出异常。以下是满足这个需求的实现。

```
@Aspect
public class ConcurrentOperationExecutor implements Ordered {
```

① 网址参见 https://www.eclipse.org/aspectj/doc/released/progguide/index.html。

```
    private static final int DEFAULT_MAX_RETRIES = 2;

    private int maxRetries = DEFAULT_MAX_RETRIES;
    private int order = 1;

    public void setMaxRetries(int maxRetries) {
        this.maxRetries = maxRetries;
    }

    public int getOrder() {
        return this.order;
    }

    public void setOrder(int order) {
        this.order = order;
    }

    @Around("com.xyz.myapp.SystemArchitecture.businessService()")
    public Object doConcurrentOperation(ProceedingJoinPoint pjp) throws
Throwable {
        int numAttempts = 0;
        PessimisticLockingFailureException lockFailureException;
        do {
            numAttempts ++;
            try {
                return pjp.proceed();
            }
            catch(PessimisticLockingFailureException ex) {
                lockFailureException = ex;
            }
        } while(numAttempts <= this.maxRetries);
        throw lockFailureException;
    }

}
```

注意：该 aspect 实现了 Ordered 接口，因此可以将 aspect 的优先级设置为高于 transaction advice（每次重试时都需要一个新的事务）。maxRetries 和 order 属性都将由 Spring 配置。正在将重试逻辑应用于所有 businessService() 方法。我们会持续尝试，直到耗尽了所有预设的重试次数（maxRetries）。

相应的 Spring 的配置如下。

```
<aop:aspectj-autoproxy/>

<bean id="concurrentOperationExecutor" class="com.xyz.myapp.service.
impl.ConcurrentOperationExecutor">
    <property name="maxRetries" value="3"/>
    <property name="order" value="100"/>
</bean>
```

第3章
Spring MVC 及常用 MediaType

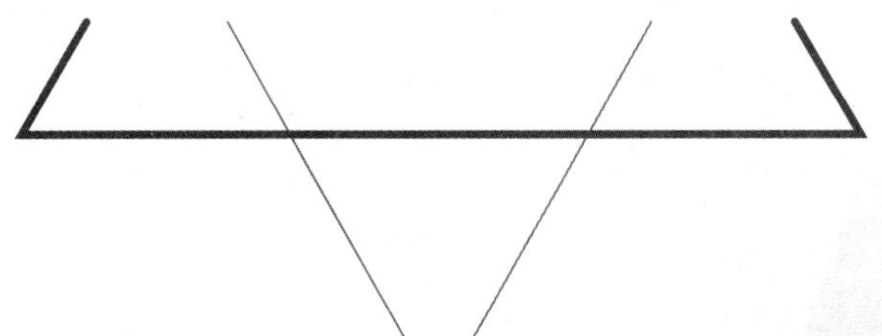

3.1 Spring MVC 简介

Spring MVC 实现了 Web 开发中的经典的 MVC（ Model-View-Controller）模式。MVC 由以下三部分组成。

（1）模型（Model）：应用程序的核心功能，管理这个模块中用到的数据和值。

（2）视图（View）：视图提供模型的展示，管理模型如何显示给用户，它是应用程序的外观。

（3）控制器（Controller）：对用户的输入做出反应，管理用户和视图的交互，是连接模型和视图的枢纽。

Spring MVC 使用 @Controller 或 @RestController 注解的 bean 来处理传入的 HTTP 请求。使用 @RequestMapping 注解将 HTTP 请求映射到相应的控制器中的方法。

以下是 @RestController 用于提供 JSON 数据的典型示例。

```
@RestController
@RequestMapping(value="/users")
public class MyRestController {

    @RequestMapping(value="/{user}", method=RequestMethod.GET)
    public User getUser(@PathVariable Long user) {
        // ...
    }

    @RequestMapping(value="/{user}/customers", method=RequestMethod.
GET)
    List<Customer> getUserCustomers(@PathVariable Long user) {
        // ...
    }

    @RequestMapping(value="/{user}", method=RequestMethod.DELETE)
    public User deleteUser(@PathVariable Long user) {
        // ...
    }

}
```

3.1.1 MVC 与三层架构的差异

用户可能认为 MVC 的任何实现都可以被自动地视为三层架构的实现，但其实不是这样的。三层架构一般定义为以下层次。

（1）表示层（Presentation Layer）：提供与用户交互的界面。GUI（图形用户界面）和 Web 页面是表示层的两个典型的例子。

（2）业务层（Business Layer）：也称为业务逻辑层，用于实现各种业务逻辑，如处理数据验证，根据特定的业务规则和任务来响应特定的行为。

（3）数据访问层（Data Access Layer）：也称为数据持久层，负责存放和管理应用的持久性业务数据。

图 3-1 所示为 MVC 各个组成部分所处的位置，以及与三层架构之间的差异。

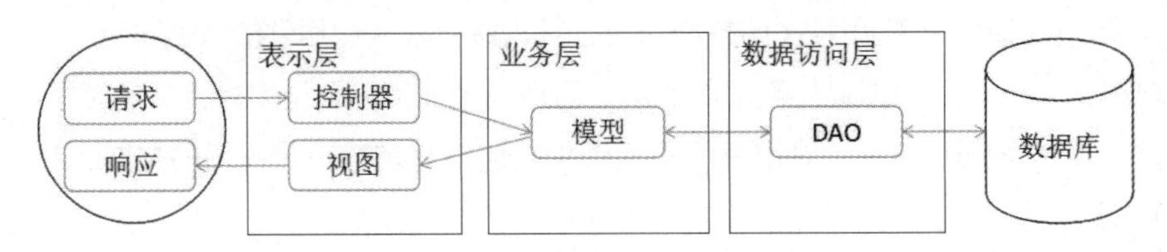

图3-1　三层架构与 MVC 的差异

在每个 MVC 的实现中，有一个经常会犯的简单但基本的错误，就是数据库连接的地方。在三层架构中，在接收到来自业务层的请求时，与数据库的所有通信（包括打开连接）都在数据访问层内完成。表示层没有与数据库的任何通信，它只能通过业务层与其通信。对于初学者来说，在使用 MVC 框架时，经常会将数据库连接放在控制器内，将连接对象传递给模型，然后在必要时使用它。这是一个错误！在控制器中会打开连接到数据库服务器实例，以便与数据库通信。这些连接被传递到不实际使用该连接的不同的 Model 组件，这会导致连接资源的浪费。

这种实现另外一个经常犯的错误是，认为数据验证（有时称为数据过滤）应该在控制器中传递到模型之前执行。这不适合三层架构的定义，其中规定这样的数据验证逻辑，连同业务逻辑和任务特定的行为，应仅存在于业务层或模型组件内。将数据验证从模型中取出并将其放入控制器中，会导致控制器和模型之间紧耦合，因为控制器不能与不同的模型一起使用，而紧耦合的缺点在于更新一个模块的结果导致其他模块的结果变化，难以重用特定的关联模块。松耦合将使控制器可以被多个模型类共享，而不是固定为一个。

正确的实现方式是，使用三层架构开发的应用程序应该将其所有逻辑（数据验证、业务规则和任务特定行为）局限于业务层。在表示层和数据访问层中不应该有应用逻辑，这样改变这两个层中的任意一个，也不会影响任何应用逻辑。

3.1.2 Spring MVC 中的自动配置

Spring Boot 提供了适用于大多数应用程序的 Spring MVC 的自动配置。

自动配置在 Spring 的默认值之上添加以下功能。

（1）包含 ContentNegotiatingViewResolver bean 和 BeanNameViewResolver bean。

（2）支持静态资源的服务，包括对 WebJars 的支持。

（3）自动注册 Converter、GenericConverter、Formatter 等 bean。

（4）支持 HttpMessageConverters。

（5）自动注册 MessageCodesResolver。

（6）支持静态 index.html。

（7）支持自定义 Favicon。

（8）自动使用 ConfigurableWebBindingInitializer bean。

1. HttpMessageConverter

Spring MVC 使用 HttpMessageConverter 接口来转换 HTTP 请求和响应。其默认值提供了开箱即用的功能，例如，对象可以自动转换为 JSON（使用 Jackson 库）或 XML（如果 Jackson XML 扩展不可用，则使用 JAXB）。字符串默认使用 UTF-8 进行编码。

如果需要添加或自定义转换器，可以使用 Spring Boot 的 HttpMessageConverters 类。

```
import org.springframework.boot.autoconfigure.web.HttpMessageConverters;
import org.springframework.context.annotation.*;
import org.springframework.http.converter.*;

@Configuration
public class MyConfiguration {

    @Bean
    public HttpMessageConverters customConverters() {
        HttpMessageConverter<?> additional = ...
        HttpMessageConverter<?> another = ...
        return new HttpMessageConverters(additional, another);
    }

}
```

2. MessageCodesResolver

MessageCodesResolver 是 Spring MVC 中用于生成错误代码的策略，可以从绑定错误中提取错误消息。可以在 Spring Boot 中设置 MessageCodesResolver 的格式。其设置属性为 spring.mvc.message-codes-resolver.format，值可以是 PREFIX_ERROR_CODE 或 POSTFIX_ERROR_CODE。

3. 静态内容

默认情况下，Spring Boot 将从类路径或 ServletContext 的根中名为 /static、/public、/resources、/META-INF/resources 的目录下加载静态内容，它使用 Spring MVC 中的 ResourceHttpRequestHandler 来实现，当然，也可以通过添加自己的 WebMvcConfigurerAdapter 并覆盖 addResourceHandlers 方法来修改该行为。

默认情况下，资源映射到 /** 路径，但也可以通过 spring.mvc.static-path-pattern 配置来调整。例如，将所有资源重定位到 /resources/** 路径，可以通过以下配置方式实现。

```
spring.mvc.static-path-pattern=/resources/**
```

还可以使用 spring.resources.static-locations（使用目录位置列表替换默认值）来自定义静态资源位置。如果这样做，默认的欢迎页面检测将切换到自定义位置，因此，如果在启动时任何位置中有一个 index.html，它将是应用程序的主页。

除了上述“标准”静态资源位置外，还提供了一个特殊情况用于 WebJars 内容。如果使用 WebJars 格式打包，则在 /webjars/** 路径中的资源可以从 Jar 文件中提供。

4. ConfigurableWebBindingInitializer

Spring MVC 使用 WebBindingInitializer 为特定请求初始化 WebDataBinder。如果创建自己的 ConfigurableWebBindingInitializer 类型的 @Bean，Spring Boot 将自动配置 Spring MVC 以使用它。

5. 模板引擎

对于动态 HTML 内容的展示，模板引擎必不可少。Spring MVC 支持 Thymeleaf 、FreeMarker 和 JSP 等多种技术。对于 Spring Boot 而言，它支持 FreeMarker、Groovy、Thymeleaf、Mustache 等引擎的自动配置功能。

由于 Thymeleaf “原型即页面”的特点适用于快速开发，因此本书在后面也会着重对 Thymeleaf 进行讲解。

3.2 JSON 类型的处理

JSON（JavaScript Object Notation）是一种轻量级的数据交换格式，具有良好的可读和便于快速编写的特性。JSON 是基于 JavaScript 编程语言的一个子集。由于 JavaScript 的流行，JSON 也开始被越来越多的人接受。JSON 可在不同平台之间进行数据交换。JSON 采用兼容性很高的、完全独立于语言文本格式，同时也具备类似于 C 语言的习惯（包括 C、 C++、 C#、Java、JavaScript、Perl、 Python 等）体系的行为。这些特性使 JSON 成为理想的数据交换语言。

本节将演示如何使用 Spring MVC 来处理 JSON 格式的数据，在 hello-world 项目源码的基础上创建了一个 media-type-json 项目作为演示案例。

3.2.1 控制器及实体

下面将创建 media-type-json 项目的控制器及实体。

1. 创建实体 User

创建 com.waylau.spring.boot.blog.domain 包，用于存放领域对象。

创建 com.waylau.spring.boot.blog.domain.User 的实体类，用于返回用户信息。User 是一个 POJO。

User.java 代码如下。

```
public class User {
```

```
    private Long id; // 实体的唯一标识

    private String name;

    private String email;

    public User() { // 无参数默认构造器
    }

    public User(Long id, String name, String email) {
        this.id = id;
        this.name = name;
        this.email = email;
    }

    public Long getId() {
        return id;
    }

    public void setId(Long id) {
        this.id = id;
    }

    public String getName() {
        return name;
    }

    public void setName(String name) {
        this.name = name;
    }

    public String getEmail() {
        return email;
    }

    public void setEmail(String email) {
        this.email = email;
    }
}
```

2. 创建 UserController

在 com.waylau.spring.boot.blog.controller.HelloController 类的基础上，创建了一个 com.waylau.spring.boot.blog.controller.UserController 类来作为本例的处理不同 MediaType 的控制器。

UserController.java 代码如下。

```
@RestController
@RequestMapping("/users")
```

```
public class UserController {

    @RequestMapping("/{id}")
    public User getUser(@PathVariable("id") Long id) {
        return new User(id,"waylau", "waylau521@gmail.com");
    }
}
```

3.2.2 返回 JSON 类型

上面的代码就已经实现了一个能够处理 JSON 类型数据响应的 API 了，将项目运行起来，看看实际的效果。

1. 运行项目

运行项目后，在浏览器里访问 http://localhost:8080/users/1 接口。

2. 查看返回数据

发现页面上已经显示了 JSON 文本。也就是说，请求返回的是 JSON 类型数据。

```
{"id":1,"name":"waylau","email":"waylau521@gmail.com"}
```

为什么没有做任何特殊设置，数据可以被自动转成 JSON 格式？

还记得项目的依赖 spring-boot-starter-web 吗？如果打开这个 spring-boot-starter-web 项目的 pom.xml，会发现该项目依赖了 spring-boot-starter-json 这个专门用于处理 JSON 的 Starter。

```
<dependencies>
    <dependency>
        <groupId>org.springframework.boot</groupId>
        <artifactId>spring-boot-starter</artifactId>
    </dependency>
    <dependency>
        <groupId>org.springframework.boot</groupId>
        <artifactId>spring-boot-starter-json</artifactId>
    </dependency>
    <dependency>
        <groupId>org.springframework.boot</groupId>
        <artifactId>spring-boot-starter-tomcat</artifactId>
    </dependency>
    <dependency>
        <groupId>org.hibernate</groupId>
        <artifactId>hibernate-validator</artifactId>
    </dependency>
    <dependency>
        <groupId>org.springframework</groupId>
        <artifactId>spring-web</artifactId>
    </dependency>
    <dependency>
```

```
        <groupId>org.springframework</groupId>
        <artifactId>spring-webmvc</artifactId>
    </dependency>
</dependencies>
```

如果再往下看 spring-boot-starter-json 的 pom.xml 描述，可以看到 spring-boot-starter-json 主要依赖的是 Jackson2 的库。

```
<dependencies>
    <dependency>
        <groupId>com.fasterxml.jackson.core</groupId>
        <artifactId>jackson-databind</artifactId>
    </dependency>
    <dependency>
        <groupId>com.fasterxml.jackson.datatype</groupId>
        <artifactId>jackson-datatype-jdk8</artifactId>
    </dependency>
    <dependency>
        <groupId>com.fasterxml.jackson.datatype</groupId>
        <artifactId>jackson-datatype-jsr310</artifactId>
    </dependency>
    <dependency>
        <groupId>com.fasterxml.jackson.module</groupId>
        <artifactId>jackson-module-parameter-names</artifactId>
    </dependency>
</dependencies>
```

Jackson2 库是非常流行的处理 JSON 的类库。只要将 Jackson2 放在 classpath 上，Spring Boot 应用程序中的任何使用了 @RestController 注解的类，都会默认呈现 JSON 格式数据的响应。

3.2.3 Web API 常用调试方式

在上面的例子中，直接在浏览器中输入 API 的地址就能看到数据的返回。但这种测试方式有一定的局限性，就是只能测试 GET 方法。如果有一个请求是 POST、PUT 或者是 DELETE 方法，那么这种测试方式就无能为力了。

1. Firefox 安装 REST 客户端插件

为了方便测试 REST API，需要一款 REST 客户端来协助。由于这里用 Firefox 浏览器居多，因此推荐安装 RESTClient 或者是 HttpRequester 插件。当然，可以根据个人喜好来安装其他软件。

在 Firefox 安装插件的界面中输入关键字“restclient”，就能看到这两款插件的信息。单击“安装”按钮即可，如图 3-2 所示。

图3-2 Firefox 安装 REST 客户端插件

2. 用 HttpRequester 来测试

运行程序后，可以对 http://localhost:8080/users/1 接口进行测试。

在 HttpRequester 中的请求 URL 中填写接口地址，而后单击“Submit”按钮来提交测试请求。在右侧响应中，能看到返回的 JSON 数据。图 3-3 所示为 HttpRequester 的使用过程。

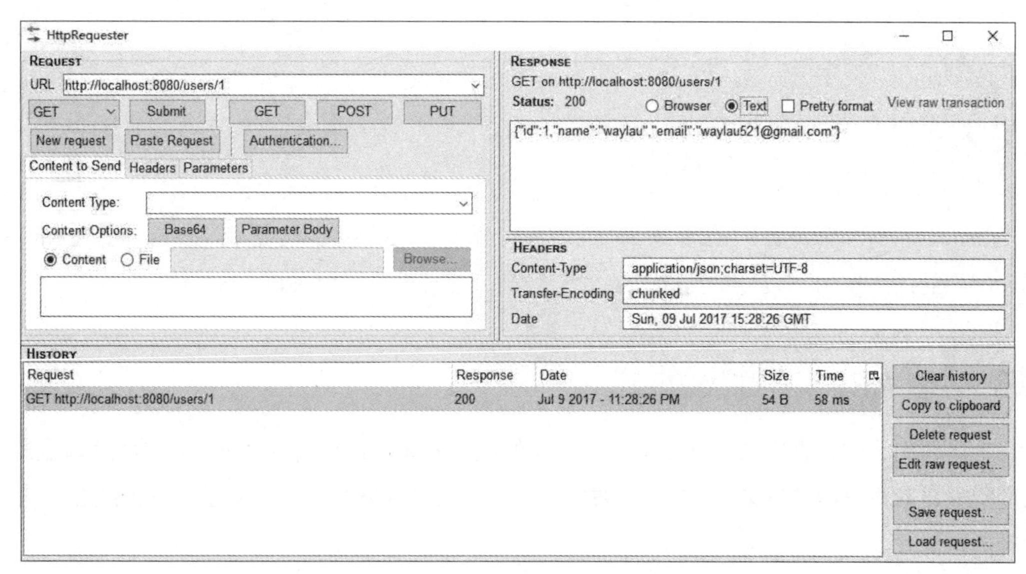

图3-3 HttpRequester 的使用

3.2.4 示例源码

本节示例源码在 media-type-json 目录下。

3.3 XML 类型的处理

XML（eXtensible Markup Language）是 Web Services 的标准数据交换格式。虽然在当前的互联网应用中，越来越倾向于使用 JSON 格式，但大部分历史遗留项目，或者与第三方系统进行对接时，仍然有机会会采用 XML，所以这里仍然将 XML 类型的处理做相关的讲解。在 media-type-json 项目源码的基础上，创建了一个 media-type-xml 项目作为演示案例。

3.3.1 返回 XML 类型

项目的依赖 spring-boot-starter-web 并没有默认对 XML 格式进行处理。所以，要支持 XML 格式的数据响应，还需要添加相关的支持。

1. 添加 JAXB

修改 User，添加 JAXB（Java Architecture for XML Binding）技术 。JAXB 是一个业界的标准，是一项可以根据 XML Schema 产生 Java 类的技术。该过程中，JAXB 也提供了将 XML 实例文档反向生成 Java 对象树的方法，并能将 Java 对象树的内容重新写到 XML 实例文档。从另一方面来讲，JAXB 提供了快速而简便的方法将 XML 模式绑定到 Java 表示，从而使 Java 开发者在 Java 应用程序中能方便地结合 XML 数据和处理函数。

JAXB 常用的注解包括 @XmlRootElement、@XmlElement 等。在 User 类上添加 @XmlRootElement 注解，以便将 User 映射为 XML。

JAXB 是 JDK 原生提供了 API，所以无须添加其他类库。

2. 修改 User 类

修改 User.java 代码如下。

```
import javax.xml.bind.annotation.XmlRootElement;

@XmlRootElement // 类转为XML
public class User {
    ...
}
```

3. 用 HttpRequester 来测试

运行程序后，可以对 http://localhost:8080/users/1 接口进行测试。

在 HttpRequester 中的请求 URL 中填写接口地址，并在 Headers 里添加额外的参数“Accept”值为“text/xml”，这样就能告诉后台接口想要的是 XML 数据。而后单击“Submit”按钮来提交测试请求。在右侧响应中能看到返回的 XML 数据。图 3-4 所示为 HttpRequester 的使用过程。

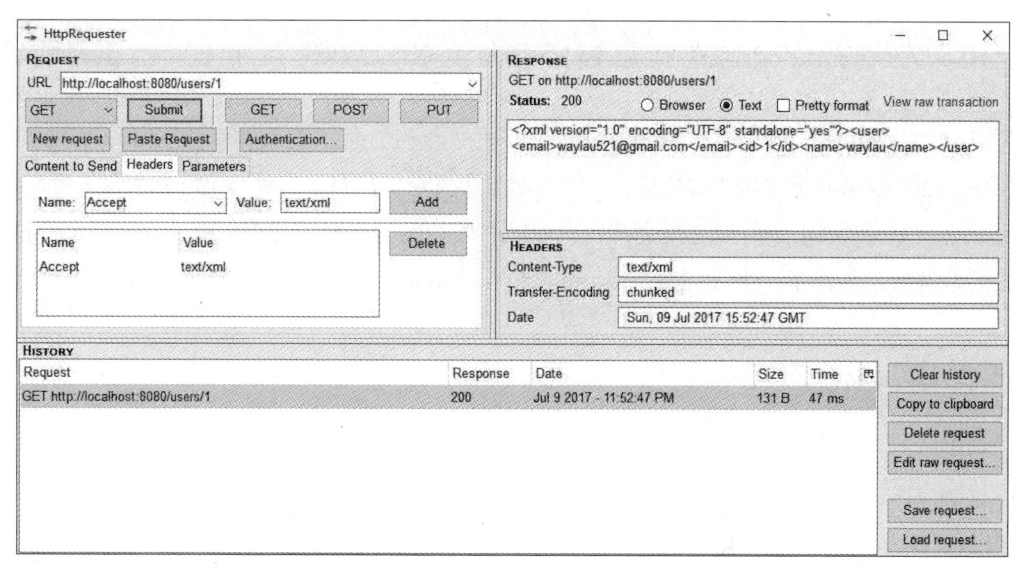

图3-4 使用 HttpRequester 进行测试

3.3.2 第三方 XML 框架

除了 JAXB 外，很多第三方框架也提供了 XML 的支持。例如，如果在类路径中具有 Jackson XML 扩展名（jackson-dataformat-xml），则将使用它来渲染 XML 响应，这样，就可以无须对代码做任何修改，直接用于 3.3.1 节中的 media-type-json 的例子了。要使用它，需要在项目中添加以下依赖。

```
<dependency>
    <groupId>com.fasterxml.jackson.dataformat</groupId>
    <artifactId>jackson-dataformat-xml</artifactId>
</dependency>
```

另一个比较不错的类库是 Woodstox，同样是一个快速开源且符合 STAX（STreaming Api for Xml processing）规范的 XML 处理器，并且还添加了打印支持和改进的命名空间处理。要使用它，需要在项目中添加以下依赖。

```
<dependency>
    <groupId>org.codehaus.woodstox</groupId>
    <artifactId>woodstox-core-asl</artifactId>
</dependency>
```

3.3.3 示例源码

本节示例源码在 media-type-xml 目录下。

3.4 文件上传的处理

Spring 框架内置支持处理 Web 应用程序中的文件上传（multipart）的功能。可以使用 org.springframework.web.multipart 包中定义的可插拔 MultipartResolver 对象来启用此功能。

3.4.1 MultipartResolver

MultipartResolver 的实现方式主要有以下两种。

1. CommonsMultipartResolver

CommonsMultipartResolver 主要依赖于 Apache 提供的 Commons FileUpload 库。

下面演示了如何使用 CommonsMultipartResolver。

```
<bean id="multipartResolver"
    class="org.springframework.web.multipart.commons.CommonsMultipar-
tResolver">

    <!-- 单个文件的最大值（单位：字节） -->
    <property name="maxUploadSize" value="100000"/>

</bean>
```

当然，为了要使用 CommonsMultipartResolver 来解析，需要确保 commons-fileupload.jar 放置在类路径中。

当 Spring DispatcherServlet 检测到文件上传的请求时，它将激活已在上下文中声明的解析器来处理请求。然后，解析器将当前的 HttpServletRequest 包装到支持文件上传的 MultipartHttpServlet Request 中。

2. StandardServletMultipartResolver

StandardServletMultipartResolver 主要是依赖于 Servlet 3.0 multipart 提供的解析，所以，要想使用 StandardServletMultipartResolver，需要在 web.xml 中将 DispatcherServlet 标记为“multipart-config”，或者在编程式 Servlet 注册中使用 javax.servlet.MultipartConfigElement，或者在定制 Servlet 类的情况下，可以在 Servlet 类上使用 javax.servlet.annotation.MultipartConfig 注解。

一旦使用上述方法之一启用 Servlet 3.0 multipart 解析，就可以将 StandardServletMultipart Resolver 添加到 Spring 配置中。

```
<bean id="multipartResolver"
    class="org.springframework.web.multipart.support.StandardServlet-
MultipartResolver">
</bean>
```

需要注意的是，Servlet 3.0 不允许从 MultipartResolver 中来设置上传文件的大小等参数，所以，需要在 Servlet 注册级别来设置。

3.4.2 通过 Form 表单来上传文件

通过 Form 表单来上传文件主要有以下几个步骤。

1. 创建带有文件输入的 Form 表单

创建一个带有文件输入的 Form 表单，来允许用户上传表单。设置编码属性（enctype="multipart/form-data"），让浏览器知道如何将表单编码为 multipart 请求。

```
<html>
    <head>
        <title>上传文件</title>
    </head>
    <body>
        <h1>请上传文件</h1>
        <form method="post" action="/form" enctype="multipart/form-da-
ta">
                <input type="text" name="name"/>
                <input type="file" name="file"/>
                <input type="submit"/>
        </form>
    </body>
</html>
```

2. 创建处理文件请求的控制器

像平常的控制器一样，这个控制器也要有 @Controller 注解，同时，在方法的参数里使用 MultipartHttpServletRequest 或者 MultipartFile 参数。

```
@Controller
public class FileUploadController {

    @PostMapping("/form")
    public String handleFormUpload(@RequestParam("name") String name,
                    @RequestParam("file") MultipartFile file) {

        if (!file.isEmpty()) {
                byte[] bytes = file.getBytes();
                // 省略存储字节的步骤
                return "redirect:uploadSuccess";
        }

        return "redirect:uploadFailure";
    }

}
```

在这个例子中，笔者没有给出 byte[] 后期的操作步骤。在实际的项目中，读者可以考虑将其保存在数据库中，或将其存储在文件系统上。

当使用 Servlet 3.0 multipart 解析时，还可以使用 javax.servlet.http.Part 来作为方法参数。

```
@Controller
public class FileUploadController {

    @PostMapping("/form")
    public String handleFormUpload(@RequestParam("name") String name,
                    @RequestParam("file") Part file) {

        InputStream inputStream = file.getInputStream();
        // 省略存储字节的步骤

        return "redirect:uploadSuccess";
    }

}
```

3.4.3 RESTful API 的文件上传

随着 RESTful API 的流行，很多 REST 客户端都会有上传文件的需要。这些 REST 客户端也包括了非浏览器的设备。与通常提交单表来上传文件的浏览器不同，REST 客户端还可以发送特定内容类型的更复杂的数据，例如，除了上传文件的请求外，第二部分还使用了 JSON 格式的数据。

```
POST /someUrl
Content-Type: multipart/mixed

--edt7Tfrdusa7r3lNQc79vXuhIIMlatb7PQg7Vp
Content-Disposition: form-data; name="meta-data"
Content-Type: application/json; charset=UTF-8
Content-Transfer-Encoding: 8bit

{
    "name": "value"
}
--edt7Tfrdusa7r3lNQc79vXuhIIMlatb7PQg7Vp
Content-Disposition: form-data; name="file-data"; filename="file.proper-
ties"
Content-Type: text/xml
Content-Transfer-Encoding: 8bit
... File Data ...
```

这里为了处理方法参数使用了 @RequestPart 注解，而不是 @RequestParam 注解。@RequestPart 注解允许通过 HttpMessageConverter 来传递请求头中带有“Content-Type”参数的特定内容。

```
@PostMapping("/upload")
public String onSubmit(@RequestPart("meta-data") MetaData metadata,
                @RequestPart("file-data") MultipartFile file) {
    // 省略处理逻辑
}
```

在本例中，@RequestPart("meta-data") MetaData metadata 方法参数将根据其“Content-Type”头的标识读取为 JSON 内容，并在 MappingJackson2HttpMessageConverter 的帮助下进行转换。

3.4.4 完整示例

本节仅仅介绍了 Spring MVC 处理文件上传的基本流程和概念，有关文件上传的完整示例会在第 15 章中做详细讲解。

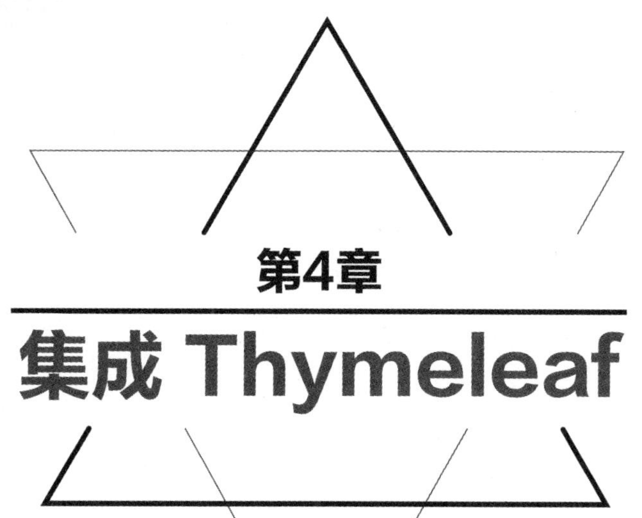

第4章 集成 Thymeleaf

4.1 常用 Java 模板引擎

对于动态 HTML 内容的展示，模板引擎必不可少。Java 主要通过 Servlet 来支持动态内容的请求和响应。Spring MVC 支持 Thymeleaf 、FreeMarker 和 JSP 等多种技术。对于 Spring Boot 而言，它支持 FreeMarker、Groovy、Thymeleaf、Mustache 等引擎的自动配置功能。

实际上，Java 模板引擎品种繁多，各有各的优势和缺点。

4.1.1 关于性能

模板引擎的性能是很多用户选择模板引擎的重要参考指标。Jeroen Reijn 对各个 Java 模板引擎做了性能分析，并将结果发表在了其个人网站上。下面来看一下性能指标的对比数据。

Jeroen Reijn 选取了以下模板作为测试的版本，这些模板引擎都是结合 Spring MVC 来工作。

- JSP + JSTL - v1.2。
- Freemarker - v2.3.25。
- Velocity - v1.7。
- Thymeleaf - v3.0.2.RELEASE。
- Mustache - v1.13。
- Scalate - v1.7.1。
- Jade - v1.2.5。
- HTTL - v1.0.11。
- Pebble - v2.2.3。
- Handlebars - v4.0.6。
- jtwig - v3.1.1。

在处理 25 000 个并发级别为 25 的请求时所需的总时间，测试结果如下所示（数值越小越好）。

```
jTwig                        4.709 seconds
Thymeleaf                    4.147 seconds
Scalate - Scaml              3.479 seconds
Handlebars                   2.936 seconds
Jade4j                       2.735 seconds
Freemarker                   2.637 seconds
HTTL                         2.531 seconds
Pebble                       2.512 seconds
Velocity                     2.491 seconds
Mustache (JMustache)         2.326 seconds
JSP                          2.227 seconds
```

从测试结果中可以看到，JSP 的性能是最高的，而 jTwig 和 Thymeleaf 相对较差。

4.1.2 为什么选择 Thymeleaf

既然 JSP 是公认的性能最好的模板引擎，为什么这些年来新的模板引擎层出不穷？为什么在实际项目中反而更加推荐使用性能相对较差的 Thymeleaf？

其实，对于开发者而言，除了从性能来考量一门语言或者一个工具外，还要考虑人的因素，即开发人员本身的开发效率，这就好比是用汇编语言和 Java 语言做比较，很显然，汇编语言在程序处理的性能上肯定会比 Java 高很多，但从开发人员的角度来看，显然 Java 语言更加符合面向对象的思维，更加容易被开发人员所掌握，并且能够最快地实现功能，推出产品。所以在衡量一款工具优劣的时候，往往需要从整体来看。

下面来对比一下 JSP 的代码及 Thymeleaf 的代码。下面是 JSP 编写的页面代码。

```
<%@ taglib prefix="sf" uri="http://www.springframework.org/tags/form" %>
<%@ taglib prefix="s" uri="http://www.springframework.org/tags" %>
<%@ taglib prefix="c" uri="http://java.sun.com/jsp/jstl/core" %>
<%@ page contentType="text/html; charset=UTF-8" pageEncoding="UTF-8"%>
<!DOCTYPE html>

<html>

  <head>
    <title>Spring MVC view layer: Thymeleaf vs. JSP</title>
    <meta http-equiv="Content-Type" content="text/html; charset=UTF-8"
/>
    <link rel="stylesheet" type="text/css" media="all" href="<s:url val-
ue='/css/thvsjsp.css' />"/>
  </head>

  <body>

    <h2>This is a JSP</h2>

    <s:url var="formUrl" value="/subscribejsp" />
    <sf:form modelAttribute="subscription" action="${formUrl}">

      <fieldset>

        <div>
          <label for="email"><s:message code="subscription.email" />:
</label>
          <sf:input path="email" />
        </div>
        <div>
          <label><s:message code="subscription.type" />: </label>
          <ul>
            <c:forEach var="type" items="${allTypes}" varStatus="-
typeStatus">
```

```
                <li>
                  <sf:radiobutton path="subscriptionType" value="${type}"
/>
                  <label for="subscriptionType${typeStatus.count}">
                    <s:message code="subscriptionType.${type}" />
                  </label>
                </li>
              </c:forEach>
            </ul>
          </div>

          <div class="submit">
            <button type="submit" name="save"><s:message code="subscrip-
tion.submit" /></button>
          </div>

        </fieldset>

      </sf:form>

    </body>

</html>
```

实现相同的功能，下面采用 Thymeleaf 来实现。

```
<!DOCTYPE html>

<html xmlns:th="http://www.thymeleaf.org">

  <head>
    <title>Spring MVC view layer: Thymeleaf vs. JSP</title>
    <meta http-equiv="Content-Type" content="text/html; charset=UTF-8"
/>
    <link rel="stylesheet" type="text/css" media="all"
      href="../../css/thvsjsp.css" th:href="@{/css/thvsjsp.css}"/>
  </head>

  <body>

    <h2>This is a Thymeleaf template</h2>

    <form action="#" th:object="${subscription}" th:action="@{/subscri-
beth}">

      <fieldset>

        <div>
          <label for="email" th:text="#{subscription.email}">Email: </
label>
```

```
          <input type="text" th:field="*{email}" />
        </div>
        <div>
          <label th:text="#{subscription.type}">Type: </label>
          <ul>
            <li th:each="type : ${allTypes}">
              <input type="radio" th:field="*{subscriptionType}" th:val-
ue="${type}" />
              <label th:for="${#ids.prev('subscriptionType')}"
                th:text="#{|subscriptionType.${type}|}">First type</
label>
            </li>
            <li th:remove="all"><input type="radio" /> <label>Second
Type</label></li>
          </ul>
        </div>

        <div class="submit">
          <button type="submit" name="save" th:text="#{subscription.
submit}">Subscribe me!</button>
        </div>

      </fieldset>

    </form>

  </body>

</html>
```

对比 JSP 的代码及 Thymeleaf 的代码可以看出以下几点。

（1）Thymeleaf 比 JSP 的代码更加接近 HTML，没有奇怪的标签，只是增加了一些有意义的属性。

（2）Thymeleaf 支持 HTML5 标准；JSP 如果要支持 HTML5 标准，需要新版的 Spring 框架来支持。

（3）JSP 需要部署到 Servlet 开发服务器上，并启动服务器。如果服务器不启动，JSP 页面不会渲染；而 Thymeleaf 即使不部署，也能直接在浏览器中打开它。

虽然，Thymeleaf 的性能不是最好的，但由于 Thymeleaf “原型即页面”的特点，非常适用于快速开发，符合 Spring Boot 开箱即用的原则。所谓“原型即页面”，是指 Thymeleaf 页面无须部署到 Servlet 开发服务器上，直接通过浏览器就能打开它。这种特点非常适合用于系统的界面原型设计。前端开发人员或者美工将原型设计好之后，提交给 Java 开发人员，只需要在原型的基础上增加少量的 Thymeleaf 表达式语句，就能转化为系统的页面了。

界面的设计与实现相分离，这就是 Thymeleaf 广为流行的原因。

4.1.3 Thymeleaf的概念

Thymeleaf 是面向 Web 和独立环境的现代服务器端 Java 模板引擎，能够处理 HTML、XML、JavaScript、CSS 甚至纯文本。类似的产品还有 JSP、Freemarker 等。

Thymeleaf 的主要目标是提供一个优雅和高度可维护的创建模板的方式。为了实现这一点，它建立在自然模板（Natural Templates）的概念上，将其逻辑注入模板文件中，不会影响模板被用作设计原型。这改善了设计的沟通，弥合了设计和开发团队之间的差距。

Thymeleaf 的设计从一开始就遵从 Web 标准，特别是 HTML5，这样就能创建完全符合验证的模板。

Thymeleaf 的语法优雅易懂。Thymeleaf 使用 OGNL（Object Graph Navigation Language），它是一种功能强大的表达式语言，通过它简单一致的表达式语法，可以存取对象的任意属性，调用对象的方法，遍历整个对象的结构图，实现字段类型转化等功能。它使用相同的表达式去存取对象的属性，这样可以更好地取得数据。在与 Spring 应用集成过程中使用 SpringEL，而 OGNL 与 SpringEL 在语法上都极其类似。

Thymeleaf 3 使用一个名为 AttoParser 2 （http://www.attoparser.org）的新解析器。AttoParser 是一个新的、基于事件（不符合 SAX 标准）的解析器，由 Thymeleaf 的作者开发，符合 Thymeleaf 的风格。

4.1.4 Thymeleaf 处理模板

Thymeleaf 能处理以下 6 种类型的模板，称为模板模式（Template Mode）。

- HTML。
- XML。
- TEXT。
- JAVASCRIPT。
- CSS。
- RAW。

其中包含有两种标记模板模式（HTML 和 XML）、三种文本模板模式（TEXT、JAVASCRIPT 和 CSS）和一个无操作模板模式（RAW）。

HTML 模板模式将允许任何类型的 HTML 输入，包括 HTML5、HTML4 和 XHTML。将不执行验证或对格式进行严格检查，这样就能尽可能地将模板代码 / 结构进行输出。

XML 模板模式将允许 XML 输入。在这种情况下，代码预期格式是良好的，不存在未关闭的标签和没有引用的属性等。如果找到未符合 XML 格式的要求，解析器将抛出异常。需要注意的是，该模式不会针对 DTD 或 XML 架构去执行验证。

TEXT 模板模式将允许对非标记性质的模板使用特殊语法，此类模板的示例可能是文本电子邮件或模板文档。注意，HTML 或 XML 模板也可以作为 TEXT 处理，在这种情况下，它们将不会被解析为标记，并且每个标签 DOCTYPE、注释等将被视为纯文本。

JAVASCRIPT 模板模式将允许在 Thymeleaf 应用程序中处理 JavaScript 文件，这意味着，JAVASCRIPT 模板模式不仅可以像 HTML 一样的使用模型数据，而且拥有额外的特定于 JavaScript 的集成功能，如专门的转义或自然脚本（Natural Scripting）。

由于 JAVASCRIPT 模板模式也是文本模式，因此使用与 TEXT 模板模式相同的语法。

CSS 模板模式将允许处理涉及 Thymeleaf 应用程序的 CSS 文件。与 JAVASCRIPT 模式类似，CSS 模板模式也是文本模式，并使用 TEXT 模板模式下的特殊处理语法。

RAW 模板模式根本不会处理模板，它用于将未经修改的资源（文件、URL 响应等）插入正在处理的模板中。例如，HTML 格式的外部不受控制的资源可以包含在应用程序模板中，这些资源可能包含的任何 Thymeleaf 代码将不会被执行。

4.1.5 标准方言

Thymeleaf 是一个非常可扩展的模板引擎，实际上它更像是一个模板引擎框架（Template Engine Framework），允许开发者定义和自定义自己的模板。

将一些逻辑应用于标记工件（如标签、某些文本、注释或只有占位符）的一个对象称为处理器（Processor）。方言（Dialect）通常包括这些处理器的集合及一些额外的工件。Thymeleaf 的核心库提供了一种称为标准方言（Standard Dialect）的方言，提供给用户开箱即用的功能。

当然，如果用户希望在利用库的高级功能的同时定义自己的处理逻辑，用户也可以创建自己的方言（甚至扩展标准的方言），也可以将 Thymeleaf 配置为同时使用几种方言。

官方的 thymeleaf-spring3 和 thymeleaf-spring4 集成包都定义了一种称为“SpringStandard Dialect”的方言，与标准方言大致相同，但是对于 Spring 框架中的某些功能则更加友好，例如，想通过使用 Spring Expression Language 或 SpringEL，而不是 OGNL。所以如果开发者是一个 Spring MVC 用户，这里的所有东西都能够在自己的 Spring 应用程序中使用。

标准方言的大多数处理器是属性处理器。这样，即使在处理之前，浏览器也可以正确地显示 HTML 模板文件，因为它们将简单地忽略其他属性。对比 JSP，在浏览器中会直接显示的代码片断如下。

```
<form:inputText name="userName" value="${user.name}" />
```

Thymeleaf 标准方言将允许开发者实现与以下代码相同的功能。

```
<input type="text" name="userName" value="James Carrot" th:value="${us-
er.name}" />
```

浏览器不仅可以正确显示这些信息，而且还可以（可选）在浏览器中静态打开原型时显示的值（可选）指定一个值属性（在这种情况下为“James Carrot”），将在模板处理期间由 ${user.name} 的评估得到的值代替。

这有助于设计师和开发人员处理相同的模板文件，并减少将静态原型转换为工作模板文件所需的工作量。这样的功能称为“自然模板”（Natural Templating）的功能。

4.2 Thymeleaf 标准方言

Thymeleaf 的可扩展性很强，它允许用户自定义模板属性集（或事件标签）、表达式、语法及应用逻辑，它更像一个模板引擎框架（Template Engine Framework），可以方便地基于该引擎来自定义模板。

当然，Thymeleaf 的优秀之处并不仅限于此。秉着“开箱即用”的原则，Thymeleaf 提供了满足大多数使用情况的默认实现——标准方言（Standard Dialects），涵盖了命名为 Standard 和 SpringStandard 的两种方言。在模板中，可以很容易地识别出这些被使用的标准方言，因为它们都以 th 属性开头，例如，

```
<span th:text="...">
```

值得注意的是，Standard 和 SpringStandard 方言在用法上几乎相同，不同之处在于 SpringStandard 包括了 Spring MVC 集成的具体特征（如用 Spring Expression Language 来替代 OGNL）。

通常，在谈论 Thymeleaf 的标准方言时，一般引用的是 Standard， 而不涉及特例。

本节的内容大多引用自笔者的开源书《Thymeleaf 教程》，有兴趣的读者可以自行参阅，书籍地址为 https://github.com/waylau/thymeleaf-tutorial。

4.2.1 Thymeleaf 标准表达式语法

大多数 Thymeleaf 属性允许设值或者包含表达式（Expressions），由于它们使用方言的关系，因此被称为标准表达式（Standard Expressions）。这些标准表达式语法主要有以下几种。

1. 简单表达式

（1）Variable Expressions（变量表达式）：${...}。

（2）Selection Expressions（选择表达式）：*{...}。

（3）Message (i18n) Expressions（消息表达式）：#{...}。

（4）Link (URL) Expressions（链接表达式）：@{...}。

（5）Fragment Expressions（分段表达式）：~{...}。

2. 字面量

（1）文本：'one text' 'Another one!' 等。

（2）数值：0、34、3.0、12.3 等。

（3）布尔：true、false。

（4）null：null。

（5）Literal Token（字面量标记）: one、sometext、 main 等。

3. 文本操作

（1）字符串拼接：+。

（2）文本替换：|The name is ${name}|。

4. 算术操作

（1）二元运算符：+、–、 *、/、%。

（2）减号（单目运算符）：–。

5. 布尔操作

（1）二元运算符：and、or。

（2）布尔否定（一元运算符）：!、not。

6. 比较和等价

（1）比较：>、<、>=、<=（gt、lt、ge、le）。

（2）等价：==、!=（eq、ne）。

7. 条件运算符

（1）If-then：(if) ? (then)。

（2）If-then-else：(if) ? (then) : (else)。

（3）Default：(value) ?: (defaultvalue)。

8. 特殊标记

No-Operation（无操作）：_。

下面的这个示例涵盖了上述大部分表达式。

```
'User is of type ' + (${user.isAdmin()} ? 'Administrator' : (${user.
type} ?: 'Unknown'))
```

4.2.2 消息表达式（Message (i18n) Expressions）

消息表达式（通常称为文本外化、国际化或 i18n）允许开发者从外部源（.properties 文件）检索特定于语言环境的消息，通过 key 引用它们（可选）应用一组参数。

在 Spring 应用程序中，这将自动与 Spring 的 MessageSource 机制集成。

```
#{main.title}

#{message.entrycreated(${entryId})}
```

在模板中的应用如下。

```
<table>
  ...
  <th th:text="#{header.address.city}">...</th>
  <th th:text="#{header.address.country}">...</th>
  ...
</table>
```

注意：如果希望消息 key 由上下文变量的值确定，或者要将变量指定为参数，则可以在消息表达式中使用变量表达式。

```
#{${config.adminWelcomeKey}(${session.user.name})}
```

4.2.3 变量表达式（Variable Expressions）

变量表达式可以是 OGNL 表达式或者 Spring EL，如果集成了 Spring，就可以在上下文变量（Context Variables）中执行。

有关 OGNL 语法和功能的详细信息，请阅读《OGNL 语言指南》（http://commons.apache.org/proper/commons-ognl/）。在 Spring MVC 启用的应用程序中，OGNL 将被替换为 SpringEL，但其语法与 OGNL 非常相似（实际上，在大多数常见情况下完全相同）。

在 Spring 术语中，变量表达式也称为模型属性（Model Attributes），它们看起来像这样。

```
${session.user.name}
```

它们作为属性值或作为属性的一部分。

```
<span th:text="${book.author.name}">
```

上面的表达式在 OGNL 和 SpringEL 中等价于：

```
((Book)context.getVariable("book")).getAuthor().getName()
```

这些变量表达式不仅涉及输出，还包括更复杂的处理，如条件判断、迭代等。

```
<li th:each="book : ${books}">
```

这里 ${books} 从上下文中选择名为 books 的变量，并将其评估为可在 th:each 循环中使用的迭代器（iterable）。

更多 OGNL 的功能如下。

```
/*
 * 使用点(.)来访问属性，等价于调用属性的getter
 */
${person.father.name}

/*
 * 访问属性也可以使用([])块
 */
${person['father']['name']}

/*
 * 如果对象是一个map，则点和块语法等价于调用其get(...)方法
 */
${countriesByCode.ES}
${personsByName['Stephen Zucchini'].age}

/*
 * 在块语法中也可以通过索引来访问数组或者集合
 */
${personsArray[0].name}

/*
 * 可以调用方法同时也支持参数
 */
${person.createCompleteName()}
${person.createCompleteNameWithSeparator('-')}
```

4.2.4 表达式基本对象（Expression Basic Objects）

当对上下文变量评估 OGNL 表达式时，某些对象可用于表达式以获得更高的灵活性，这些对象将被引用（按照 OGNL 标准），从#符号开始。

- #ctx：上下文对象。
- #vars：上下文变量。
- #locale：上下文区域设置。
- #request：HttpServletRequest 对象（仅在 Web 上下文中）。
- #response：HttpServletResponse 对象（仅在 Web 上下文中）。
- #session：HttpSession 对象（仅在 Web 上下文中）。
- #servletContext：ServletContext 对象（仅在 Web 上下文中）。

所以可以这样做。

```
Established locale country: <span th:text="${#locale.country}">US</span>.
```

有关表达式基本对象的完整内容可以参考 4.6 节的内容。

4.2.5 表达式工具对象（Expression Utility Objects）

除了上面这些基本的对象之外，Thymeleaf 提供了一组工具对象，这些对象将帮助我们在表达式中执行常见任务。

- #execInfo：模板执行的信息。
- #messages：在变量内获取外部消息的方法表达式，与使用 # {...} 语法获得的方式相同。
- #uris：用于转义 URL/URI 部分的方法。
- #conversions：执行已配置的 conversion service。
- #dates：java.util.Date 对象的方法，如格式化、组件提取等。
- #calendars：类似于 # dates，但是对应于 java.util.Calendar 对象。
- #numbers：格式化数字对象的方法。
- #strings：String 对象的方法，包括 contains、startsWith、prepending/appending 等。
- #objects：对象通常的方法。
- #bools：布尔判断的方法。
- #arrays：array 方法。
- #lists：list 方法。
- #sets：set 方法。
- #maps：map 方法。
- #aggregates：在数组或集合上创建聚合的方法。
- #ids：用于处理可能重复的 id 属性的方法（如作为迭代的结果）。

下面是一个格式化日期的例子。

```
<p>
  Today is: <span th:text="${#calendars.format(today,'dd MMMM yyyy')}">
13 May 2011</span>
</p>
```

有关表达式工具对象的完整内容可以参考“附录 C：Thymeleaf 表达式工具对象”部分。

4.2.6 选择表达式（Selection Expressions）

选择表达式与变量表达式很像，区别在于，它们是在当前选择的对象上执行的，而不是在整个上下文变量映射上执行的，它们看起来像这样。

```
*{customer.name}
```

它们所作用的对象由 th:object 属性指定。

```
<div th:object="${book}">
  ...
  <span th:text="*{title}">...</span>
  ...
</div>
```

这等价于：

```
{
  // th:object="${book}"
  final Book selection = (Book) context.getVariable("book");
  // th:text="*{title}"
  output(selection.getTitle());
}
```

4.2.7 链接表达式（Link (URL) Expressions）

链接表达式旨在构建URL并向其添加有用的上下文和会话信息（通常称为URL重写的过程）。因此，对于部署在 Web 服务器的 /myapp 上下文中的 Web 应用程序，可以使用以下表达式。

```
<a th:href="@{/order/list}">...</a>
```

可以转成：

```
<a href="/myapp/order/list">...</a>
```

cookie 没有启用下，如果需要保持会话，可以这样。

```
<a href="/myapp/order/list;jsessionid=23fa31abd41ea093">...</a>
```

URL 是可以携带参数的。

```
<a th:href="@{/order/details(id=${orderId},type=${orderType})}">...</a>
```

最终的结果如下。

```
<a href="/myapp/order/details?id=23&amp;type=online">...</a>
```

链接表达式可以是相对的，在这种情况下，应用程序上下文将不会作为 URL 的前缀。

```
<a th:href="@{../documents/report}">...</a>
```

也可以是服务器相对（同样，没有应用程序上下文前缀）。

```
<a th:href="@{~/contents/main}">...</a>
```

可以是协议相对（就像绝对 URL，但浏览器将使用在显示的页面中相同的 HTTP 或 HTTPS 协议）。

```
<a th:href="@{//static.mycompany.com/res/initial}">...</a>
```

当然，Link 表达式可以是绝对的。

```
<a th:href="@{http://www.mycompany.com/main}">...</a>
```

在绝对（或协议相对）的 URL 等中 Thymeleaf 链接表达式添加的是什么值？答案是，可能是由响应过滤器定义的 URL 重写。在基于 Servlet 的 Web 应用程序中，对于每个输出的 URL（上下文相对、相对、绝对……）Thymeleaf 将总是在显示 URL 之前调用 HttpServletResponse.encodeUrl(...) 机制，这意味着过滤器可以通过包装 HttpServletResponse 对象（通常使用的机制）来为应用程序执行定制的 URL 重写。

4.2.8 分段表达式（Fragment Expressions）

分段表达式是 Thymeleaf 3.x 版本新增的内容。

分段表达式是一种表示标记片段并将其移动到模板周围的简单方法，正是由于这些表达式，片段可以被复制，或者作为参数传递给其他模板等。

最常见的用法是使用 th:insert 或 th:replace 插入片段。

```
<div th:insert="~{commons :: main}">...</div>
```

但是它们可以在任何地方使用，就像任何其他变量一样。

```
<div th:with="frag=~{footer :: #main/text()}">
  <p th:insert="${frag}">
</div>
```

分段表达式是可以有参数的。

分段表达式相关内容可以参考 4.5 节的内容。

4.2.9 字面量（Literals）

Thymeleaf 有一组可用的字面量和操作。

1. 文本

文本文字只是在单引号之间指定的字符串，它们可以包含任何字符，但应该避免其中的任何单引号使用。

```
<p>
  Now you are looking at a <span th:text="'working web application'">
```

```
template file</span>.
</p>
```

2. 数值

数值文字就是数字。

```
<p>The year is <span th:text="2013">1492</span>.</p>
<p>In two years, it will be <span th:text="2013 + 2">1494</span>.</p>
```

3. 布尔

布尔文字为“true”和“false”。例如，

```
<div th:if="${user.isAdmin()} == false"> ...
```

在这个例子中，“== false”写在大括号之外，Thymeleaf 会做处理。如果是写在大括号内，那就是由 OGNL/SpringEL 引擎负责处理。

```
<div th:if="${user.isAdmin() == false}"> ...
```

4. null

null 字面量使用如下。

```
<div th:if="${variable.something} == null"> ...
```

5. 字面量标记

数字、布尔和 null 字面实际上是“字面量标记”（Literal Token）的特殊情况。这些标记允许在标准表达式中进行一点简化。它们工作与文本文字（'...'）完全相同，但只允许使用字母（A-Z 和 a-z）、数字（0-9）、括号（[和]）、点（.）、连字符（-）与下画线（_），所以没有空白、逗号等。

标记无须任何引号，所以可以这样做。

```
<div th:class="content">...</div>
```

用来代替：

```
<div th:class="'content'">...</div>
```

6. 附加文本

无论是文字，还是评估变量或消息表达式的结果，都可以使用“+”操作符轻松地附加文本。

```
<span th:text="'The name of the user is ' + ${user.name}">
```

7. 字面量替换

字面量替换允许容易地格式化包含变量值的字符串，而不需要使用‘...’+‘...’附加文字。

这些替换必须被“|”包围，例如，

```
<span th:text="|Welcome to our application, ${user.name}!|">
```

其等价于：

```
<span th:text="'Welcome to our application, ' + ${user.name} + '!'">
```

字面量替换可以与其他类型的表达式相结合。

```
<span th:text="${onevar} + ' ' + |${twovar}, ${threevar}|">
```

|...| 字面量替换只允许使用变量表达式或消息表达式（${...}, *{...}, #{...}），其他字面量（'...'）、布尔或数字标记、条件表达式等是不允许的。

4.2.10 算术运算

支持算术运算：+, –, *, / 和 %。

```
<div th:with="isEven=(${prodStat.count} % 2 == 0)">
```

注意：这些运算符也可以在 OGNL 变量表达式本身中应用（在这种情况下将由 OGNL 执行，而不是 Thymeleaf 标准表达式引擎）。

```
<div th:with="isEven=${prodStat.count % 2 == 0}">
```

注意：其中一些运算符存在文本别名：div（/）、mod（%）。

4.2.11 比较与相等

表达式中的值可以与 >、<、>= 和 <= 进行比较，并且可以使用 == 和 != 运算符来检查是否相等。

注意：< 和 > 不应该在 XML 属性值中使用，因此它们应被替换为 < 和 >。

```
<div th:if="${prodStat.count} &gt; 1">
<span th:text="'Execution mode is ' + ( (${execMode} == 'dev')? 'Devel-
opment' : 'Production')">
```

一个更简单的替代方案可能是使用一些这些运算符存在的文本别名：gt (>)、lt (<)、ge (>=)、le (<=)、not (!). eq (==)、neq/ne (!=)。

4.2.12 条件表达式

条件表达式仅用于评估两个表达式中的一个，这取决于评估条件（本身就是另一个表达式）的结果。

下面来看一个示例 th:class 片段 。

```
<tr th:class="${row.even}? 'even' : 'odd'">
  ...
```

```
</tr>
```

条件表达式（condition、then 和 else）的 3 个部分都是自己的表达式，这意味着它们可以是变量（${...}, *{...}），消息（#{...}），（@{...}）或字面量（'...'）。

条件表达式也可以使用括号嵌套。

```
<tr th:class="${row.even}? (${row.first}? 'first' : 'even') : 'odd'">
  ...
</tr>
```

else 表达式也可以省略，在这种情况下，如果条件为 false，则返回 null 值。

```
<tr th:class="${row.even}? 'alt'">
  ...
</tr>
```

4.2.13 默认表达式（Elvis Operator）

默认表达式（Default Expression）是一种特殊的条件值，没有 then 部分，它相当于某些语言中的“Elvis operator”，如 Groovy。指定两个表达式，如果第一个不是 null，则使用第二个。

查看如下示例。

```
<div th:object="${session.user}">
  ...
  <p>Age: <span th:text="*{age}?: '(no age specified)'">27</span>.</p>
</div>
```

这相当于：

```
<p>Age: <span th:text="*{age != null}? *{age} : '(no age speci-
fied)'">27</span>.</p>
```

与条件表达式一样，它们之间可以包含嵌套表达式。

```
<p>
  Name:
  <span th:text="*{firstName}?: (*{admin}? 'Admin' : #{default.user-
name})">Sebastian</span>
</p>
```

4.2.14 无操作标记

无操作标记由下画线（_）表示。表示什么也不做，这允许开发人员使用原型中的文本默认值。例如，

```
<span th:text="${user.name} ?: 'no user authenticated'">...</span>
```

可以直接使用“no user authenticated”作为原型文本，这样代码从设计的角度看起来很简洁。

```
<span th:text="${user.name} ?: _">no user authenticated</span>
```

4.2.15 数据转换及格式化

Thymeleaf 的双大括号为变量表达式（$ {...}）和选择表达式（* {...}）提供了数据转换服务，它看上去是这样的。

```
<td th:text="${{user.lastAccessDate}}">...</td>
```

注意到了“$ {{...}} ”里面的双括号吗？这意味着 Thymeleaf 可以通过转换服务将结果转换为 String。

假设 user.lastAccessDate 类型为 java.util.Calendar，如果转换服务（IStandardConversionService 接口的实现）已经被注册并且包含有效的 Calendar - > String 的转换，则它将被应用。

IStandardConversionService 的默认实现类为 StandardConversionService，只需在转换为“String”的任何对象上执行 .toString()。

4.2.16 表达式预处理

表达式预处理（Expression Preprocessing）被定义在下画线（_）之间。

```
#{selection.__${sel.code}__}
```

变量表达式 ${sel.code} 将先被执行，假如结果是“ALL”，那么“_”之间的值“ALL”将被看作表达式的一部分被执行，在这里会变成 selection.ALL。

4.3 Thymeleaf 设置属性值

本节将介绍如何在标记中设置（或修改）属性值的方式。

th:attr 用于设置任意属性。

```
<form action="subscribe.html" th:attr="action=@{/subscribe}">
  <fieldset>
    <input type="text" name="email" />
    <input type="submit" value="Subscribe!" th:attr="value=#{subscribe.
submit}"/>
```

```
  </fieldset>
</form>
```

th:attr 会将表达式的结果设置到相应的属性中。上面模板的结果如下。

```
<form action="/gtvg/subscribe">
  <fieldset>
    <input type="text" name="email" />
    <input type="submit" value="¡Suscríbe!"/>
  </fieldset>
</form>
```

也能同时设置多个属性值。

```
<img src="../../images/gtvglogo.png"
     th:attr="src=@{/images/gtvglogo.png},title=#{logo},alt=#{logo}" />
```

输出如下。

```
<img src="/gtgv/images/gtvglogo.png" title="Logo de Good Thymes" alt=
"Logo de Good Thymes" />
```

4.3.1 设置值到指定的属性

使用 th:attr 来设置属性的好处是可以设置任意命名的属性，这样方便自定义属性。但是，缺点也很明显，如果使用任何属性都是采用自定义的方式，那么，属性的维护有时就会变得很困难。考虑下面的示例。

```
<input type="submit" value="Subscribe!" th:attr="value=#{subscribe.
submit}"/>
```

上面可以指定一个自定义属性 value 的值，这在语法上完全没有问题，但这并不是最佳的方式。因为 Thymeleaf 已经提供了属性名为 value 的属性。通常，将使用其他的 th：* 属性设置 Thymeleaf 特定的标签属性（而不仅仅是像 th:attr 这样的任意属性）。

例如，要设置 value 属性，可以使用 th:value。

```
<input type="submit" value="Subscribe!" th:value="#{subscribe.submit}"/>
```

要设置 action 属性，使用 th:action。

```
<form action="subscribe.html" th:action="@{/subscribe}">
```

Thymeleaf 提供了很多属性，每个都针对特定的 HTML5 属性，如 th:abbr、th:action、th:background、th:form、th:height、th:style 等。详细的 Thymeleaf 属性列表可以参阅本书后面的附录 B 部分。

4.3.2 同时设置多个值

th:alt-title 和 th:lang-xmllang 是两个特殊的属性，可以同时设置同一个值到两个属性。

（1）th:alt-title 用于设置 alt 和 title。

（2）th:lang-xmllang 用于设置 lang 和 xml:lang。

观察下面的示例。

```
<img src="../../images/gtvglogo.png"
    th:attr="src=@{/images/gtvglogo.png},title=#{logo},alt=#{logo}" />
```

这个例子等价于：

```
<img src="../../images/gtvglogo.png"
    th:src="@{/images/gtvglogo.png}" th:alt-title="#{logo}" />
```

两者最终的结果都是：

```
<img src="../../images/gtvglogo.png"
    th:src="@{/images/gtvglogo.png}" th:title="#{logo}" th:alt="#
{logo}" />
```

4.3.3 附加和添加前缀

th:attrappend 和 th:attrprepend 用于附加和添加前缀属性。例如，

```
<input type="button" value="Do it!" class="btn" th:attrappend="class=${''
 + cssStyle}" />
```

执行模板， cssStyle 变量设置为 “warning” 时，输出如下。

```
<input type="button" value="Do it!" class="btn warning" />
```

同时，th:classappend 和 th:styleappend 用于设置 CSS 的 class 和 style。例如，

```
<tr th:each="prod : ${prods}" class="row" th:classappend="${prodStat.
odd}? 'odd'">
```

4.3.4 固定值布尔属性

HTML 具有布尔属性的概念，没有值的属性意味着该值为“true”。 在 XHTML 中，这些属性只取一个值，即它本身。

例如，属性 checked 的用法。

```
<input type="checkbox" name="option2" checked /> <!-- HTML -->
<input type="checkbox" name="option1" checked="checked" /> <!-- XHTML
```

```
-->
```

标准方言包括允许用户通过评估条件来设置这些属性，如果评估为 true，则该属性将被设置为其固定值，如果评估为 false，则不会设置该属性。

```
<input type="checkbox" name="active" th:checked="${user.active}" />
```

标准方言中存在以下固定值布尔属性。

th:async

th:autofocus

th:autoplay

th:checked

th:controls

th:declare

th:default

th:defer

th:disabled

th:formnovalidate

th:hidden

th:ismap

th:loop

th:multiple

th:novalidate

th:nowrap

th:open

th:pubdate

th:readonly

th:required

th:reversed

th:scoped

th:seamless

th:selected

4.3.5 默认属性处理器

Thymeleaf 提供“默认属性处理器”（Default Attribute Processor），当标准方言没有提供的属

性时，也可以设置其属性。例如，

```
<span th:whatever="${user.name}">...</span>
```

th:whatever 并不是标准方言中提供的属性，但仍然可以正确对属性进行赋值。最终输出如下。

```
<span whatever="John Apricot">...</span>
```

默认属性处理器与 th:attr 设置任意属性有着异曲同工之妙。

4.3.6 支持对 HTML5 友好的属性及元素名称

data-{prefix}-{name} 语法是 HTML5 中编写自定义属性的标准方式，不需要开发人员使用任何命名空间的名称（如 th: * ）。Thymeleaf 支持多种语法，自动提供所有的方言（而不只是标准方言）。

考虑下面的例子。

```
<table>
    <tr data-th-each="user : ${users}">
        <td data-th-text="${user.login}">...</td>
        <td data-th-text="${user.name}">...</td>
    </tr>
</table>
```

其实完全等价于：

```
<table>
    <tr th:each="user : ${users}">
        <td th:each="${user.login}">...</td>
        <td th:each="${user.name}">...</td>
    </tr>
</table>
```

如果是一个对 HTML5 语法有“强迫症”的开发人员，那么可以放心地通过 data-{prefix}-{name} 语法来使用 Thymeleaf 元素。

4.4 Thymeleaf 迭代器与条件语句

本节将介绍 Thymeleaf 迭代器与条件语句，这两者在实际开发中经常被使用。

4.4.1 迭代器

迭代器是程序中常见的设计模式，是可在容器上遍访元素的接口。

1. 基本的迭代

Thymeleaf 的 th:each 将循环 array 或 list 中的元素并重复打印一组标签，语法相当于 Java foreach 表达式。

```
<li th:each="book : ${books}" th:text="${book.title}">En las Orillas
del Sar</li>
```

可以使用 th：each 属性进行遍历的对象如下。

（1）任何实现 java.util.Iterable 的对象。

（2）任何实现 java.util.Enumeration 的对象。

（3）任何实现 java.util.Iterator 的对象，其值将被迭代器返回，而不需要在内存中缓存所有的值。

（4）任何实现 java.util.Map 的对象。迭代映射时，迭代变量 将是 java.util.Map.Entry 类。

（5）任何数组。

（6）任何其他对象将被视为包含对象本身的单值列表。

2. 状态变量

Thymeleaf 提供“状态变量”（Status Variable）来跟踪迭代器的状态。

th:each 属性中定义了如下状态变量。

（1）index 属性是当前“迭代器索引”（Iteration Index），从 0 开始。

（2）count 属性是当前“迭代器索引”（Iteration Index），从 1 开始。

（3）size 属性是迭代器元素的总数。

（4）current 是当前“迭代变量”（Iter Variable）。

（5）even/odd 判断当前迭代器是否为 even 或 odd。

（6）first 判断当前迭代器是否为第一个。

（7）last 判断当前迭代器是否为最后。

看下面的例子。

```
<table>
  <tr>
    <th>NAME</th>
    <th>PRICE</th>
    <th>IN STOCK</th>
  </tr>
  <tr th:each="prod,iterStat : ${prods}" th:class="${iterStat.odd}?
'odd'">
    <td th:text="${prod.name}">Onions</td>
    <td th:text="${prod.price}">2.41</td>
    <td th:text="${prod.inStock}? #{true} : #{false}">yes</td>
  </tr>
</table>
```

状态变量（在本示例中为“iterStat”）在 th：each 中定义了。

下面来看看模板处理后的结果。

```
<!DOCTYPE html>

<html>

  <head>
    <title>Good Thymes Virtual Grocery</title>
    <meta content="text/html; charset=UTF-8" http-equiv="Content-Type"/>
    <link rel="stylesheet" type="text/css" media="all" href="/gtvg/css/
gtvg.css" />
  </head>

  <body>

    <h1>Product list</h1>

    <table>
      <tr>
        <th>NAME</th>
        <th>PRICE</th>
        <th>IN STOCK</th>
      </tr>
      <tr class="odd">
        <td>Fresh Sweet Basil</td>
        <td>4.99</td>
        <td>yes</td>
      </tr>
      <tr>
        <td>Italian Tomato</td>
        <td>1.25</td>
        <td>no</td>
      </tr>
      <tr class="odd">
        <td>Yellow Bell Pepper</td>
        <td>2.50</td>
        <td>yes</td>
      </tr>
      <tr>
        <td>Old Cheddar</td>
        <td>18.75</td>
        <td>yes</td>
      </tr>
    </table>

    <p>
      <a href="/gtvg/" shape="rect">Return to home</a>
    </p>
```

```
  </body>

</html>
```

注意：从输出结果可以看到，迭代状态变量能够良好地运行，并且已经在奇数行上创建了具有“odd”样式的标签。

如果没有明确设置状态变量，则 Thymeleaf 将始终创建一个状态变量，可以通过后缀“Stat”获取到迭代变量的名称。

```
<table>
  <tr>
    <th>NAME</th>
    <th>PRICE</th>
    <th>IN STOCK</th>
  </tr>
  <tr th:each="prod : ${prods}" th:class="${prodStat.odd}? 'odd'">
    <td th:text="${prod.name}">Onions</td>
    <td th:text="${prod.price}">2.41</td>
    <td th:text="${prod.inStock}? #{true} : #{false}">yes</td>
  </tr>
</table>
```

4.4.2 条件语句

条件语句是用来判断给定的条件是否满足，并根据判断的结果决定执行的语句。

1. if 和 unless

th:if 属性用法如下。

```
<a href="comments.html"
   th:href="@{/product/comments(prodId=${prod.id})}"
   th:if="${not #lists.isEmpty(prod.comments)}">view</a>
```

注意：th:if 属性不仅能够判断布尔条件，它还能判断以下表达式。

- 如果值不为 null：
  - 如果值为布尔值，则为 true；
  - 如果值是数字，并且不为零；
  - 如果值是一个字符且不为零；
  - 如果 value 是 String，而不是“false”“off”或“no”；
  - 如果值不是布尔值、数字、字符或字符串。
- 如果值为 null，则 th:if 将为 false。

另外，th:if 有一个相反的属性 th:unless，前面的例子改为：

```
<a href="comments.html"
   th:href="@{/comments(prodId=${prod.id})}"
   th:unless="${#lists.isEmpty(prod.comments)}">view</a>
```

2. switch 语句

switch 语句使用 th:switch 与 th:case 属性集合来实现。

```
<div th:switch="${user.role}">
  <p th:case="'admin'">User is an administrator</p>
  <p th:case="#{roles.manager}">User is a manager</p>
</div>
```

注意：只要一个 th:case 属性被评估为 true，每个其他同一个 switch 语句中的 th:case 属性将被评估为 false。

用 th:case="*" 来设置默认选项。

```
<div th:switch="${user.role}">
  <p th:case="'admin'">User is an administrator</p>
  <p th:case="#{roles.manager}">User is a manager</p>
  <p th:case="*">User is some other thing</p>
</div>
```

4.5 Thymeleaf 模板片段

本节将介绍如何来使用 Thymeleaf 模板片段。

Thymeleaf 模板片段的目的是让模板片段可以在多个页面实现重用，从而减少代码量，也使整个页面更加“模块化”。

4.5.1 定义和引用片段

在自己的模板中经常需要从其他模板中添加 html 页面片段，如页脚、标题、菜单等，这些页面片段由于在各个页面都会被引用到，因此设置为页面片段，从而实现页面片段的重用。

为了做到这一点，Thymeleaf 需要人们来定义这些“片段”，可以使用 th:fragment 属性来完成。

以定义 /WEB-INF/templates/footer.html 页面作为例子。

```
<!DOCTYPE html>

<html xmlns:th="http://www.thymeleaf.org">

  <body>
```

```
    <div th:fragment="copy">
      &copy; 2017 <a href="https://waylau.com">waylau.com</a>
    </div>

  </body>

</html>
```

如果想引用这个 copy 代码片段，可以用 th:insert 或 th:replace 属性（th:include 也可以实现类似功能，但自 Thymeleaf 3.0 以来就不再推荐使用了）。

```
<body>

  ...

  <div th:insert="~{footer :: copy}"></div>

</body>
```

注意： th:insert 需要一个“片段表达式”（~{...}）。在上面的例子中，“片段表达式”（~{, }）是完全可选的，所以上面的代码将等效于：

```
<body>

  ...

  <div th:insert="footer :: copy"></div>

</body>
```

4.5.2 Thymeleaf 片段规范语法

以下是 Thymeleaf 片段规范语法。

- “~{templatename::selector}”：名为 templatename 的模板上的指定标记选择器。selector 可以只是一个片段名。
- “~{templatename}”：包含完整的模板 templatename。
- ~{::selector}” 或 “~{this::selector}”：指相同模板中的代码片段。

4.5.3 不使用 th:fragment

不使用 th:fragment 也可以引用 HTML 片段，例如，

```
...
```

```
<div id="copy-section">
  &copy; 2017 <a href="https://waylau.com">waylau.com</a>
</div>
...
```

通过 id 也可以引用到页面片段。

```
<body>

  ...

  <div th:insert="~{footer :: #copy-section}"></div>

</body>
```

4.5.4 th:insert、 th:replace 、th:include三者的区别

th:insert、th:replace、th:include 三者都能实现片段的引用，但在最终的实现效果上还是存在差异。其中：

- th:insert 是最简单的，它将简单地插入指定的片段作为正文的主标签。
- th:replace 用指定实际片段来替换其主标签。
- th:include 类似于 th:insert，但不是插入片段，它只插入此片段的“内容”。

所以考虑下面的例子。

```
<footer th:fragment="copy">
  &copy; 2017 <a href="https://waylau.com">waylau.com</a>
</footer>
```

3 种方式同时引用该片段。

```
<body>

  ...

  <div th:insert="footer :: copy"></div>

  <div th:replace="footer :: copy"></div>

  <div th:include="footer :: copy"></div>

</body>
```

结果为：

```
<body>
```

```
  ...

  <div>
    <footer>
      &copy; 2017 <a href="https://waylau.com">waylau.com</a>
    </footer>
  </div>

  <footer>
    &copy; 2017 <a href="https://waylau.com">waylau.com</a>
  </footer>

  <div>
    &copy; 2017 <a href="https://waylau.com">waylau.com</a>
  </div>

</body>
```

4.6 Thymeleaf 表达式基本对象

在 Thymeleaf 中，一些对象和变量 map 总是可以被调用，这些对象称为“表达式基本对象”，一起来看看它们。

4.6.1 基本对象

（1）#ctx：上下文对象，是 org.thymeleaf.context.IContext 或者 org.thymeleaf.context.IWebContext 的实现，取决于环境是桌面程序或者是 Web 程序。

需要注意的是，#vars 和 #root 是同一个对象的同义词，但建议使用 #ctx。

```
${#ctx.locale}
${#ctx.variableNames}

${#ctx.request}
${#ctx.response}
${#ctx.session}
${#ctx.servletContext}
```

（2）#locale：直接访问与 java.util.Locale 关联的当前的请求。

```
${#locale}
```

4.6.2 Web 上下文命名空间

Thymeleaf 在 Web 环境中有一系列的快捷方式用于访问请求参数、会话属性等应用属性。需要注意的是，这些不是“上下文对象”（Context Objects），但有 map 添加到上下文作为变量，这样就能访问它们而无须 #。它们类似于“命名空间”（Namespaces）。

（1）param：用于检索请求参数。${param.foo} 是一个使用 foo 请求参数的值 String[]，所以 ${param.foo[0]} 将会通常用于获取第一个值。

```
${param.foo}
${param.size()}
${param.isEmpty()}
${param.containsKey('foo')}
...
```

（2）session：用于检索会话属性。

```
${session.foo}
${session.size()}
${session.isEmpty()}
${session.containsKey('foo')}
...
```

（3）application：用于检索应用及上下文属性。

```
${application.foo}
${application.size()}
${application.isEmpty()}
${application.containsKey('foo')}
...
```

需要注意的是，没有必要指定访问请求属性的命名空间，因为所有请求属性都会自动添加到上下文中作为上下文根中的变量。

```
${myRequestAttribute}
```

4.6.3 Web 上下文对象

在 Web 环境，下列对象可以直接访问（注意它们是对象，而非 map 或者是命名空间）。

（1）#request：直接访问与当前请求关联的 javax.servlet.http.HttpServletRequest 对象。

```
${#request.getAttribute('foo')}
${#request.getParameter('foo')}
${#request.getContextPath()}
${#request.getRequestName()}
...
```

（2）#session：直接访问与当前请求关联的 javax.servlet.http.HttpSession 对象。

```
${#session.getAttribute('foo')}
${#session.id}
${#session.lastAccessedTime}
...
```

（3）#servletContext：直接访问与当前请求关联的 javax.servlet.ServletContext 对象。

```
${#servletContext.getAttribute('foo')}
${#servletContext.contextPath}
...
```

4.7 Thymeleaf 与 Spring Boot 集成

本节将演示如何把 Thymeleaf 技术框架集成到 Spring Boot 项目中。在 media-type-json 项目基础上构建了一个新的项目 thymeleaf-in-action。

4.7.1 所需环境

本例子采用的开发环境如下。

- Thymeleaf 3.0.6.RELEASE。
- Thymeleaf Layout Dialect 2.2.2。

4.7.2 build.gradle

开发人员需要添加 Thymeleaf 的依赖。Spring Boot 已经提供了相关的 Starter 来实现 Thymeleaf 开箱即用的功能，所以只需要在 build.gradle 文件中添加 Thymeleaf Starter 的库即可。

```
// 依赖关系
dependencies {
    //...

    // 添加Thymeleaf的依赖
    compile('org.springframework.boot:spring-boot-starter-thymeleaf')

    //...
}
```

spring-boot-starter-thymeleaf 库默认使用的是 Thymeleaf 3.0.6.RELEASE 版本，目前也是 Thymeleaf 最新的版本。spring-boot-starter-thymeleaf 库主要依赖了以下 3 个库。

```
<dependencies>
    <dependency>
        <groupId>org.springframework.boot</groupId>
        <artifactId>spring-boot-starter</artifactId>
    </dependency>
    <dependency>
        <groupId>org.thymeleaf</groupId>
        <artifactId>thymeleaf-spring5</artifactId>
    </dependency>
    <dependency>
        <groupId>org.thymeleaf.extras</groupId>
        <artifactId>thymeleaf-extras-java8time</artifactId>
    </dependency>
</dependencies>
```

上述依赖，可以参见 spring-boot-starter-thymeleaf 库的 pom.xml 说明。

4.7.3 如何自定义版本

spring-boot-starter-thymeleaf 库默认使用的是 Thymeleaf 最新的版本 3.0.6.RELEASE。有时，出于学习或者工作的需要，例如，想升级到最新的类库或者想降级使用旧版本，此时就需要替换 spring-boot-starter-thymeleaf 库中的默认版本。下面演示了如何使用 Thymeleaf 旧版本来替换 Spring Boot 依赖库中的 Thymeleaf 和 Thymeleaf Layout Dialect 的版本号，用法如下。

```
buildscript {
    //...

    // 自定义Thymeleaf和Thymeleaf Layout Dialect的版本
    ext['thymeleaf.version'] = '3.0.3.RELEASE'
    ext['thymeleaf-layout-dialect.version'] = '2.2.0'

    //...
}
```

4.8 Thymeleaf 实战

在 4.7 节构建了一个新的项目 thymeleaf-in-action。在本节中将通过该项目，采用 Thymeleaf 来实现一个最简单的“用户管理”功能。

“用户管理”可以实现对用户的查询、新增、删除和修改。为了简便，没有使用数据库管理系统，而是将数据直接保存在了内存。需要注意的是，只要应用重启，数据就会丢失。

4.8.1 修改 application.properties

application.properties 文件在 Spring Boot 项目中是用于配置项目的配置文件。在 thymeleaf-in-action 项目中修改 application.properties，增加下面几项配置。

```
# Thymeleaf编码
spring.thymeleaf.encoding=UTF-8
# 热部署静态文件
spring.thymeleaf.cache=false
# 使用HTML5标准
spring.thymeleaf.mode=HTML5
```

上面的配置中的注释已经非常清楚地说明了各个配置的意图。其中 spring.thymeleaf.cache=false 是指不对 Thymeleaf 模板进行缓存，这样在开发阶段修改了 Thymeleaf 模板后，就能在浏览器中及时看到修改后的页面效果。

4.8.2 后台编码

按照习惯，先写好后台的接口，而后再实现前端页面的编码。当然，也可根据自己的习惯来进行编码。

1. 创建资源库

建包 com.waylau.spring.boot.blog.repository，用于存放资源库。在该包下创建 UserRepository 接口，用于处理用户的资源库。

```
public interface UserRepository {
    /**
     * 新增或者修改用户
     * @param user
     * @return
     */
    User saveOrUpateUser(User user);

    /**
     * 删除用户
     * @param id
     */
    void deleteUser(Long id);

    /**
     * 根据用户id获取用户
     * @param id
     * @return
     */
    User getUserById(Long id);
```

```
    /**
     * 获取所有用户的列表
     * @return
     */
    List<User> listUser();
}
```

在项目的包下，新增一个 UserRepositoryImpl 类作为该类的实现类。

```
@Repository
public class UserRepositoryImpl implements UserRepository {

    private static AtomicLong counter = new AtomicLong();

    private final ConcurrentMap<Long, User> userMap = new ConcurrentHash-
Map<Long, User>();

    @Override
    public User saveOrUpateUser(User user) {
        Long id = user.getId();
        if (id == null) {
            id = counter.incrementAndGet();
            user.setId(id);
        }
        this.userMap.put(id, user);
        return user;
    }

    @Override
    public void deleteUser(Long id) {
        this.userMap.remove(id);
    }

    @Override
    public User getUserById(Long id) {
        return this.userMap.get(id);
    }

    @Override
    public List<User> listUser() {
        return new ArrayList<User>(this.userMap.values());
    }

}
```

其中，用 ConcurrentMap<Long, User> userMap 来模拟数据的存储， AtomicLong counter 用来生成一个递增的 id，作为用户的唯一编号。@Repository 注解用于标识 UserRepositoryImpl 类是一个可注入的 bean。

2. 修改控制器

修改 com.waylau.spring.boot.blog.controller.UserController 用于处理界面的请求。

```
@RestController
@RequestMapping("/users")
public class UserController {

    @Autowired
    private UserRepository userRepository;

    /**
     * 查询所有用户
     * @param model
     * @return
     */
    @GetMapping
    public ModelAndView list(Model model) {
        model.addAttribute("userList", userRepository.listUser());
        model.addAttribute("title", "用户管理");
        return new ModelAndView("users/list", "userModel", model);
    }

    /**
     * 根据id查询用户
     * @param id
     * @param model
     * @return
     */
    @GetMapping("{id}")
    public ModelAndView view(@PathVariable("id") Long id, Model model)
{
        User user = userRepository.getUserById(id);
        model.addAttribute("user", user);
        model.addAttribute("title", "查看用户");
        return new ModelAndView("users/view", "userModel", model);
    }

    /**
     * 获取创建表单页面
     * @param model
     * @return
     */
    @GetMapping("/form")
    public ModelAndView createForm(Model model) {
        model.addAttribute("user", new User());
        model.addAttribute("title", "创建用户");
        return new ModelAndView("users/form", "userModel", model);
    }
```

```
    /**
     * 保存用户
     * @param user
     * @return
     */
    @PostMapping
    public ModelAndView saveOrUpateUser(User user) {
        user = userRepository.saveOrUpateUser(user);
        return new ModelAndView("redirect:/users");
    }

    /**
     * 删除用户
     * @param id
     * @return
     */
    @GetMapping(value = "delete/{id}")
    public ModelAndView delete(@PathVariable("id") Long id) {
        userRepository.deleteUser(id);
        return new ModelAndView("redirect:/users");
    }

    /**
     * 获取修改用户的界面
     * @param id
     * @param model
     * @return
     */
    @GetMapping(value = "modify/{id}")
    public ModelAndView modifyForm(@PathVariable("id") Long id, Model
model) {
        User user = userRepository.getUserById(id);

        model.addAttribute("user", user);
        model.addAttribute("title", "修改用户");
        return new ModelAndView("users/form", "userModel", model);
    }

}
```

整体的 API 设计如下。

（1）GET /users： 返回用于展现用户列表的 list.html 页面。

（2）GET /users/{id}： 返回用于展现用户的 view.html 页面。

（3）GET /users/form： 返回用于新增或者修改用户的 form.html 页面。

（4）POST /users： 用于新增或者修改用户请求处理。成功处理后重定向到 list.html 页面。

（5）GET /users/delete/{id}： 根据 id 删除相应的用户数据。成功处理后重定向到 list.html 页面。

（6）GET /users/modify/{id}： 根据 id 获取相应的用户数据，返回 form.html 页面用来执行修改。

需要注意的是，本 API 仅为展现 Spring MVC 的功能，并非完全符合 REST 风格。如果需要了解详细的 REST 风格架构，可以参考笔者的另外一本开源书《REST 实战》（https://github.com/waylau/rest-in-action）。

4.8.3 编写前台页面

页面主要采用 Thymeleaf 引擎来开发。本节内容为了专注于 Thymeleaf 核心功能，故不涉及 CSS 样式和 JS 脚本的编写。

在项目的 templates 目录下新建一个 users 目录，来归档“用户管理”功能相关的页面。

其中：list.html 用于展现用户列表；form.html 用于新增或者修改用户的资料；view.html 用于查看某个用户的资料。

在 templates 目录下新建一个 fragments 页面，来归档页面共用的部分。

其中：header.html 为共用的头部页面；footer.html 为共用的底部页面。

1. list.html

list.html 的完整代码如下。

```
<!DOCTYPE html>
<html xmlns:th="http://www.thymeleaf.org"
    xmlns:layout="http://www.ultraq.net.nz/thymeleaf/layout">

<head>
    <meta charset="UTF-8">
    <title>Thymeleaf in action</title>
</head>

<body>
    <div th:replace="~{fragments/header :: header}"></div>
    <h3 th:text="${userModel.title}">waylau</h3>
    <div>
        <a href="/users/form.html" th:href="@{/users/form}">创建用户</a>
    </div>
    <table border="1">
        <thead>
            <tr>
                <td>ID</td>
                <td>Email</td>
                <td>Name</td>
            </tr>
        </thead>
        <tbody>
            <tr th:if="${userModel.userList.size()} eq 0">
                <td colspan="3">没有用户信息！</td>
            </tr>
            <tr th:each="user : ${userModel.userList}">
```

```
                <td th:text="${user.id}"></td>
                <td th:text="${user.email}"></td>
                <td><a th:href="@{'/users/'+${user.id}}" th:text="${us-
er.name}"></a></td>
            </tr>
        </tbody>
    </table>
    <div th:replace="~{fragments/footer :: footer}"></div>
</body>

</html>
```

2. form.html

form.html 的完整代码如下。

```
<!DOCTYPE html>
<html xmlns:th="http://www.thymeleaf.org"
    xmlns:layout="http://www.ultraq.net.nz/thymeleaf/layout">

<head>
    <meta charset="UTF-8">
    <title>Thymeleaf in action</title>
</head>

<body>
    <div th:replace="~{fragments/header :: header}"></div>
    <h3 th:text="${userModel.title}">waylau</h3>
    <form action="/users" th:action="@{/users}" method="POST" th:ob-
ject="${userModel.user}">
        <input type="hidden" name="id" th:value="*{id}"> 名称:
        <br>
        <input type="text" name="name" th:value="*{name}">
        <br> 邮箱:
        <br>
        <input type="text" name="email" th:value="*{email}">
        <input type="submit" value="提交">
    </form>
    <div th:replace="~{fragments/footer :: footer}"></div>
</body>

</html>
```

3. view.html

view.html 的完整代码如下。

```
<!DOCTYPE html>
<html xmlns:th="http://www.thymeleaf.org"
    xmlns:layout="http://www.ultraq.net.nz/thymeleaf/layout">
```

```
<head>
    <meta charset="UTF-8">
    <title>Thymeleaf in action</title>
</head>

<body>
    <div th:replace="~{fragments/header :: header}"></div>
    <h3 th:text="${userModel.title}">waylau</h3>
    <div>
        <P><strong>ID:</strong><span th:text="${userModel.user.id}"></
span></P>
        <P><strong>Name:</strong><span th:text="${userModel.user.
name}"></span></P>
        <P><strong>Email:</strong><span th:text="${userModel.user.
email}"></span></P>
    </div>
    <div>
        <a th:href="@{'/users/delete/'+${userModel.user.id}}">删除</a>
        <a th:href="@{'/users/modify/'+${userModel.user.id}}">修改</a>
    </div>
    <div th:replace="~{fragments/footer :: footer}"></div>
</body>

</html>
```

4. header.html

header.html 的完整代码如下。

```
<!DOCTYPE html>
<html xmlns:th="http://www.thymeleaf.org"
    xmlns:layout="http://www.ultraq.net.nz/thymeleaf/layout">

<head>
    <meta charset="UTF-8">
    <title>Thymeleaf in action</title>
</head>

<body>
    <div th:fragment="header">
        <h1>Thymeleaf in action</h1>
        <a href="/users" th:href="@{~/users}">首页</a>
    </div>
</body>

</html>
```

5. footer.html

footer.html 的完整代码如下。

```
<!DOCTYPE html>
<html xmlns:th="http://www.thymeleaf.org"
    xmlns:layout="http://www.ultraq.net.nz/thymeleaf/layout">

<head>
    <meta charset="UTF-8">
    <title>Thymeleaf in action</title>
</head>

<body>
    <div th:fragment="footer">
        <a href="https://waylau.com">Welcome to waylau.com</a>
    </div>
</body>

</html>
```

上述代码较为简单，在此不再赘述。

4.8.4 运行

启动 thymeleaf-in-action 项目后，在浏览器访问 http://localhost:8080/users，可以看到项目的运行效果。其中，图 4-1 展示的是 list.html 页面，用于实现查看用户列表的功能；图 4-2 展示的是 form.html 页面，用于实现新增用户的功能；图 4-3 展示的是 view.html 页面，用于实现查看用户的功能。

图4-1 查看用户列表

图4-2 新增用户

图4-3 查看用户

4.8.5 示例源码

本节示例源码在 thymeleaf-in-action 目录下。

第5章

数据持久化

5.1 JPA 概述

JPA（Java Persistence API）是用于管理 Java EE 和 Java SE 环境中的持久化，以及对象 / 关系映射的 Java API。

JPA 最新规范为 JSR 338: JavaTM Persistence 2.1（https://jcp.org/en/jsr/detail?id=338）。目前，市面上实现该规范的常见 JPA 框架有 EclipseLink（http://www.eclipse.org/eclipselink）、Hibernate（http://hibernate.org/orm）、Apache OpenJPA（http://openjpa.apache.org/）等。本书主要介绍以 Hibernate 为实现的 JPA。

本节只对 JPA 做简单的介绍。读者如果要了解详细的 JPA 用法，可以参见笔者的另外一本开源书《Java EE 编程要点》（https://github.com/waylau/essential-javaee）中的“数据持久化”相关内容。

5.1.1 JPA 的产生背景

在 JPA 产生之前，围绕如何简化数据库操作的相关讨论已经是层出不穷，众多厂商和开源社区也都提供了持久层框架的实现，其中 ORM 框架最为开发人员所关注。

ORM（Object Relational Mapping，对象关系映射）是一种用于实现面向对象编程语言里不同类型系统的数据之间的转换的程序技术。由于面向对象数据库系统（OODBS）的实现，在技术上还存在难点，目前，市面上流行的数据库还是以关系型数据库为主。

由于关系型数据库使用的 SQL 是一种非过程化的面向集合的语言，而目前许多应用仍然是由高级程序设计语言（如 Java）来实现的，但是高级程序设计语言是过程化的，而且是面向单个数据的，这使得 SQL 与它之间存在着不匹配，这种不匹配称为“阻抗失配”。由于“阻抗失配”的存在，开发人员在使用关系型数据库时不得不花很多功夫去完成两种语言之间的相互转化。

而 ORM 框架的产生，正是为了简化这种转化操作。使用 ORM，在编程语言中，就可以使用面向对象的方式来完成数据库的操作。

ORM 框架的出现，使直接存储对象成为可能，它们将对象拆分成 SQL 语句，从而来操作数据库。但是不同的 ORM 框架在使用上存在比较大的差异，这也导致开发人员需要学习各种不同的 ORM 框架，增加了技术学习的成本。

而 JPA 规范就是为了解决这个问题：规范 ORM 框架，使用 ORM 框架统一的接口和用法。这样，在采用了面向接口编程的技术中，即便是更换了不同的 ORM 框架，也无须变更业务逻辑。

最早的 JPA 规范是由 Java 官方提出的，随 Java EE 5 规范一同发布。

5.1.2 实体

实体是轻量级的持久化域对象。通常，实体表示关系数据库中的表，并且每个实体实例对应于该表中的行。实体的主要编程工件是实体类，尽管实体可以使用辅助类。

在 EJB 3 之前，EJB 主要包含 3 种类型：会话 Bean、消息驱动 Bean、实体 Bean。但自 EJB 3.0 开始，实体 Bean 被单独分离出来，形成了新的规范：JPA。所以，JPA 完全可以脱离 EJB 3 来使用。实体是 JPA 中的核心概念。

实体的持久状态通过持久化字段或持久化属性来表示，这些字段或属性使用对象 / 关系映射注解将实体和实体关系映射到基础数据存储中的关系数据。

与实体在概念上比较接近的另外一个领域对象是值对象。实体是可以被跟踪的，通常会有一个主键（唯一标识），来追踪其状态。而值对象则没有这种标识，用户只关心值对象的属性。

1. 实体类的要求

一个实体类需满足以下条件。

（1）类必须用 javax.persistence.Entity 注解。

（2）类必须有一个 public 或 protected 的无参数的构造函数。该类可以具有其他构造函数。

（3）类不能声明为 final 。没有方法的或持久化实例变量必须声明为 final。

（4）如果实体实例被当作值以分离对象方式进行传递（如通过会话 Bean 的远程业务接口），则该类必须实现 Serializable 接口。

（5）实体可以扩展实体类或者是非实体类，并且非实体类可以扩展实体类。

（6）持久化实例变量必须声明为 private、protected 或 package-private，并且只能通过实体类的方法直接访问。客户端必须通过访问器或业务方法访问实体的状态。

以下是一个用户（User）的实体例子。

```
@Entity // 实体
public class User {

    @Id // 主键
    @GeneratedValue(strategy=GenerationType.IDENTITY) // 自增长策略
    private Long id; // 实体一个唯一标识
    private String name;
    private String email;

    protected User() { // 无参构造函数；设为 protected 防止直接使用
    }

    public User(Long id, String name, String email) {
        this.id = id;
        this.name = name;
        this.email = email;
    }

    // 省略getter/setter方法
}
```

2. 实体类中的持久化字段和属性

可以通过实体的实例变量或属性访问实体的持久状态。字段或属性必须是以下 Java 语言类型。

- Java 基本数据类型。
- java.lang.String。
- 其他可序列化类型如下。
  - Java 基本数据类型的包装器
  - java.math.BigInteger
  - java.math.BigDecimal
  - java.util.Date
  - java.util.Calendar
  - java.sql.Date
  - java.sql.Time
  - java.sql.TimeStamp
  - 用户定义的可序列化类型
  - byte[]
  - Byte[]
  - char[]
  - Character[]
- 枚举类型。
- 其他实体和（或）实体集合。
- 可嵌入类。

实体可以使用持久化字段、持久化属性或两者的组合。如果映射注解应用于实体的实例变量，则实体使用持久化字段；如果映射注解应用于实体的 JavaBean 风格属性的 getter 方法，则实体使用持久化属性。

如果实体类使用持久化字段，则 Persistence 运行时直接访问实体类实例变量。所有未使用 javax.persistence.Transient 注解的字段或未标记为 Java transient 将被持久化到数据存储中。对象 / 关系映射注解必须应用于实例变量。

如果实体使用持久化属性，实体必须遵循 JavaBean 组件的方法约定。 JavaBean 风格的属性使用 getter 方法和 setter 方法，它们通常以实体类的实例变量名称命名。对于实体的每个持久化属性，都有一个 getter 方法和 setter 方法。如果属性是布尔值，可以使用 isProperty，而不是 getProperty。例如，Customer 实体使用持久化属性并具有称为 firstName 的专用实例变量，则该类定义用于检索和设置 firstName 实例变量的状态的 getFirstName 方法和 setFirstName 方法。

单值持久化属性的方法签名如下。

```
Type getProperty()
void setProperty(Type type)
```

持久化属性的对象 / 关系映射注解必须应用于 getter 方法。映射注解不能应用于注解为 @Transient 或标记为 transient 的字段或属性。

3. 在实体字段和属性中使用集合

集合作为持久化字段和属性，必须使用受 Java 集合支持的接口，而不管实体是否使用持久化字段或属性，可以使用以下集合接口。

- java.util.Collection。
- java.util.Set。
- java.util.List。
- java.util.Map。

如果实体类使用持久化字段，则上述方法签名中的类型必须是这些集合类型之一。也可以使用这些集合类型的泛型。例如，如果它具有包含一组电话号码的持久化属性，那么 Customer 实体将具有以下方法。

```
Set<PhoneNumber> getPhoneNumbers() { ... }
void setPhoneNumbers(Set<PhoneNumber>) { ... }
```

如果实体的字段或属性由基本类型或可嵌入类的集合组成，请在字段或属性上使用 javax.persistence.ElementCollection 注解。

@ElementCollection 的两个属性是 targetClass 和 fetch。 targetClass 属性指定基本类或可嵌入类的类名，如果字段或属性是使用 Java 编程语言泛型定义的，则它是可选的。可选的 fetch 属性用于分别使用 LAZY 或 EAGER 的 javax.persistence.FetchType 常量来指定是否应“懒加载”或“急加载”地检索集合。默认情况下，系统会用 LAZY。

以下实体 Person 具有持久化字段，它是抓取 String 类的集合的时候使用了“急加载”。 targetClass 元素不是必需的，因为它使用泛型定义字段。

```
@Entity
public class Person {
    ...
    @ElementCollection(fetch=EAGER)
    protected Set<String> nickname = new HashSet();
    ...
}
```

实体元素和关系的集合可以由 java.util.Map 表示。 Map 由键和值组成。

使用 Map 元素或关系时，以下规则适用。

（1）Map 键或值可以是基本 Java 编程语言类型、可嵌入类或实体。

（2）当 Map 值是可嵌入类或基本类型时，使用 @ElementCollection 注解。

（3）当 Map 值是实体时，使用 @OneToMany 或 @ManyToMany 注解。

（4）仅在双向关系的一侧使用 Map 类型。

如果 Map 的键类型是 Java 编程语言基本类型，请使用注解 javax.persistence.MapKeyColumn 来设置键的列映射。默认情况下，@MapKeyColumn 的 name 属性的格式为 RELATIONSHIP-FIELD/PROPERTY-NAME_KEY。例如，如果引用关系字段名称为 image，则默认名称属性为 IMAGE_KEY。

如果 Map 的键类型是实体，请使用 javax.persistence.MapKeyJoinColumn 注解。如果需要多个列来设置映射，使用注解 javax.persistence.MapKeyJoinColumns 来包含多个 @MapKeyJoinColumn 注解。如果不存在 @MapKeyJoinColumn，则映射列名称默认设置为 RELATIONSHIP-FIELD/PROPERTY-NAME_KEY。例如，如果关系字段名称为 employee，那么默认名称属性为 EMPLOYEE_KEY。

如果 Java 编程语言通用类型未在关系字段或属性中使用，则必须使用 javax.persistence.MapKeyClass 注解显式设置键类。

如果 Map 键是主键，或者实体的持久字段或属性是 Map 值，请使用 javax.persistence.MapKey 注解。@MapKeyClass 和 @MapKey 注解不能在同一个字段或属性上使用。

如果 Map 值是 Java 编程语言基本类型或可嵌入类，它将被映射为底层数据库中的集合表。如果不使用通用类型，则 @ElementCollection 注解的 targetClass 属性必须设置为 Map 值的类型。

如果 Map 值是一个实体，并且是多对多或一对多单向关系的一部分，则它将被映射为底层数据库中的连接表。使用映射的单向一对多关系也可以使用 @JoinColumn 注解映射。

如果实体是一对多或多对一双向关系的一部分，则它将被映射到表示 Map 的值的实体的表中。如果不使用通用类型，则 @OneToMany 和 @ManyToMany 注解的 targetEntity 属性必须设置为 Map 值的类型。

4. 验证持久化字段和属性

Java API for JavaBeans Validation（Bean Validation）提供了一种验证应用程序数据的机制。Bean Validation 集成到了 Java EE 容器中，允许在企业应用程序的任何层中使用相同的验证逻辑。

Bean Validation 约束可以应用于持久化实体类、可嵌入类和映射的超类。默认情况下，Persistence 提供程序将在 PrePersist、PreUpdate 和 PreRemove 生命周期事件之后立即自动对具有持久字段或以 Bean Validation 约束注解的属性的实体执行验证。

Bean Validation 约束是应用于 Java 编程语言类的字段或属性的注解。Bean Validation 提供了一组约束及用于定义自定义约束的 API。自定义约束可以是默认约束的特定组合，或不使用默认约束的新约束。每个约束与至少一个验证器类相关联，验证器类验证约束字段或属性的值。自定义约束开发人员还必须为约束提供一个验证器类。

Bean Validation 约束应用于持久化类的持久化字段或属性。当添加 Bean Validation 约束时，使用与持久类相同的访问策略。也就是说，如果持久化类使用字段访问，则在类的字段上应用 Bean Validation 约束注解。如果类使用属性访问，应用对 getter 方法的约束。

Bean Validation 提供了很多内置约束注解，例如，

```
@Max(10)
int quantity;
```

@Max(10) 约束了字段或者属性最大值不能大于 10。

又如，

```
@NotNull
String username;
```

@NotNull 约束了字段或者属性值不能为空。

本书的最后部分“附录 D：Bean Validation 内置约束”中列出了 Bean Validation 常见的内置约束，供读者查阅。这些约束都在 javax.validation.constraints 包中做了定义。以下实体类 Contact 具有应用于其持久化字段的 Bean Validation 约束。

```
@Entity
public class Contact implements Serializable {
    @Id
    @GeneratedValue(strategy = GenerationType.AUTO)
    private Long id;
    @NotNull
    protected String firstName;
    @NotNull
    protected String lastName;
    @Pattern(regexp = "[a-z0-9!#$%&'*+/=?^_`{|}~-]+(?:\\."
            + "[a-z0-9!#$%&'*+/=?^_`{|}~-]+)*@"
            + "(?:[a-z0-9](?:[a-z0-9-]*[a-z0-9])?\\.)+[a-z0-9]"
            + "(?:[a-z0-9-]*[a-z0-9])?",
            message = "{invalid.email}")
    protected String email;
    @Pattern(regexp = "^\\(?(\\d{3})\\)?[- ]?(\\d{3})[- ]?(\\d{4})$",
            message = "{invalid.phonenumber}")
    protected String mobilePhone;
    @Pattern(regexp = "^\\(?(\\d{3})\\)?[- ]?(\\d{3})[- ]?(\\d{4})$",
            message = "{invalid.phonenumber}")
    protected String homePhone;
    @Temporal(javax.persistence.TemporalType.DATE)
    @Past
    protected Date birthday;
    ...
}
```

5.1.3 实体中的主键

每个实体都有唯一的对象标识符。例如，客户实体可以通过客户号码来标识。唯一标识符或主键（Primary Key）使客户端能够定位特定实体实例。每个实体都必须有一个主键。实体可以具有简单主键或复合主键。

简单主键使用 javax.persistence.Id 注解来表示主键属性或字段。

当主键由多个属性组成时，使用复合主键，该属性对应于一组单个持久化属性或字段。复合主键必须在主键类中定义。复合主键使用 javax.persistence.EmbeddedId 和 javax.persistence.IdClass 注解来表示。

主键或复合主键的属性或字段必须是以下 Java 语言类型之一。

- Java 基本数据类型。
- Java 基本数据类型的包装类型。
- java.lang.String。
- java.util.Date（时间类型应为 DATE）。
- java.sql.Date。
- java.math.BigDecimal。
- java.math.BigInteger。

不应在主键中使用浮点类型。如果使用生成的主键，则只有整型类型是可移植的。

主键类必须满足以下要求。

（1）类的访问控制修饰符必须是 public。

（2）如果使用基于属性的访问，主键类的属性必须为 public 或 protected。

（3）该类必须有一个公共默认构造函数。

（4）类必须实现 hashCode() 方法和 equals(Object other) 方法。

（5）类必须是可序列化的。

（6）复合主键必须被表示并映射到实体类的多个字段或属性，或者必须被表示并映射为可嵌入类。

（7）如果类映射到实体类的多个字段或属性，则主键类中的主键字段或属性的名称和类型必须与实体类的名称和类型匹配。

以下主键类是组合键，customerOrder 和 itemId 字段一起唯一标识一个实体。

```
public final class LineItemKey implements Serializable {
    private Integer customerOrder;
    private int itemId;

    public LineItemKey() {}
```

```
    public LineItemKey(Integer order, int itemId) {
        this.setCustomerOrder(order);
        this.setItemId(itemId);
    }

    @Override
    public int hashCode() {
        return ((this.getCustomerOrder() == null
                ? 0 : this.getCustomerOrder().hashCode())
                ^ ((int) this.getItemId()));
    }

    @Override
    public boolean equals(Object otherOb) {
        if (this == otherOb) {
            return true;
        }
        if (!(otherOb instanceof LineItemKey)) {
            return false;
        }
        LineItemKey other = (LineItemKey) otherOb;
        return ((this.getCustomerOrder() == null
                ? other.getCustomerOrder() == null : this.getCustomer-
Order()
                .equals(other.getCustomerOrder()))
                && (this.getItemId() == other.getItemId()));
    }

    @Override
    public String toString() {
        return "" + getCustomerOrder() + "-" + getItemId();
    }
    /* Getters and setters */
}
```

5.1.4 实体间的关系

实体之间一般具有如下类型的关系。

（1）一对一（One to One）：每个实体实例与另一个实体的单个实例相关。例如，要建模其中每个存储仓包含单个窗口小部件的物理仓库，StorageBin 和 Widget 将具有一对一的关系。一对一关系使用对应的持久性属性或字段上的 javax.persistence.OneToOne 注解。

（2）一对多（One to Many）：实体实例可以与其他实体的多个实例相关。例如，销售订单可以有多个订单项。在订单应用程序中，CustomerOrder 将与 LineItem 具有一对多关系。一对多关系使用对应的持久性属性或字段上的 javax.persistence.OneToMany 注解。

（3）多对一（Many to One）：实体的多个实例可以与另一个实体的单个实例相关。这种多重性与一对多关系相反。在刚刚提到的示例中，从 LineItem 的角度来看，与 CustomerOrder 的关系是多对一的。多对一关系在对应的持久性属性或字段上使用 javax.persistence.ManyToOne 注解。

（4）多对多（Many to Many）：实体实例可以与彼此的多个实例相关。例如，每个大学课程有很多学生，每个学生可以采取几个课程。因此，在注册申请中，课程和学生将具有多对多关系。多对多关系使用对应的持久化属性或字段上的 javax.persistence.ManyToMany 注解。

实体关系中的方向可以是双向的或单向的。双向关系具有拥有方（Owner Side）和被拥有方（Inverse Side）。单向关系只有一个拥有方。关系的拥有方决定 Persistence 运行时如何更新数据库中的关系。

1. 双向关系

在双向关系中，每个实体都有一个引用另一个实体的关系字段或属性。通过关系字段或属性，实体类的代码可以访问其相关对象。如果实体具有相关字段，则该实体称为“知道”其相关对象。例如，如果 CustomerOrder 知道它具有什么 LineItem 实例，并且如果 LineItem 知道它属于哪个 CustomerOrder，则它们具有双向关系。

双向关系必须遵循以下规则。

（1）双向关系的反面必须通过使用 @OneToOne、@OneToMany 或 @ManyToMany 注解的 mappedBy 元素引用其拥有方。mappedBy 元素指定实体中作为关系所有者的属性或字段。

（2）多对一双向关系的许多方面不能定义 mappedBy 元素。许多方面总是关系的拥有方。

（3）对于一对一双向关系，拥有侧对应于包含相应外键的一侧。

（4）对于多对多双向关系，任一侧可以是拥有侧。

2. 单向关系

在单向关系中，只有一个实体具有引用另一个实体的关系字段或属性。例如，LineItem 将具有标识产品的关系字段，但产品不具有 LineItem 的关系字段或属性。换句话说，LineItem 知道产品，但产品不知道哪些 LineItem 实例引用它。

3. 查询和关系的方向

Java Persistence 查询语言和 Criteria API 查询通常在关系之间导航。关系的方向确定查询是否可以从一个实体导航到另一个实体。例如，查询可以从 LineItem 导航到 Product，但不能以相反的方向导航。对于 CustomerOrder 和 LineItem，查询可以在两个方向上导航，因为这两个实体具有双向关系。

4. 级联操作和关系

使用关系的实体通常依赖于关系中另一个实体的存在。例如，订单项是订单的一部分；如果订单被删除，订单项也应该被删除，这称为级联删除关系。

javax.persistence.CascadeType 枚举类型定义在各种级联操作。表 5-1 列出了实体的级联操作。

表5-1 实体的级联操作

级联操作	描述
ALL	ALL 级联操作将应用于父实体的相关实体。所有等同于指定 cascade={DETACH, MERGE, PERSIST, REFRESH, REMOVE}
DETACH	如果父实体与持久化上下文分离，则相关实体也将被分离
MERGE	如果父实体被合并到持久化上下文中，则相关实体也将被合并
PERSIST	如果父实体被持久化到持久化上下文中，则相关实体也将被持久化
REFRESH	如果父实体在当前持久化上下文中被刷新，相关实体也将被刷新
REMOVE	如果父实体从当前持久化上下文中删除，相关实体也将被删除

级联删除关系使用 @OneToOne 和 @OneToMany 关系的 cascade=REMOVE 元素指定。例如，

```
@OneToMany(cascade=REMOVE, mappedBy="customer")
public Set<CustomerOrder> getOrders() { return orders; }
```

5. 删除关系中的“孤儿”

当从关系中去除一对一或一对多关系中的目标实体时，通常希望将删除操作级联到目标实体。这样的目标实体被认为是“孤儿”，并且 orphanRemoval 属性可以用于指定应该移除孤立的实体。例如，如果订单有许多订单项，其中一个订单项从订单中删除，则删除的订单项将被视为“孤儿”。如果 orphanRemoval 设置为 true，则在从订单中删除订单项时，订单项实体将被删除。

@OneToMany 和 @oneToOne 中的 orphanRemoval 属性采用布尔值，默认值为 false。

以下示例将在删除客户实体时把删除操作级联到“孤儿”订单实体。

```
@OneToMany(mappedBy="customer", orphanRemoval="true")public List
<CustomerOrder> getOrders() { ... }
```

5.1.5 实体中的可嵌入类

可嵌入类用于表示实体的状态，但不具有它们自己的持久性标识，这点与实体类不同。可嵌入类的实例共享拥有它的实体的身份。可嵌入类仅作为另一个实体的状态存在。实体可以具有单值或集合值可嵌入类属性。

可嵌入类与实体类具有相同的规则，但是使用 javax.persistence.Embeddable 注解，而不是 @Entity 进行注解。

以下可嵌入类 ZipCode 具有字段 zip 和 plusFour。

```
@Embeddable
public class ZipCode {
    String zip;
```

```
    String plusFour;
    ...
}
```

以下是可嵌入类在 Address 实体中的使用。

```
@Entity
public class Address {
    @Id
    protected long id
    String street1;
    String street2;
    String city;
    String province;
    @Embedded
    ZipCode zipCode;
    String country;
    ...
}
```

拥有可嵌入类作为其持久状态的一部分的实体可以使用 javax.persistence.Embedded 来注解字段或属性，但不是必须这样做的。

可嵌入类本身可以使用其他可嵌入类来表示它们的状态，它们还可以包含基本 Java 编程语言类型或其他可嵌入类的集合。可嵌入类也可以包含与其他实体或实体集合的关系，如果可嵌入类有这样的关系，则关系是从目标实体或实体集合到拥有可嵌入类的实体。

5.1.6 实体继承

实体支持类继承、多态关联和多态查询。实体类可以扩展非实体类，非实体类可以扩展实体类。实体类可以是抽象的和具体的。

1. 抽象实体

抽象类可以通过用 @Entity 装饰类来声明一个实体。抽象实体就像具体实体，但不能实例化。

抽象实体可以像具体实体一样被查询。 如果抽象实体是查询的目标，则查询对抽象实体的所有具体子类进行操作。

```
@Entity
public abstract class Employee {
    @Id
    protected Integer employeeId;
    ...
}
@Entity
public class FullTimeEmployee extends Employee {
    protected Integer salary;
```

```
    ...
}
@Entity
public class PartTimeEmployee extends Employee {
    protected Float hourlyWage;
}
```

3. 映射超类

实体可以从包含持久状态和映射信息但不是实体的超类继承。也就是说，超类不使用 @Entity 注解进行修饰，并且不由 Java Persistence 提供程序映射为实体。当具有多个实体类共用的状态和映射信息时，最常使用这些超类。

映射的超类通过用注解 javax.persistence.MappedSuperclass 装饰类来指定。

```
@MappedSuperclass
public class Employee {
    @Id
    protected Integer employeeId;
    ...
}
@Entity
public class FullTimeEmployee extends Employee {
    protected Integer salary;
    ...
}
@Entity
public class PartTimeEmployee extends Employee {
    protected Float hourlyWage;
    ...
}
```

映射的超类不能被查询，不能在 EntityManager 或 Query 操作中使用。必须在 EntityManager 或 Query 操作中使用映射超类的实体子类。映射的超类不能是实体关系的目标，映射的超类可以是抽象的或具体的。

映射的超类在底层数据存储中没有任何对应的表，从映射的超类继承的实体定义表映射。例如，在上面的代码示例中，基础表将是 FULLTIMEEMPLOYEE 和 PARTTIMEEMPLOYEE，但没有 EMPLOYEE 表。

3. 非实体超类

实体可以具有非实体超类，并且这些超类可以是抽象的或具体的。非实体超类的状态是非持久的，并且通过实体类从非实体超类继承的任何状态是非持久的。非实体超类不能在 EntityManager 或 Query 操作中使用，将忽略非实体超类中的任何映射或关系注解。

4. 实体继承映射策略

javax.persistence.Inheritance 注解用来指定对象继承关系持久化的方式。以下映射策略用于将实体数据映射到底层数据库。

（1）每个类层次结构的单个表。

（2）每个具体实体类的表。

（3）“连接”（Join）策略，其中特定于子类的字段或属性被映射到与父类共同的字段或属性不同的表。

策略通过将 @Inheritance 的 strategy 元素设置为 javax.persistence.InheritanceType 枚举类型中定义的选项之一来配置。

```
public enum InheritanceType {
    SINGLE_TABLE,
    JOINED,
    TABLE_PER_CLASS
};
```

如果未在实体层次结构的根类上指定 @Inheritance 注解，则使用默认策略 InheritanceType.SINGLE_TABLE。

5. 连接子类策略

在连接子类策略中，对应于 InheritanceType.JOINED，类层次结构的根由单个表表示，每个子类具有单独的表，其中仅包含特定于该子类的字段。也就是说，子类表不包含用于继承字段或属性的列。子类表还具有表示其主键的一个或多个列，其是超类表的主键的外键。

连接子类策略为多态关系提供良好的支持，但需要在实例化实体子类时执行一个或多个连接操作，这可能导致广泛的类层次结构的性能差。类似地，覆盖整个类层次结构的查询需要子类表之间的连接操作，从而导致性能降低。

某些 Java Persistence API 提供程序（包括 GlassFish Server 中的默认提供程序）在使用连接的子类策略时，需要有一个与根实体对应的鉴别器列（Discriminator Column）。如果未在应用程序中使用自动表创建，请确保针对标识符列默认值正确设置数据库表，或使用 @DiscriminatorColumn 注解来匹配数据库模式。

5.1.7 管理实体

实体由实体管理器管理，它由 javax.persistence.EntityManager 实例表示。 每个 EntityManager 实例与持久上下文相关联（存在于特定数据存储中的一组被管实体实例）。 持久化上下文定义了创建、持久化和删除特定实体实例的范围。 EntityManager 接口定义用于与持久性上下文进行交互的方法，可以用于创建和删除持久实体实例，通过实体的主键查找实体，并允许在实体上运行查询。

1. 容器管理的实体管理器

对于容器管理的实体管理器，EntityManager 实例的持久化上下文由容器自动传播到在单个

Java Transaction API（JTA）事务中使用 EntityManager 实例的所有应用程序组件。

JTA 事务通常涉及跨应用程序组件的调用。要完成 JTA 事务，这些组件通常需要访问单个持久性上下文。当 EntityManager 通过 javax.persistence.PersistenceContext 注解注入应用程序组件时，会发生这种情况。持久化上下文自动地与当前 JTA 事务一起传播，并且映射到相同持久性单元的 EntityManager 引用提供对该事务内的持久性上下文的访问。通过自动传播持久化上下文，应用程序组件不需要将对 EntityManager 实例的引用彼此传递，以便在单个事务中进行更改。Java EE 容器管理着实体管理器的生命周期。

要获取 EntityManager 实例，请将实体管理器注入应用程序组件。

```
@PersistenceContext
EntityManager em;
```

2. 应用程序管理的实体管理器

另外，对于应用程序管理的实体管理器，持久化上下文不会传播到应用程序组件，并且 EntityManager 实例的生命周期由应用程序管理。

当应用程序需要访问不通过特定持久性单元中的 EntityManager 实例使用 JTA 事务传播的持久性上下文时，使用应用程序管理的实体管理器。在这种情况下，每个 EntityManager 创建一个新的、隔离的持久上下文。 EntityManager 及其相关联的持久化上下文由应用程序显式创建和销毁，它们也用于直接注入 EntityManager 实例时无法完成，因为 EntityManager 实例不是线程安全的，EntityManagerFactory 实例是线程安全的。在这种情况下，应用程序使用 javax.persistence.EntityManagerFactory 的 createEntityManager 方法创建 EntityManager 实例。

要获取 EntityManager 实例，首先必须通过 javax.persistence.PersistenceUnit 注解将 EntityManager Factory 实例注入应用程序组件中。

```
@PersistenceUnit
EntityManagerFactory emf;
```

从 EntityManagerFactory 获得 EntityManager。

```
EntityManager em = emf.createEntityManager();
```

应用程序管理的实体管理器不会自动传播 JTA 事务上下文。 当执行实体操作时，这样的应用程序需要手动地获得对 JTA 事务管理器的访问并且添加事务划分信息。 javax.transaction.User Transaction 接口定义用于开始、提交和回滚事务的方法。 通过创建用 @Resource 注解的实例变量来注入 UserTransaction 的实例。

```
@Resource
UserTransaction utx;
```

要开始事务，请调用 UserTransaction.begin 方法。当所有实体操作完成时，调用 UserTransaction.

commit 方法提交事务。UserTransaction.rollback 方法用于回滚当前事务。

以下示例显示如何在使用应用程序管理的实体管理器的应用程序中管理事务。

```
@PersistenceUnit
EntityManagerFactory emf;
EntityManager em;
@Resource
UserTransaction utx;
...
em = emf.createEntityManager();
try {
    utx.begin();
    em.persist(SomeEntity);
    em.merge(AnotherEntity);
    em.remove(ThirdEntity);
    utx.commit();
} catch (Exception e) {
    utx.rollback();
}
```

3. 使用 EntityManager 查找实体

EntityManager.find 方法用于通过实体的主键在数据存储中查找实体。

```
@PersistenceContext
EntityManager em;
public void enterOrder(int custID, CustomerOrder newOrder) {
    Customer cust = em.find(Customer.class, custID);
    cust.getOrders().add(newOrder);
    newOrder.setCustomer(cust);
}
```

4. 管理实体实例的生命周期

通过 EntityManager 实例调用实体上的操作来管理实体实例。实体实例处于 4 种状态之一：新建（New）、受管（Managed）、分离（Detached）或删除（Removed）。

（1）新建实体实例没有持久性标识，并且尚未与持久性上下文相关联。

（2）受管实体实例具有持久性标识，并与持久性上下文相关联。

（3）分离的实体实例具有持久性标识，并且当前不与持久性上下文相关联。

（4）删除的实体具有持久性标识，与持久上下文相关联，并且被调度为从数据存储中移除。

5. 持久化实体实例

新建实体实例通过调用 persist 方法或通过从关系注解中设置了 cascade=PERSIST 或 cascade=ALL 元素的相关实体调用的级联 persist 操作来成为管理和持久性，这意味着当与 persist 操作相关联的事务完成时，实体的数据被存储到数据库。如果实体已经被管理，则 persist 操作将被忽略，尽管 persist 操作将级联到在关系注解中具有级联元素设置为 PERSIST 或 ALL 的相关实体。

如果在删除的实体实例上调用 persist，则该实体将成为受管实体。如果实体被分离，则 persist 将抛出 IllegalArgumentException，否则事务提交将失败。以下方法执行 persist 操作。

```
@PersistenceContext
EntityManager em;
...
public LineItem createLineItem(CustomerOrder order, Product product,
        int quantity) {
    LineItem li = new LineItem(order, product, quantity);
    order.getLineItems().add(li);
    em.persist(li);
    return li;
}
```

persist 操作传播到与关系注解中的 cascade 元素设置为 ALL 或 PERSIST 的调用实体相关的所有实体。

```
@OneToMany(cascade=ALL, mappedBy="order")
public Collection<LineItem> getLineItems() {
    return lineItems;
}
```

6. 删除实体实例

通过调用 remove 方法或通过从关系注解中设置了 cascade=REMOVE 或 cascade=ALL 元素的相关实体调用的级联 remove 操作来删除托管实体实例。如果对新实体调用 remove 方法，则忽略 remove 操作，尽管 remove 将级联到在关系注解中将 cascade 元素设置为 REMOVE 或 ALL 的相关实体。如果在分离的实体上调用 remove，则 remove 将抛出 IllegalArgumentException，否则事务提交将失败。如果在已删除的实体上调用，则会忽略 remove。当事务完成或作为 flush 操作的结果时，实体的数据将从数据存储中移除。

在以下示例中，与订单关联的所有 LineItem 实体也被删除，因为 CustomerOrder.getLineItems 在关系注解中设置了 cascade=ALL。

```
public void removeOrder(Integer orderId) {
    try {
        CustomerOrder order = em.find(CustomerOrder.class, orderId);
        em.remove(order);
    }
...
```

7. 将实体数据同步到数据库

当与实体相关联的事务提交时，持久实体的状态被同步到数据库。如果被管实体与另一个被管实体具有双向关系，则数据将基于关系的所有者侧被持久化。

要强制将受管实体同步到数据存储，请调用 EntityManager 实例的 flush 方法。如果实体与另一个实体相关，并且关系注解具有设置为 PERSIST 或 ALL 的 cascade 元素，则当调用 flush 时，相关实体的数据将与数据存储器同步。

如果实体被删除，调用 flush 将从数据存储中删除实体数据。

8. 持久单元

持久单元（Persistence Uni）定义由应用程序中的 EntityManager 实例管理的所有实体类的集合，这组实体类表示包含在单个数据存储中的数据。

持久单元由 persistence.xml 配置文件定义。 以下是 persistence.xml 文件的示例。

```
<persistence>
    <persistence-unit name="OrderManagement">
        <description>This unit manages orders and customers.
            It does not rely on any vendor-specific features and can
            therefore be deployed to any persistence provider.
        </description>
        <jta-data-source>jdbc/MyOrderDB</jta-data-source>
        <jar-file>MyOrderApp.jar</jar-file>
        <class>com.widgets.CustomerOrder</class>
        <class>com.widgets.Customer</class>
    </persistence-unit>
</persistence>
```

此文件定义名为 OrderManagement 的持久单元，该单元使用支持 JTA 的数据源 jdbc/MyOrderDB。jar-file 和 class 元素指定管理的持久类：实体类、可嵌入类和映射超类。jar-file 元素指定对打包持久单元可见的 JAR 文件，打包持久单元包含了受管理的持久类。而 class 元素则显示地指定受管理的持久类。

jta-data-source（用于 JTA 感知的数据源）和 non-jta-data-source（用于非 JTA 感知的数据源）元素指定要由容器使用的数据源的全局 JNDI 名称。

META-INF 目录包含 persistence.xml 的 JAR 文件或目录称为持久单元的根。持久单元的范围由持久单元的根确定。每个持久单元必须使用对持久单元范围唯一的名称标识。

持久单元可以打包为 WAR 或 EJB JAR 文件的一部分，也可以打包为 JAR 文件，然后将其包含在 WAR 或 EAR 文件中。

（1）如果将持久单元打包为 EJB JAR 文件中的一组类，那么 persistence.xml 应放在 EJB JAR 的 META-INF 目录中。

（2）如果将持久单元打包为 WAR 文件中的一组类，persistence.xml 应位于 WAR 文件的 WEBINF/classes/META-INF 目录中。

（3）如果将持久单元打包到将包含在 WAR 或 EAR 文件中的 JAR 文件中，则 JAR 文件应位于 WAR 的 WEB-INF/lib 目录或者 EAR 文件的库目录。

注意：在 Java Persistence API 1.0 中，JAR 文件可以位于 EAR 文件的根位置，作为持久单元的根，这个在新版本中不再支持。便携式应用程序应该使用 EAR 文件的库目录作为持久单元的根。

5.1.8 查询实体

（1）Java Persistence API 提供以下用于查询实体的方法。

（2）Java 持久化查询语言（Java Persistence Query Language，JPQL）是一种简单的基于字符串的语言，类似于用于查询实体及其关系的 SQL。

Criteria API 用于使用 Java 编程语言 API 来创建类型安全查询，以查询实体及其关系。

JPQL 和 Criteria API 都有优点和缺点。JPQL 查询通常比 Criteria 查询更简捷和更可读。熟悉 SQL 的开发人员会发现很容易学习 JPQL 的语法。JPQL 命名查询可以在实体类中使用 Java 编程语言注解或在应用程序的部署描述符中定义。但是，JPQL 查询不是类型安全的，并且在从实体管理器检索查询结果时需要转换，这意味着在编译时可能不会捕获类型转换错误。JPQL 查询不支持开放式参数。

Criteria 查询允许开发人员在应用程序的业务层中定义查询。尽管使用 JPQL 动态查询也是可能的，但是 Criteria 查询提供更好的性能，因为每次调用 JPQL 动态查询时都必须解析。Criteria 查询是类型安全的，因此不需要转换，就像 JPQL 查询那样。Criteria API 只是另一种 Java 编程语言 API，不需要开发人员学习另一种查询语言的语法。Criteria 查询通常比 JPQL 查询更详细，并且需要开发人员在向实体管理器提交查询之前创建多个对象并对这些对象执行操作。

5.1.9 数据库模式创建

持久化提供程序可以配置为在应用程序部署期间使用应用程序部署描述符中的标准属性自动创建数据库表、将数据加载到表中及删除表，这些任务通常在发布的开发阶段使用，而不是针对生产数据库。

以下是 persistence.xml 部署描述符的示例，它指定提供程序应使用提供的脚本删除所有数据库工件、使用提供的脚本创建工件，以及在部署应用程序时从提供的脚本加载数据。

```
<?xml version="1.0" encoding="UTF-8"?>
<persistence version="2.1" xmlns="http://xmlns.jcp.org/xml/ns/persis-
tence"
 xmlns:xsi="http://www.w3.org/2001/XMLSchema-instance"
 xsi:schemaLocation="http://xmlns.jcp.org/xml/ns/persistence
 http://xmlns.jcp.org/xml/ns/persistence/persistence_2_1.xsd">
  <persistence-unit name="examplePU" transaction-type="JTA">
    <jta-data-source>java:global/ExampleDataSource</jta-data-source>
    <properties>
        <property name="javax.persistence.schema-generation.database.
action"
```

```
                    value="drop-and-create"/>
          <property name="javax.persistence.schema-generation.create-
source"
                    value="script"/>
          <property name="javax.persistence.schema-generation.create-
script-source"
                    value="META-INF/sql/create.sql" />
          <property name="javax.persistence.sql-load-script-source"
                    value="META-INF/sql/data.sql" />
          <property name="javax.persistence.schema-generation.drop-source"

                    value="script" />
          <property name="javax.persistence.schema-generation.drop-script-
source"
                    value="META-INF/sql/drop.sql" />
      </properties>
    </persistence-unit>
</persistence>
```

1. 配置应用程序以创建或删除数据库表

javax.persistence.schema-generation.database.action 属性用于指定部署应用程序时持久化提供程序所采取的操作。如果未设置属性，持久化提供程序将不会创建或删除任何数据库工件，如表 5-2 所示。

表5-2　设置属性参数

设置	描述
none	不会创建或删除模式
create	提供程序将在应用程序部署上创建数据库工件。应用程序重新部署后，工件将保持不变
drop-and-create	数据库中的任何工件将被删除，并且提供程序将在部署时创建数据库工件
drop	应用程序部署时将删除数据库中的任何工件

在此示例中，持久化提供程序将删除任何剩余的数据库工件，然后在部署应用程序时创建工件。

```
<property name="javax.persistence.schema-generation.database.action"
          value="drop-and-create"/>
```

默认情况下，持久单元中的对象 / 关系元数据用于创建数据库工件。还可以提供供应商用来创建和删除数据库工件的脚本。javax.persistence.schema-generation.create-source 和 javax.persistence.schema-generation.drop-source 属性控制提供程序将如何创建或删除数据库工件，如表 5-3 所示。

表5-3　属性参数的设置及描述

设置	描述
metadata	使用应用程序中的对象/关系元数据创建或删除数据库工件
script	使用提供的脚本创建或删除数据库工件
metadata-then-script	使用对象/关系元数据的组合，然后使用用户提供的脚本创建或删除数据库工件
script-then-metadata	使用用户提供的脚本，然后使用对象/关系元数据来创建和删除数据库工件

在此示例中，持久化提供程序将使用在应用程序中打包的脚本来创建数据库工件。

```
<property name="javax.persistence.schema-generation.create-source"
          value="script"/>
```

如果在 create-source 或 drop-source 中指定脚本，请使用 javax.persistence.schema-generation.create-script-source 或 javax.persistence.schema-generation.drop-script-source 属性指定脚本的位置。脚本的位置是相对于持久单元的根。

```
<property name="javax.persistence.schema-generation.create-script-
source"
          value="META-INF/sql/create.sql" />
```

在上面的示例中，create-script-source 设置为 META-INF/sql 目录中名为 create.sql 的 SQL 文件，相对于持久单元的根。

2. 使用SQL脚本加载数据

如果要在应用程序加载之前使用数据填充数据库表，请在 javax.persistence.sql-load-script-source 属性中指定加载脚本的位置，此属性中指定的位置是相对于持久单元的根。

5.2 Spring Data JPA

Spring Data JPA 是更大的 Spring Data 家族（http://projects.spring.io/spring-data）的一部分，从而使得轻松实现基于 JPA 的存储库变得更容易，该模块用于处理对基于 JPA 的数据访问层的增强支持。它使得更容易构建基于使用 Spring 数据访问技术栈的应用程序。

通过上一节的学习，知道了 JPA 是一套规范，这样在使用不同 ORM 实现时，可以只需要关注 JPA 中的 API，而无须关注具体的实现。但同时，JPA 也提供了 EntityManager 接口来管理实体。

然而，Spring Data JPA 对于 JPA 的支持则是更近一步。使用 Spring Data JPA 开发者无须过多关注 EntityManager 的创建、事务处理等 JPA 相关的处理，这基本上也是作为一个开发框架而言所能做到的极限了，甚至 Spring Data JPA 让开发者连实现持久层业务逻辑的工作都省了，唯一要做的只是声明持久层的接口，其他都交给 Spring Data JPA 来完成。

Spring Data JPA 就是这么强大，让数据持久层开发工作简化，只需声明一个接口。例如，开发者声明了一个 findUserById() 方法，Spring Data JPA 就能判断出这是根据给定条件的 ID 查询出满足条件的 User 对象，而其中的实现过程开发者无须关心，这一切都交予 Spring Data JPA 来完成。

5.2.1 Spring Data 的含义

Spring Data 是一个用于简化数据库访问，并支持云服务的开源框架，其主要目标是使对数据的访问变得方便快捷，并支持 map-reduce 框架和云计算数据服务①。Spring Data 包含多个子项目。

（1）Spring Data Commons：提供共享的基础框架，适合各个子项目使用，支持跨数据库持久化。

（2）Spring Data JPA：简化创建 JPA 数据访问层和跨存储的持久层功能。

（3）Spring Data Hadoop：基于 Spring 的 Hadoop 作业配置和一个 POJO 编程模型的 MapReduce 作业。

（4）Spring Data KeyValue：集成了 Redis 和 Riak，提供多个常用场景下的简单封装。

（5）Spring Data JDBC Extensions：支持 Oracle RAD、高级队列和高级数据类型。

本书中也用到了 Spring Data 家族中的 Spring Data Elasticsearch（http://projects.spring.io/spring-data-elasticsearch/）、Spring Data Mongodb（http://projects.spring.io/spring-data-mongodb/）等，这些都是针对 Elasticsearch、Mongodb 等 NoSQL 提供了数据访问层框架。

简而言之，Spring Data 旨在统一包括数据库系统和 NoSQL 数据存储在内不同持久化存储的访问方式，让开发者通过统一的接口进行功能的实现。

5.2.2 Spring Data JPA的特性

Spring Data JPA 是对 JPA 规范的实现。

对于普通开发者而言，自己实现应用程序的数据访问层是一件极其烦琐的过程。开发者必须编写太多的样板代码来执行简单查询、分页和审计。Spring Data JPA 旨在通过将工作量减少到实际需要的量来显著改进数据访问层的实现。作为开发人员，只需要编写存储库的接口，包括自定义查询方法，而这些接口的实现，Spring Data JPA 将会自动提供。

Spring Data JPA 包含如下特征。

（1）基于 Spring 和 JPA 来构建复杂的存储库。

（2）支持 Querydsl（http://www.querydsl.com）谓词，因此支持类型安全的 JPA 查询。

---

① 有关 map-reduce 框架和云计算数据服务的内容，可参阅笔者所著的《分布式系统常用技术及案例分析》一书。

（3）域类的透明审计。

（4）具备分页支持、动态查询执行、集成自定义数据访问代码的能力。

（5）在引导时验证 @Query 带注解的查询。

（6）支持基于 XML 的实体映射。

（7）通过引入 @EnableJpaRepositories 来实现基于 JavaConfig 的存储库配置。

5.2.3 如何使用 Spring Data JPA

在项目中使用 Spring Data JPA 的推荐方法是使用依赖关系管理系统。下面是使用 Gradle 构建的示例。

```
dependencies {
    compile 'org.springframework.data:spring-data-jpa:2.0.0.M4'
}
```

在代码中只需声明继承自 Spring Data JPA 中的接口。

```
import org.springframework.data.jpa.repository.JpaRepository;
...

public interface UserRepository extends JpaRepository<User, Long>{

      List<User> findByNameLike(String name);

}
```

在这个例子中代码继承了 Spring Data JPA 中的 JpaRepository 接口，而后声明相关的方法即可。例如，声明 findByNameLike，就能自动实现通过名称来模糊查询的方法。

5.2.4 核心概念

Spring Data 存储库抽象中的中央接口是 Repository，它将域类及域类的 id 类型作为类型参数进行管理，此接口主要作为标记接口捕获要使用的类型，并帮助开发者发现扩展此接口。而 Crud Repository 为受管理的实体类提供复杂的 CRUD 功能。

```
public interface CrudRepository<T, ID extends Serializable>
    extends Repository<T, ID> {

    <S extends T> S save(S entity);   // (1)

    T findOne(ID primaryKey);         // (2)

    Iterable<T> findAll();            // (3)
```

```
    Long count();                    // (4)

    void delete(T entity);           // (5)

    boolean exists(ID primaryKey);   // (6)

    ...// 省略更多方法
}
```

CrudRepository 接口中的方法含义如下。

- 保存给定实体。
- 返回由给定 id 标识的实体。
- 返回所有实体。
- 返回实体的数量。
- 删除给定的实体。
- 指示是否存在具有给定 ID 的实体。

同时还提供其他特定的持久化技术的抽象，如 JpaRepository 或 MongoRepository，这些接口扩展了 CrudRepository。

在 CrudRepository 的顶部有一个 PagingAndSortingRepository 抽象，它增加了额外的方法来简化对实体的分页访问。

```
public interface PagingAndSortingRepository<T, ID extends Serializable>
  extends CrudRepository<T, ID> {

  Iterable<T> findAll(Sort sort);

  Page<T> findAll(Pageable pageable);
}
```

例如，想访问用户的第二页的页面大小为 20，可以简单地做这样的事情。

```
PagingAndSortingRepository<User, Long> repository = // … 获取 beanPage
<User> users = repository.findAll(new PageRequest(1, 20));
```

除了查询方法外，还可以使用计数和删除查询。

派生计数查询。

```
public interface UserRepository extends CrudRepository<User, Long> {

  Long countByLastname(String lastname);
}
```

派生删除查询。

```
public interface UserRepository extends CrudRepository<User, Long> {
```

```
  Long deleteByLastname(String lastname);

  List<User> removeByLastname(String lastname);

}
```

5.2.5 查询方法

对于底层数据存储的管理，通常使用标准 CRUD 功能的资源库来实现。使用 Spring Data 声明这些查询将会变得简单，只需要 4 步过程。

1. 声明扩展Repository或其子接口之一的接口

声明接口并输入将处理的域类和 ID 类型。

interface PersonRepository extends Repository<Person, Long> {...}

2. 在接口上声明查询方法

```
interface PersonRepository extends Repository<Person, Long> {
  List<Person> findByLastname(String lastname);
}
```

3. 为这些接口创建代理实例

可以通过 JavaConfig 的方式。

```
interface PersonRepository extends Repository<Person, Long> {
  List<Person> findByLastname(String lastname);
}
```

或通过 XML 配置方式。

```
<?xml version="1.0" encoding="UTF-8"?>
<beans xmlns="http://www.springframework.org/schema/beans"
   xmlns:xsi="http://www.w3.org/2001/XMLSchema-instance"
   xmlns:jpa="http://www.springframework.org/schema/data/jpa"
   xsi:schemaLocation="http://www.springframework.org/schema/beans
     http://www.springframework.org/schema/beans/spring-beans.xsd
     http://www.springframework.org/schema/data/jpa
     http://www.springframework.org/schema/data/jpa/spring-jpa.xsd">

   <jpa:repositories base-package="com.waylau.repositories"/>

</beans>
```

在此示例中使用了 JPA 命名空间。如果使用任何其他存储库的存储库抽象，则需要将其更改为存储模块的相应命名空间。

另外，请注意，JavaConfig 变量不会明确配置包，因为默认情况下使用注解类的包。如果自定

义要扫描的程序包，请使用数据存储特定存储库的 @Enable... 注解。例如，

```
@EnableJpaRepositories(basePackages = "com.waylau.repositories.jpa")
@EnableMongoRepositories(basePackages = "com.waylau.repositories.mongo")
interface Configuration { }
```

4. 获取注入的存储库实例并使用它

```
public class SomeClient {

  @Autowired
  private PersonRepository repository;

  public void doSomething() {
    List<Person> persons = repository.findByLastname("Lau");
  }
}
```

5.2.6 定义资源库的接口

首先需要定义实体类的接口，接口必须继承资源库并且输入实体类型和 ID 类型，如果需要用到 CRUD 方法，可以使用 CrudRepository 来替代 Repository。

1. 自定义接口

通常，存储库接口将会扩展 Repository、CrudRepository 或 PagingAndSortingRepository 等这些 Spring Data 接口。另外，如果不想继承 Spring Data 接口，还可以在接口上添加 @Repository Definition 注解，用以声明这是一个 Repository 接口。扩展 CrudRepository 将会公开一套完整的方法来操作开发者的实体。 如果开发者喜欢其中的方法来调用，也可以简单地复制 CrudRepository 中的部分方法到自己的 repository。

这允许开发者自定义数据库的功能的抽象。

下面是一个有选择地公开 CRUD 方法的例子。

```
@NoRepositoryBean
interface MyBaseRepository<T, ID extends Serializable> extends Repository<T,
ID> {

  T findOne(ID id);

  T save(T entity);
}

interface UserRepository extends MyBaseRepository<User, Long> {
  User findByEmailAddress(EmailAddress emailAddress);
}
```

第一步定义了一个公共基础的接口 findOne(...) 和 save(...) 方法，这些方法将会引入 Spring Data 的实现类中，如 SimpleJpaRepository，因为它们匹配 CrudRepository 的方法签名，所以 User Repository 将会具备 save(...) 的功能和 findOne(...) 的功能，当然也具备 findByEmailAddress 的功能。

注意：如果中间的资源库接口添加了 @NoRepositoryBean 注解，这样运行时，Spring Data 将不会创建拥有该注解的实例。

2. 使用 Spring Data 多模块来创建资源库

使用单个 Spring Data 模块在应用中非常简单，但有时需要多个 Spring Data 模块，例如，需要定义的资源库需要去区分两种不同的持久化技术，如果在 classpath 中发现多个资源库时， Spring Data 会进行严格的配置限制，确保每个资源库或实体决定绑定哪个 Spring Data 模块。

（1）如果资源库定义了继承特定的资源库，那么它是一个特定的 Spring Data 模块。

（2）如果实体注解了一个特定的声明，它是一个特定的 Spring Data 模块。Spring Data 模块可以接纳第三方的声明，如 JPA 的 @Entity，或者提供来自 Spring Data MonggoDB/Spring Data Elasticsearch 的 @Document 。

下面是自定义特定模块接口的资源库。

```
interface MyRepository extends JpaRepository<User, Long> { }

@NoRepositoryBean
interface MyBaseRepository<T, ID extends Serializable> extends JpaRe-
pository<T, ID> {
  …
}

interface UserRepository extends MyBaseRepository<User, Long> {
  …
}
```

MyRepository 和 UserRepository 继承于 JpaRepository，在这个层级中是对 Spring Data JPA 模块的合法替代。

使用一般的接口定义的资源库如下。

```
interface AmbiguousRepository extends Repository<User, Long> {
 …
}

@NoRepositoryBean
interface MyBaseRepository<T, ID extends Serializable> extends CrudRe-
pository<T, ID> {
  …
}

interface AmbiguousUserRepository extends MyBaseRepository<User, Long>
```

```
{
  …
}
```

AmbiguousRepository 和 AmbiguousUserRepository 仅在它们的层级来继承 Repository 和 CrudRepostory，当它们使用单个 Spring Data 模块的时候是完美的，但是当使用多模块 Spring Data 时， Spring 将无法区分每个资源库的范围。

下面的例子是使用实体类注解来定义资源库的使用范围。

```
interface PersonRepository extends Repository<Person, Long> {
 …
}

@Entity
public class Person {
  …
}

interface UserRepository extends Repository<User, Long> {
 …
}

@Document
public class User {
  …
}
```

PersonRepository 所引用的 Person 使用了 @Entity 注解，所以这个仓库清晰地使用了 Sping Data JPA。 UserRepository 所引用的 User 声明了 @Document，表明这个仓库将使用 Spring Data MongoDB 模块。

下面是使用混合的注解来定义资源库的例子。

```
interface JpaPersonRepository extends Repository<Person, Long> {
 …
}

interface MongoDBPersonRepository extends Repository<Person, Long> {
 …
}

@Entity
@Document
public class Person {
  …
}
```

这个例子中实体类 Person 使用了 JPA 和 Spring Data MongoDB 两种注解，表明这个实体类既可以用于 JpaPersonRepository，也可以用于 MongoDBPersonRepository，Spring Data 因不能区分类型而导致未定义的行为。

在同一个域类型上使用多个持久化技术特定的注释可以跨多个持久性技术重用域类型，但是 Spring Data 不再能够确定一个唯一的模块来绑定存储库。

最后，还可以使用包路径来区分不同的仓库类型。不同的包路径下的仓库使用不同的仓库类型，通过在配置类 Configuration 中声明注解来实现，也可以通过 XML 配置来定义。

通过注解来实现不同包路径下使用不同的仓库。

```
@EnableJpaRepositories(basePackages = "com.waylau.repositories.jpa")
@EnableMongoRepositories(basePackages = "com.waylau.repositories.mongo")
interface Configuration { }
```

5.2.7 定义查询方法

资源库代理有两种方法去查询，可以根据方法名或者自定义查询。可用的选项取决于实际的存储。但不管如何，必须要有一个策略来决定创建什么实际查询。下面来看一下可用的选项。

1. 查询查找策略

以下策略可供查询库基础设施来解决。开发者可以配置策略名称空间，通过 query-lookup-strategy 属性的 XML 配置或通过 queryLookupStrategy 启用的属性 ${store} 库注解的 Java 配置。一些策略可能不支持特定的数据存储。

（1）CREATE 尝试从查询方法名称中来构造特定的查询语句。一般的方法是从方法名称中移除一组已知的前缀，然后解析方法的其余部分。

（2）USE_DECLARED_QUERY 试图找到一个声明查询，没有找到就抛出一个异常。查询可以定义在注解上。

（3）CREATE_IF_NOT_FOUND，如果不使用任何显式配置，它是默认选项，它结合了 CREATE 和 USE_DECLARED_QUERY。它首先查找已声明的查询，如果未找到已声明的查询，则会创建一个基于名称的自定义查询。它允许通过方法名称进行快速查询定义，但也可以通过根据需要引入已声明的查询来自定义这些查询。

2. 创建查询

内置到 Spring Data 存储库基础结构中的查询构建器机制对于在存储库的实体上构建约束查询很有用。 机制剥离前缀 find...By、read...By、query...By、count...By 和 get...By 从该方法开始解析其余部分。 引入子句可以包含其他表达式，如在要创建的查询上设置不同标志的区别。 但是，第一个 By 作为分隔符指示实际标准的开始。在非常基本的层次上，可以定义实体属性的条件，并将它们与 And 和 Or 连接。

下面是根据方法名创建查询的例子。

```
public interface PersonRepository extends Repository<User, Long> {

  List<Person> findByEmailAddressAndLastname(EmailAddress emailAddress,
String lastname);

  // 启用distinct标志
  List<Person> findDistinctPeopleByLastnameOrFirstname(String lastname,
String firstname);
  List<Person> findPeopleDistinctByLastnameOrFirstname(String lastname,
String firstname);

  // 给独立的属性启用ignore case
  List<Person> findByLastnameIgnoreCase(String lastname);

  // 给所有合适的属性启用ignore case
  List<Person> findByLastnameAndFirstnameAllIgnoreCase(String lastname,
String firstname);

  // 启用ORDER BY
  List<Person> findByLastnameOrderByFirstnameAsc(String lastname);
  List<Person> findByLastnameOrderByFirstnameDesc(String lastname);
}
```

实际结果的解析方法取决于持久化存储所创建的查询。须注意以下问题。

（1）表达式通常可以在运算符组合的属性上进行遍历。开发者可以组合属性表达式 AND 和 OR，还可获得对诸如 Between、LessThan、GreaterThan 之间的运算符的支持，对于属性表达式，受支持的操作符可能因数据存储方式而异。

（2）方法解析器支持为单个属性设置 IgnoreCase 标志 [如 findByLastnameIgnoreCase(...)] 也可以是所有属性 [通常为 String 实例，如 findByLastnameAndFirstnameAllIgnoreCase(...)]。是否支持 ignore case 写取决于具体的存储方式。

（3）可以通过将 OrderBy 子句附加到引用属性的查询方法并提供排序方向（Asc 或 Desc）来应用静态排序。

3. 属性表达式

属性表达式只能引用受管理实体的直接属性，如前面的示例。 在查询创建时，开发者已经确保已解析的属性是受管域类的属性。但是，也可以通过遍历嵌套属性来定义约束。 假设一个 Person 有一个带有 ZipCode 的 Address。在这种情况下，方法名称为：

```
List<Person> findByAddressZipCode(ZipCode zipCode);
```

解析算法首先将整个部分（AddressZipCode）解释为属性，然后检查域类是否具有该属性名称。如果算法解析成功，则使用该属性。如果未解析成功，算法将继续解析最右侧的驼峰分割的单词，此时会先分为头部和尾部（在我们的示例就是 AddressZip 和 Code），并试图找到相应的属性。如果算法找到一个头部的属性，它将从尾部那继续建立搜索树，按照刚刚描述的方式分割尾部。如果第一个分割不匹配，则算法将分割点移到左侧（即拆分为 Address 和 ZipCode）并继续上述算法。

虽然这应该适用于大多数情况下，算法可能选择错误的属性。假设 Person 类也有一个 addressZip 属性。该算法将在第一个分割循环中匹配，并且基本上选择错误的属性，最后失败（因为 addressZip 的类型可能没有 code 属性）。

要解决这种模糊性，可以在方法名称中使用下画线（_）手动定义遍历点。所以方法名称最终会这样。

```
List<Person> findByAddress_ZipCode(ZipCode zipCode);
```

由于将下画线（_）视为保留字符，强烈建议遵循标准 Java 命名约定（即不在属性名称中使用下画线，而是使用驼峰案例）。

4. 特殊参数处理

要处理查询中的参数，只需定义方法参数，如上面的示例中所示。 此外，基础设施将识别某些特定类型，如 Pageable 和 Sort，以动态地对查询应用进行分页和排序。

下面使用 Pageable、Slice 和 Sort 来查询。

```
Page<User> findByLastname(String lastname, Pageable pageable);

Slice<User> findByLastname(String lastname, Pageable pageable);

List<User> findByLastname(String lastname, Sort sort);

List<User> findByLastname(String lastname, Pageable pageable);
```

第一个方法允许在查询方法的静态定义查询中通过一个 org.springframework.data.domain.Pageable 实例来动态地添加分页。分页类知道元素的总数和可用页数，它通过基础库来触发一个统计查询计算所有的总数。由于这个查询可能对存储库消耗巨大，可以使用 Slice 来替代。Slice 仅仅知道是否有下一个 Slice 可用，这对查询大数据已经足够了。

排序选项与分页的处理方式一样，如果开发者需要排序，简单地添加一个 org.springframework.data.domain.Sort 参数到自己定义的方法即可。也正因为如此，简单地返回一个列表也是可以的，在这种情况下，将不会创建构建实际页面实例所需的附加元数据（这反过来意味着将不必发出额外的计数查询），而是简单地限制查询，以仅查找给定范围实体。

要找出在查询中有多少页，需要触发一个额外的计数查询。按照默认来说这个查询可以从实际触发查询中衍生出来。

5. 限制查询结果

查询方法的结果可以通过关键字 First 或 Top 来限制，它们可以互换使用。 可选的数字值可以追加到 top/first，以指定要返回的最大结果大小。 如果省略该数字，则假定结果大小为 1。

下面示例用 Top 和 First 查询限制结果大小。

```
User findFirstByOrderByLastnameAsc();

User findTopByOrderByAgeDesc();

Page<User> queryFirst10ByLastname(String lastname, Pageable pageable);

Slice<User> findTop3ByLastname(String lastname, Pageable pageable);

List<User> findFirst10ByLastname(String lastname, Sort sort);

List<User> findTop10ByLastname(String lastname, Pageable pageable);
```

限制表达式也支持 Distinct 关键字。对于限制查询的结果集，定义到一个实例中包装这个结果到一个 Optional 中也是被支持的。

如果分页或切片被应用到一个限制查询分页（计算多少页可用），则它也能应用于限制结果。

要注意结合通过 Sort 参数动态排序的限制结果允许表达查询的方法为“K”最小的及“K”最大的元素。

6. 流查询结果

可以通过使用 Java 8 Stream<T> 作为返回类型来递增地处理查询方法的结果。不是简单地将查询结果包装在 Stream 数据存储中，而是使用特定方法来执行流传输。

下面的例子是以 Java 8 Stream<T> 来进行查询的流处理结果。

```
@Query("select u from User u")
Stream<User> findAllByCustomQueryAndStream();
Stream<User> readAllByFirstnameNotNull();

@Query("select u from User u")
Stream<User> streamAllPaged(Pageable pageable);
```

一个数据流可能包裹底层数据存储特定资源，因此在使用后必须关闭。也可以使用 close() 方法或使用 try-with-resources 语句[①] 来关闭数据流。

下面是在 try-with-resources 块中操作一个 StreamStream<T> 的例子。

```
try (Stream<User> stream = repository.findAllByCustomQueryAndStream()) {
  stream.forEach(...);
}
```

当前不是所有的 Spring Data 模块都支持 Stream<T> 作为返回类型。

---

① 有关 try-with-resources 语句的详细介绍，可参阅笔者的博客 https://waylau.com/concise-try-with-resources-jdk9。

7. 异步查询结果

可以使用 Spring 的异步方法执行能力来异步地执行存储库查询，这意味着该方法将在调用时立即返回，并且实际的查询执行将在已经提交到 Spring 任务执行器的任务中发生。

```
@Async
Future<User> findByFirstname(String firstname);                    // (1)

@Async
CompletableFuture<User> findOneByFirstname(String firstname);   // (2)

@Async
ListenableFuture<User> findOneByLastname(String lastname);      // (3)
```

（1）使用 java.util.concurrent.Future 作为返回类型。

（2）使用 Java 8 java.util.concurrent.CompletableFuture 作为返回类型。

（3）使用 org.springframework.util.concurrent.ListenableFuture 作为返回类型。

5.2.8 创建资源实例

为存储库接口创建实例及定义 bean 的方式有两种。一种方法是 Spring 命名空间；另一种方法是 Java Config 配置方式，该配置方式也是推荐的方式。

1. XML配置

每个 Spring Data 模块都包含一个资源库元素，开发者可以简单地定义 Spring 所要扫描的基础包。

下面是通过 XML 来配置 Spring Data 资源库的例子。

```
<?xml version="1.0" encoding="UTF-8"?>
<beans:beans xmlns:beans="http://www.springframework.org/schema/beans"
  xmlns:xsi="http://www.w3.org/2001/XMLSchema-instance"
  xmlns="http://www.springframework.org/schema/data/jpa"
  xsi:schemaLocation="http://www.springframework.org/schema/beans
    http://www.springframework.org/schema/beans/spring-beans.xsd
    http://www.springframework.org/schema/data/jpa
    http://www.springframework.org/schema/data/jpa/spring-jpa.xsd">

  <repositories base-package="com.waylau.repositories" />

</beans:beans>
```

在上面的示例中，指示 Spring 扫描 com.waylau.repositories 及其所有子包，以扩展 Repository 或其子接口之一。对于找到的每个接口，基础结构注册持久化技术特定的 FactoryBean，以创建处理查询方法的调用的适当代理。每个 bean 都注册在从接口名派生的 bean 名称下，因此 UserRepository 的接口将注册在 userRepository 下。base-package 属性允许使用通配符，以便开发者可以定义扫描包的模式。

2. 使用过滤器

可以使用 <include-filter /> 或 <exclude-filter /> 来进行相关的过滤。

```
<repositories base-package="com.waylau.repositories">
  <context:exclude-filter type="regex" expression=".*SomeRepository" />
</repositories>
```

3. JavaConfig 方式

可以在 JavaConfig 类上使用 @Enable${store}Repositories 注解，来实现触发存储库基础结构，示例配置如下所示。

```
@Configuration
@EnableJpaRepositories("com.waylau.repositories")
class ApplicationConfiguration {

  @Bean
  public EntityManagerFactory entityManagerFactory() {
    // ...
  }
}
```

该示例使用 JPA 特定的注解，开发者可以根据实际使用的存储模块更改它，这同样适用于 EntityManagerFactory bean 的定义。

4. 独立使用

还可以使用 Spring 容器之外的存储库基础结构，如在 CDI 环境中。开发者仍然需要在类路径中有一些 Spring 库，但一般来说可以通过编程方式设置存储库。提供存储库支持的 Spring Data 模块提供了一个持久化技术特定的 RepositoryFactory，示例如下所示。

```
RepositoryFactorySupport factory = // ...在这里实例化工厂UserRepository
repository = factory.getRepository(UserRepository.class);
```

5.2.9 Spring Data 自定义实现

通常有必要为几个存储库方法提供自定义实现。Spring 数据存储库很容易允许开发者提供自定义存储库代码，并将其与通用 CRUD 抽象和查询方法功能集成。

1. 向单个存储库添加自定义行为

要使用自定义功能丰富存储库，首先需要定义自定义功能的接口和实现。使用提供的存储库接口来扩展自定义接口。

自定义资源库方法的接口。

```
interface UserRepositoryCustom {
  public void someCustomMethod(User user);
```

```
}
```

自定义资源库方法接口的实现。

```
class UserRepositoryImpl implements UserRepositoryCustom {

  public void someCustomMethod(User user) {
    // 自定义实现
  }
}
```

实现本身不依赖于 Spring Data，可以是普通的 Spring bean。 因此，可以使用标准依赖注入行为来注入其他 bean 的引用，如 JdbcTemplate 等方面。

```
interface UserRepository extends CrudRepository<User, Long>, UserRepos-
itoryCustom {

  // 此处用于声明查询方法
}
```

将标准存储库接口扩展自定义接口，这样做就能结合 CRUD 和自定义的功能。

如果使用命名空间配置，存储库基础架构尝试通过扫描我们发现存储库的包下面的类来自动检测自定义实现，这些类需要遵循将命名空间元素的属性 repository-impl-postfix 附加到找到的库的命名约定接口名称，此后缀默认为 Impl。

配置示例如下。

```
<repositories base-package="com.waylau.repository" />

<repositories base-package="com.waylau.repository" repository-impl-post-
fix="FooBar" />
```

第一个配置示例将尝试查找类 com.waylau.repository.UserRepositoryImpl 作为自定义存储库实现，而第二个示例将尝试查找 com.waylau.repository.UserRepositoryFooBar。

如果自定义实现仅使用基于注解的配置和自动装配，那么上面所示的方法将很有效，因为它将被视为任何其他 Spring bean。如果自定义实现 bean 需要特殊的装配，只需声明 bean 并命名它刚才描述的约定。基础设施然后将通过名称而不是创建一个自身来引用手动定义的 bean 定义。

下面是自定义实现的手动装配的例子。

```
<repositories base-package="com.waylau.repository" />

<beans:bean id="userRepositoryImpl" class="...">
  <!-- 更多配置 -->
</beans:bean>
```

2. 向所有存储库添加自定义行为

当要向所有的存储库接口添加单个方法时，上述方法是不可行的，要向所有存储库添加自定义行为，首先要添加一个中间接口来声明共享行为。

声明自定义共享行为的接口。

```
@NoRepositoryBean
public interface MyRepository<T, ID extends Serializable>
  extends PagingAndSortingRepository<T, ID> {

  void sharedCustomMethod(ID id);
}
```

现在，各个存储库接口将扩展此中间接口，而不是 Repository 接口，以包含已声明的功能。接下来，创建中间接口的实现，以扩展持久性技术特定的存储库基类，这个类将作为存储库代理的自定义基类。

下面是自定义存储库基类的例子。

```
public class MyRepositoryImpl<T, ID extends Serializable>
  extends SimpleJpaRepository<T, ID> implements MyRepository<T, ID> {

  private final EntityManager entityManager;

  public MyRepositoryImpl(JpaEntityInformation entityInformation,
                          EntityManager entityManager) {
    super(entityInformation, entityManager);

    this.entityManager = entityManager;
  }

  public void sharedCustomMethod(ID id) {
    // ...
  }
}
```

类需要具有特定于存储库的工厂实现使用的超类的构造函数。如果存储库基类具有多个构造函数，则覆盖采用 EntityInformation 加上特定于存储库的基础结构对象（如 EntityManager 或模板类）的类。

Spring <repositories /> 命名空间的默认行为是为 base-package 下的所有接口提供一个实现，这意味着如果保持在它的当前状态，MyRepository 的实现实例将由 Spring 创建。这当然不是所期望的，因为它只是作为 Repository 和要为每个实体定义的实际存储库接口之间的中介。要排除将 Repository 从实例化为存储库实例的接口，可以使用 @NoRepositoryBean（如上所示）对其进行注解，或将其移动到已配置的 base-package 之外。

最后一步是使 Spring Data 基础结构感知自定义的存储库基类。在 JavaConfig 中，这是通过使

用 @Enable...Repositories 的 repositoryBaseClass 属性来实现的。

下面的例子是使用 JavaConfig 配置自定义存储库基类。

```
@Configuration
@EnableJpaRepositories(repositoryBaseClass = MyRepositoryImpl.class)
class ApplicationConfiguration {...}
```

相应的属性在 XML 命名空间中可用。

```
<repositories base-package="com.acme.repository"
    base-class="... .MyRepositoryImpl" />
```

5.2.10 从聚合根发布事件

Spring Data 提供了 @DomainEvents 用来发布领域事件（Domain Event）。用法如下。

```
class AnAggregateRoot {

    @DomainEvents
    Collection<Object> domainEvents() {
         ... //返回想在这里发布的事件
    }

    @AfterDomainEventsPublication
    void callbackMethod() {
        ...//清理邻域事件列表
    }
}
```

5.3 Spring Data JPA 与 Hibernate、Spring Boot 集成

本节将演示如何在 Spring Boot 中集成 JPA 的功能。在 thymeleaf-in-action 项目基础上构建了一个新的项目 jpa-in-action。

5.3.1 所需环境

本例采用的开发环境如下。

（1）MySQL Community Server 5.7.17。

（2）MySQL Workbench 6.3.9。

（3）Spring Data JPA 2.0.0.M4。

- Hibernate 5.2.10.Final。
- MySQL Connector/J 6.0.5。
- H2 Database 1.4.196。

5.3.2 build.gradle

我们需要添加 Spring Data JPA 及 MySQL 连接驱动的依赖。Spring Boot 已经提供了相关的 Starter 来实现 Spring Data JPA 开箱即用的功能，所以只需要在 build.gradle 文件中添加 Spring Data JPA 的 Starter 的库即可。

```
// 依赖关系
dependencies {
    //...

    //添加Spring Data JPA 的依赖
    compile('org.springframework.boot:spring-boot-starter-data-jpa')

    //添加MySQL连接驱动的依赖
    compile('mysql:mysql-connector-java:6.0.5')

    //...
}
```

spring-boot-starter-data-jpa 库同时提供了如下依赖。

- Hibernate。
- Spring Data JPA。
- Spring ORM。

在编译过程中，如果仔细观察控制台，可以看到如下编译信息，这是下载依赖包的整个过程。

```
Download http://maven.aliyun.com/nexus/content/groups/public/mysql/
mysql-connector-java/6.0.5/mysql-connector-java-6.0.5.pom
Download https://repo.spring.io/milestone/org/springframework/boot/
spring-boot-starter-data-jpa/2.0.0.M2/spring-boot-starter-data-jpa-
2.0.0.M2.pom
Download https://repo.spring.io/milestone/org/springframework/data/
spring-data-jpa/2.0.0.M4/spring-data-jpa-2.0.0.M4.pom
Download https://repo.spring.io/milestone/org/springframework/data/
build/spring-data-parent/2.0.0.M4/spring-data-parent-2.0.0.M4.pom
Download https://repo.spring.io/milestone/org/springframework/spring-as-
pects/5.0.0.RC2/spring-aspects-5.0.0.RC2.pom
Download https://repo.spring.io/milestone/org/springframework/boot/
spring-boot-starter-jdbc/2.0.0.M2/spring-boot-starter-jdbc-2.0.0.M2.pom
```

```
Download https://repo.spring.io/milestone/org/springframework/boot/
spring-boot-starter-aop/2.0.0.M2/spring-boot-starter-aop-2.0.0.M2.pom
Download http://maven.aliyun.com/nexus/content/groups/public/org/hiber-
nate/hibernate-core/5.2.10.Final/hibernate-core-5.2.10.Final.pom
Download http://maven.aliyun.com/nexus/content/groups/public/org/as-
pectj/aspectjweaver/1.8.10/aspectjweaver-1.8.10.pom
Download http://maven.aliyun.com/nexus/content/groups/public/com/zaxx-
er/HikariCP/2.6.2/HikariCP-2.6.2.pom
Download https://repo.spring.io/milestone/org/springframework/spring-jd-
bc/5.0.0.RC2/spring-jdbc-5.0.0.RC2.pom
Download https://repo.spring.io/milestone/org/springframework/data/
spring-data-commons/2.0.0.M4/spring-data-commons-2.0.0.M4.pom
Download https://repo.spring.io/milestone/org/springframework/spring-tx-
/5.0.0.RC2/spring-tx-5.0.0.RC2.pom
Download https://repo.spring.io/milestone/org/springframework/spring-
orm/5.0.0.RC2/spring-orm-5.0.0.RC2.pom
Download https://repo.spring.io/milestone/org/springframework/boot/
spring-boot-starter-data-jpa/2.0.0.M2/spring-boot-starter-data-jpa-
2.0.0.M2.jar
Download http://maven.aliyun.com/nexus/content/groups/public/mysql/
mysql-connector-java/6.0.5/mysql-connector-java-6.0.5.jar
Download https://repo.spring.io/milestone/org/springframework/boot/
spring-boot-starter-aop/2.0.0.M2/spring-boot-starter-aop-2.0.0.M2.jar
Download https://repo.spring.io/milestone/org/springframework/boot/
spring-boot-starter-jdbc/2.0.0.M2/spring-boot-starter-jdbc-2.0.0.M2.jar
Download http://maven.aliyun.com/nexus/content/groups/public/org/hiber-
nate/hibernate-core/5.2.10.Final/hibernate-core-5.2.10.Final.jar
Download https://repo.spring.io/milestone/org/springframework/data/
spring-data-jpa/2.0.0.M4/spring-data-jpa-2.0.0.M4.jar
...
```

由于篇幅限制，这里只展示了部分下载过程，有兴趣的读者可以自己去控制台进行查看。

5.3.3 集成 H2

在未配置数据源的情况下，启动项目是失败的，报错如下。

```
2017-07-18 23:06:51.347  WARN 7912 --- [           main] ConfigServlet
WebServerApplicationContext : Exception encountered during context
initialization - cancelling refresh attempt: org.springframework.beans.
factory.UnsatisfiedDependencyException: Error creating bean with name 'org.
springframework.boot.autoconfigure.orm.jpa.HibernateJpaAutoConfiguration':
Unsatisfied dependency expressed through constructor parameter 0; nested
exception is org.springframework.beans.factory.BeanCreationException:
Error creating bean with name 'dataSource' defined in class path resource
[org/springframework/boot/autoconfigure/jdbc/DataSourceConfiguration
$Hikari.class]: Bean instantiation via factory method failed; nested
```

```
exception is org.springframework.beans.BeanInstantiationException:
Failed to instantiate [com.zaxxer.hikari.HikariDataSource]: Factory
method 'dataSource' threw exception; nested exception is org.spring-
framework.boot.autoconfigure.jdbc.DataSourceProperties$DataSourceBean-
CreationException: Cannot determine embedded database driver class for
database type NONE. If you want an embedded database please put a supported
one on the classpath. If you have database settings to be loaded from a
particular profile you may need to active it (no profiles are currently
active).
2017-07-18 23:06:51.349  INFO 7912 --- [           main] o.apache.
catalina.core.StandardService   : Stopping service [Tomcat]
2017-07-18 23:06:51.372  WARN 7912 --- [ost-startStop-1] o.a.c.loader.
WebappClassLoaderBase       : The web application [ROOT] appears to have
started a thread named [Abandoned connection cleanup thread] but has
failed to stop it. This is very likely to create a memory leak. Stack
trace of thread:
 java.lang.Object.wait(Native Method)
 java.lang.ref.ReferenceQueue.remove(Unknown Source)
 com.mysql.cj.jdbc.AbandonedConnectionCleanupThread.run(AbandonedCon-
nectionCleanupThread.java:43)
2017-07-18 23:06:51.400  INFO 7912 --- [           main] utoConfigura-
tionReportLoggingInitializer :

Error starting ApplicationContext. To display the auto-configuration
report re-run your application with 'debug' enabled.
2017-07-18 23:06:51.413 ERROR 7912 --- [           main] o.s.b.d.Log-
gingFailureAnalysisReporter   :

***************************
APPLICATION FAILED TO START
***************************

Description:

Cannot determine embedded database driver class for database type NONE

Action:

If you want an embedded database please put a supported one on the
classpath. If you have database settings to be loaded from a particular
profile you may need to active it (no profiles are currently active).
```

Spring Boot 的自动配置功能会自动检测有无指定数据库的类型（数据源）。如果没有指定，Spring Boot 会认为用户是想要配置一个内嵌的数据库，此时会扫描 classpath 有无相应的内嵌数据库。如果还是没有找到，则提示上述错误。

所以，解决方式是修改 build.gradle 文件，添加一个内嵌数据库。这里演示了添加 H2 内存数据库的依赖的过程。

```
// 依赖关系
dependencies {
    //...

    // 添加H2的依赖
    runtime('com.h2database:h2:1.4.196')

    //...
}
```

其中“runtime”表明该依赖对于运行时是必需的。

有关 H2 的相关知识，可参阅笔者所著的开源书《H2 Database 教程》（https://github.com/waylau/h2-database-doc）。

5.3.4 MySQL 安装及使用

H2 方便在开发阶段快速初始化表和数据，并且在重启项目后，会自动清理历史数据。但在真实项目中，项目部署到生产环境中，就必须要将数据放置在安全、可靠的数据库中。本书将选择 MySQL 作为关系型数据库的例子。

1. 下载 MySQL

MySQL 官方网站提供了 MySQL Community Server 的免费下载，下载地址为 https://dev.mysql.com/downloads/mysql/。如果操作系统是 Windows，则可选用 MySQL Workbench 作为 MySQL 的可视化管理工具。MySQL Workbench 同样来自 MySQL 官方，下载地址为 https://dev.mysql.com/downloads/workbench/。

如果是其他操作系统平台，读者可自行选择合适的管理工具。当然，使用命令行工具也能胜任本书讲解案例的要求。

3. 安装 MySQL

安装 MySQL Community Server 的教程现成的资料也比较多，这里就不再赘述。有需要的读者可参考笔者的相关博文 https://waylau.com/windows-install-mysql-noinstall-zip/。

5.4 数据持久化实战

如果读者对第 4 章的内容还有印象的话，应该还记得，我们用 Thymeleaf 实现了一个简单的“用户管理”功能。为了简便，并没有使用数据库管理系统，而是将数据直接保存在了内存中，这样导致的后果是只要应用重启，数据就会丢失。本节将通过 JPA 来将数据存储到关系型数据库中，这样就实现了数据的持久化。

在 5.3 节创建的 jpa-in-action 项目的基础上来实现 JPA 功能。

5.4.1 定义实体

修改 User 类，参考 JPA 的规范，将其修改成为实体。

（1）User 类上增加了 @Entity 注解，以标识其为实体。

（2）@Id 标识 id 字段为主键。

（3）@GeneratedValue(strategy=GenerationType.IDENTITY) 标识 id 字段，以便使用数据库的自增长字段为新增加的实体的标识。这种情况下需要数据库提供对自增长字段的支持，一般的数据库（如 HSQL、SQL Server、MySQL、DB2、Derby 等）都能够提供这种支持。

（4）应 JPA 的规范要求，设置无参的构造函数 protected User() {}，并设为 protected，防止直接被使用。

（5）重写 toString 方法，将 User 信息自定义输出。

```
import javax.persistence.Entity;
import javax.persistence.GeneratedValue;
import javax.persistence.GenerationType;
import javax.persistence.Id;

@Entity // 实体
public class User {

    @Id // 主键
    @GeneratedValue(strategy=GenerationType.IDENTITY) // 自增长策略
    private Long id; // 实体一个唯一标识
    private String name;
    private String email;

    protected User() { // 无参构造函数；设为 protected 防止直接使用
    }

    public User(Long id, String name, String email) {
        this.id = id;
        this.name = name;
        this.email = email;
    }

    public Long getId() {
        return id;
    }
```

```
    public void setId(Long id) {
        this.id = id;
    }
    public String getName() {
        return name;
    }
    public void setName(String name) {
        this.name = name;
    }
    public String getEmail() {
        return email;
    }
    public void setEmail(String email) {
        this.email = email;
    }

    @Override
    public String toString() {
        return String.format("User[id=%d, name='%s', email='%s']", id,
name, email);
    }
}
```

5.4.2 修改资源库

修改用户资源库的接口，继承自 CrudRepository。

```
import org.springframework.data.repository.CrudRepository;
import com.waylau.spring.boot.blog.domain.User;

public interface UserRepository extends CrudRepository<User, Long> {

}
```

由于 Spring Data JPA 已经帮助用户做了实现，因此，用户不需要做任何实现，甚至都无须在 UserRepository 中定义任何方法，就可以直接删除了之前定义的 UserRepositoryImpl 类。

5.4.3 修改控制器

UserController 也要做一些调整，将原来 UserRepositoryImpl 实现的方法全部换成 JPA 的默认实现。

```
@RestController
@RequestMapping("/users")
public class UserController {
```

```
    @Autowired
    private UserRepository userRepository;

    /**
     * 查询所有用户
     * @param model
     * @return
     */
    @GetMapping
    public ModelAndView list(Model model) {
        model.addAttribute("userList", userRepository.findAll());
        model.addAttribute("title", "用户管理");
        return new ModelAndView("users/list","userModel",model);
    }

    /**
     * 根据id查询用户
     * @param id
     * @param model
     * @return
     */
    @GetMapping("{id}")
    public ModelAndView view(@PathVariable("id") Long id, Model model)
{
        Optional<User> user = userRepository.findById(id);
        model.addAttribute("user", user.get());
        model.addAttribute("title", "查看用户");
        return new ModelAndView("users/view","userModel",model);
    }

    /**
     * 获取创建表单页面
     * @param model
     * @return
     */
    @GetMapping("/form")
    public ModelAndView createForm(Model model) {
        model.addAttribute("user", new User(null, null, null));
        model.addAttribute("title", "创建用户");
        return new ModelAndView("users/form","userModel",model);
    }

    /**
     * 保存或者修改用户
     * @param user
     * @return
     */
    @PostMapping
```

```
    public ModelAndView saveOrUpdateUser(User user) {
        userRepository.save(user);
        return new ModelAndView("redirect:/users");// 重定向到list页面
    }

    /**
     * 删除用户
     * @param id
     * @return
     */
    @GetMapping("/delete/{id}")
    public ModelAndView delete(@PathVariable("id") Long id) {
        userRepository.deleteById(id);
        return new ModelAndView("redirect:/users"); // 重定向到list页面
    }

    /**
     * 获取修改用户的界面
     * @param id
     * @param model
     * @return
     */
    @GetMapping("/modify/{id}")
    public ModelAndView modify(@PathVariable("id") Long id, Model mod
el) {
        Optional<User> user = userRepository.findById(id);
        model.addAttribute("user", user.get());
        model.addAttribute("title", "修改用户");
        return new ModelAndView("users/form","userModel",model);
    }
}
```

5.4.4 查看 H2 表结构

因为目前使用的是 H2 内存数据库，如果想看一下内存数据库的表结构，该怎么办呢？

修改 application.properties 文件，增加下面的配置。

```
# 使用H2控制台
spring.h2.console.enabled=true
```

而后启动项目，访问 http://localhost:8080/h2-console 即可。其中，设置 JDBC URL 为“jdbc:h2:mem:testdb”，如图 5-1 所示。

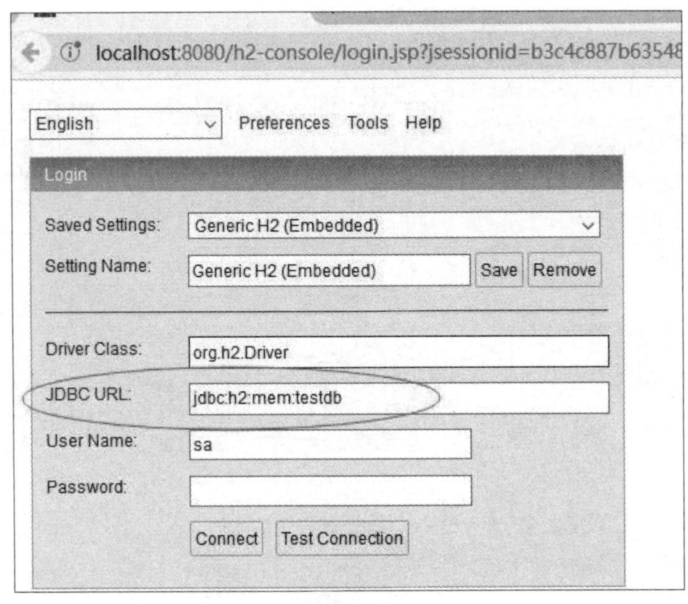

图5-1 H2设置JDBC URL

在项目中初始化相应的数据，如图 5-2 所示。

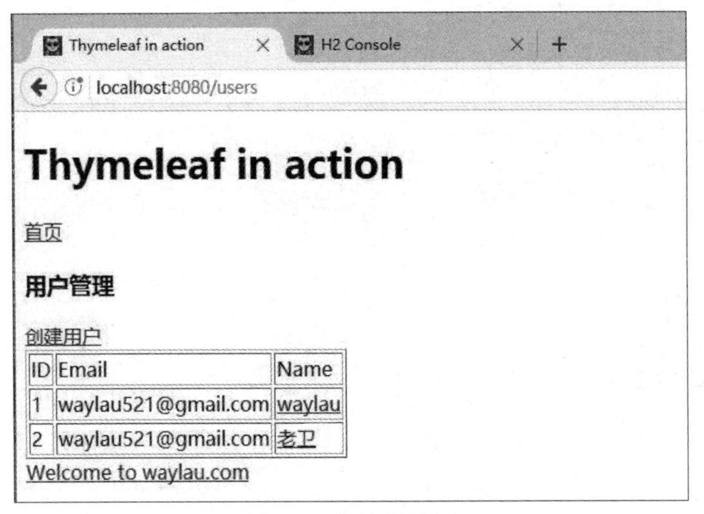

图5-2 初始化数据

而后在 H2 数据库中就能查看到相应的数据，如图 5-3 所示。

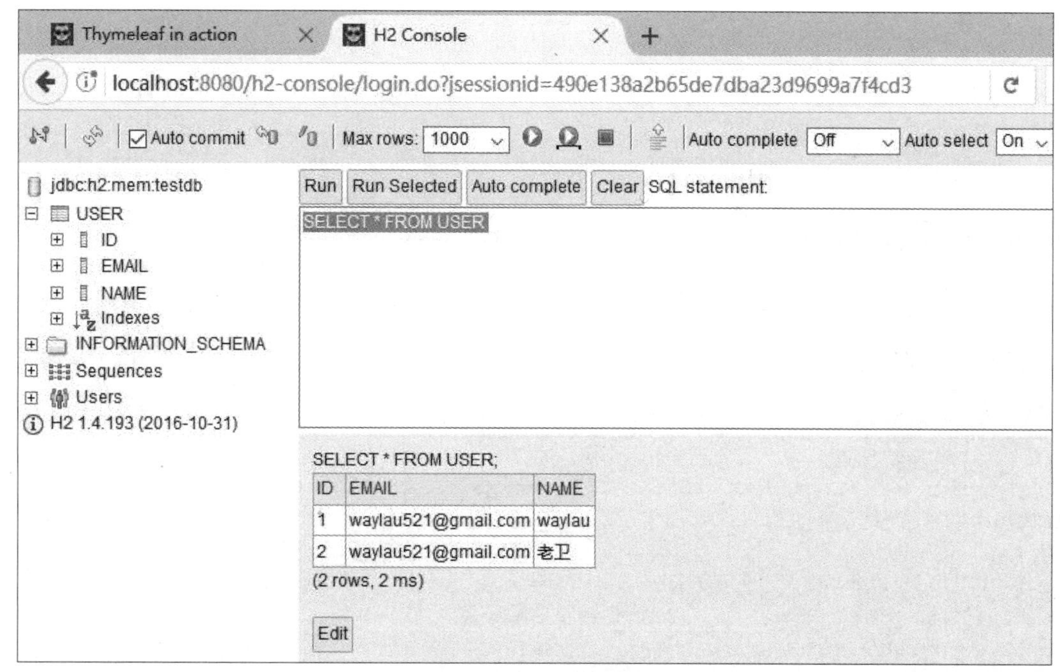

图5-3　H2查看数据

5.4.5 使用 MySQL 数据库

通过命令行使用 MySQL 客户端来操作数据库。

```
$ mysql -u root -p
```

首先，创建名为 blog 的数据库，编码为 UTF-8。

```
mysql> DROP DATABASE IF EXISTS blog;
Query OK, 9 rows affected (1.63 sec)

mysql> CREATE DATABASE blog DEFAULT CHARSET utf8 COLLATE utf8_general_
ci;
Query OK, 1 row affected (0.00 sec)
```

修改 application.properties 文件，增加下面的几项配置。

```
# DataSource
spring.datasource.url=jdbc:mysql://localhost/blog?useSSL=false&server-
Timezone=UTC&characterEncoding=utf-8
spring.datasource.username=root
spring.datasource.password=123456
spring.datasource.driver-class-name=com.mysql.cj.jdbc.Driver

# JPA
spring.jpa.show-sql = true
```

```
spring.jpa.hibernate.ddl-auto=create-drop
```

其中 spring.jpa.hibernate.ddl-auto 中的 create-drop 是指每次应用启动，都会自动删除并创建数据库表，在开发时，非常方便。

5.4.6 运行查看效果

启动项目，在控制台可以看到 Hibernate 的执行情况。

```
2017-07-19 00:18:03.276  INFO 8540 --- [           main] org.hibernate.
Version                          : HHH000412: Hibernate Core {5.2.10.Final}
2017-07-19 00:18:03.277  INFO 8540 --- [           main] org.hibernate.
cfg.Environment                  : HHH000206: hibernate.properties not found
2017-07-19 00:18:03.309  INFO 8540 --- [           main] o.hibernate.
annotations.common.Version       : HCANN000001: Hibernate Commons Annota-
tions {5.0.1.Final}
2017-07-19 00:18:03.406  INFO 8540 --- [           main] org.hibernate.
dialect.Dialect                  : HHH000400: Using dialect: org.hibernate.
dialect.MySQL5Dialect
Hibernate: drop table if exists user
Hibernate: create table user (id bigint not null auto_increment, email
varchar(255), name varchar(255), primary key (id)) engine=MyISAM
```

可以发现，Hibernate 会自动在 blog 数据库中创建表 user。

通过浏览器访问 http://localhost:8080/users ，可以看到项目的运行效果。在页面进行一些增删改查的操作。

可以在数据库看到操作后的数据。

```
mysql> show databases;
+--------------------+
| Database           |
+--------------------+
| information_schema |
| blog               |
| mysql              |
| performance_schema |
| sys                |
+--------------------+
5 rows in set (0.00 sec)

mysql> use blog;
Database changed

mysql> show tables;
+----------------+
| Tables_in_blog |
+----------------+
```

```
| user           |
+----------------+
1 row in set (0.00 sec)

mysql> select * from user;
+----+---------------------+---------+
| id | email               | name    |
+----+---------------------+---------+
|  1 | waylau521@gmail.com | Way lau |
|  2 | waylau521@163.com   | 老卫    |
+----+---------------------+---------+
2 rows in set (0.00 sec)

mysql>
```

5.4.7 相关错误解决

在配置数据的数据源时要格外小心，配置不当，可能会遇到如下错误。

1. SSL 连接未验证

```
Establishing SSL connection without server's identity verification is
not recommended. According to MySQL 5.5.45+, 5.6.26+ and 5.7.6+ require-
ments SSL connection must be established by default if explicit option
isn't set. For compliance with existing applications not using SSL the
verifyServerCertificate property is set to 'false'. You need either to
explicitly disable SSL by setting useSSL=false, or set useSSL=true and
provide truststore for server certificate verification.
```

解决方法：在数据源的 spring.datasource.url 里添加参数 useSSL=false。

2. 时区无法确认

```
java.sql.SQLException: The server time zone value '�й���ʱ��' is unrec-
ognized or represents more than one time zone. You must configure either
the server or JDBC driver (via the serverTimezone configuration proper-
ty) to use a more specifc time zone value if you want to utilize time
zone support.
```

解决方法：在数据源的 spring.datasource.url 里添加参数 serverTimezone=UTC。

5.4.8 示例源码

本节示例源码在 jpa-in-action 目录下。

第6章

全文搜索

6.1 全文搜索概述

本节将带读者理解全文搜索的概念。全文搜索也称为全文检索，它的工作原理是计算机索引程序通过扫描文章中的每一个词，对每一个词建立一个索引，指明该词在文章中出现的次数和位置，当用户查询时，检索程序就根据事先建立的索引进行查找，并将查找的结果反馈给用户的检索方式。这个过程类似于通过字典中的检索字表查字的过程。

由于在执行搜索之前，搜索的关键字已经被建立了索引，因此搜索时的速度往往很快。

6.1.1 数据结构

在理解全文搜索的概念之前，先对数据结构进行理解。数据结构主要分为结构化数据与非结构化数据。

（1）结构化数据：指具有固定格式或有限长度的数据，如数据库、元数据等。

（2）非结构化数据：指不定长或无固定格式的数据，如邮件、Word 文档等。

6.1.2 数据搜索的方式

针对不同的数据结构（结构化数据或是非结构化数据），有不同的数据搜索方式。

1. 结构化数据的搜索方式

对于结构化的数据而言，数据一般就存储在关系型数据库中，通过 SQL 语句来对数据库进行搜索。

如果是针对元数据的搜索，则可以利用操作系统本身的机制，如利用 Windows 搜索对文件名、类型、修改时间进行搜索等。

2. 非结构化数据的搜索方式

对于非结构化的数据而言，有下面两种搜索方式。

（1）顺序扫描法（Serial Scanning）：所谓顺序扫描，例如，要找内容包含某一个字符串的文件，就是一个文档一个文档地查看，对于每一个文档，从头看到尾，如果此文档包含此字符串，则此文档为要找的文件，接着看下一个文件，直到扫描完所有的文件。利用 Windows 的搜索也可以搜索到文件的内容。Linux 下的 grep 命令也是这种方式。这些方式都是操作系统提供的简单方法，但对于小数据量的文件而言，这种方法还是最直接、最方便的。但是对于大量的文件，这种方法就很慢了。

（2）全文检索（Full-text Search）：将非结构化数据中的一部分信息提取出来，重新组织，使其变得有一定结构，然后对此有一定结构的数据进行搜索，从而达到搜索相对较快的目的。这部分从非结构化数据中提取出的然后重新组织的信息，我们称为索引。以字典为例，字典的拼音表和部首检字表就相当于字典的索引，对每一个字的解释是非结构化的，如果字典没有音节表和部首检字表，要找一个字只能顺序扫描。然而字的某些信息可以提取出来进行结构化处理，如读音，就比较结构化，分声母和韵母，分别只有几种情况可以一一列举，于是将读音提取出来按一定的顺序排列，每一项读音都指向此字的详细解释的页数。搜索时按结构化的拼音搜到读音，然后按其指向的页数，便可找到非结构化数据，即对字的解释。这种先建立索引，再对索引进行搜索的过程就称为全文搜索。

有人可能会说，全文检索的确加快了搜索的速度，但是多了创建索引的过程，两者加起来不一定比顺序扫描快多少。的确，加上索引的过程，全文检索不一定比顺序扫描快，尤其是在数据量小的时候更是如此。而对大量的数据创建索引也是一个很慢的过程。

然而两者还是有区别的，顺序扫描是每次都要扫描，而创建索引的过程仅仅需要一次，以后便是一劳永逸了，每次搜索，创建索引的过程不是必需的，仅仅搜索创建好的索引就可以了。

6.1.3 全文搜索的原理

实现全文搜索一般分为以下步骤。

1. 建文本库

在开发功能之前，一个信息检索系统需要做一些准备工作，首先，必须建立一个文本数据库，用于保存所有用户可能检索的信息。根据这些信息，确定索引中的文本类型。文本类型是系统识别的信息格式。这种格式应该具有可识别和低冗余的特性。一旦确定了文本类型，就不应当对其进行大的改动。

2. 建立索引

有了文本类型之后，就应该根据数据库中的文本来建立索引。索引可以大大提高信息检索的速度。目前，建立索引的方法有很多种。使用哪种方法取决于信息检索系统的大小。大型信息检索系统，如百度、Google 等，都是采用倒排的方式来构建索引。

3. 执行搜索

在文档建立索引之后，就可以开始对其进行搜索。这时，通常由用户提交搜索请求，将请求进行分析，然后用文本操作来处理。对于一个真实的信息检索系统而言，在真正处理请求前，还可以对请求进行一些预处理，然后再将请求送到后台，并返回用户需要的信息。

4. 过滤结果

通常情况下，在信息检索系统中检索到用户需要的信息后，还要做进一步操作，即将信息按一定顺序进行排序或过滤，然后返回给用户。这一步实际上关乎到最终用户的体验。

6.1.4 全文搜索相关的技术

基于 Java 的全文搜索相关的技术，开源实现如下。

- Lucene：http://lucene.apache.org/core/。
- Elasticsearch：https://www.elastic.co/cn/products/elasticsearch。
- Solr：http://lucene.apache.org/solr/。

Lucene 是搜索引擎，而 Elasticsearch 与 Solr 都是基于 Lucene 之上而实现的全文检索系统。有关 Elasticsearch 与 Solr 的对比，总结如下。

（1）Solr 利用 Zookeeper 进行分布式管理，而 Elasticsearch 自身带有分布式协调管理功能。

（2）Solr 支持更多格式的数据，如 JSON、XML、CSV，而 Elasticsearch 仅支持 JSON 文件格式。

（3）Solr 官方提供的功能更多，而 Elasticsearch 本身更注重于核心功能，高级功能多由第三方插件提供。

（4）Solr 在传统的搜索应用中表现好于 Elasticsearch，但在处理实时搜索应用时效率明显低于 Elasticsearch。

（5）Solr 是传统搜索应用的有力解决方案，但 Elasticsearch 更适用于新兴的实时搜索应用。

更加详细的对比，可以参见 http://solr-vs-elasticsearch.com/。本书选用 Elasticsearch 作为全文搜索案例的技术框架。

6.2 Elasticsearch 核心概念

Elasticsearch 是一个高度可扩展的开源全文搜索和分析引擎，它允许用户快速地、近实时地对大数据进行存储、搜索和分析，它通常用来支撑有复杂的数据搜索需求的企业级应用。Elasticsearch 基于 Apache Lucene 构建，作为一个非常流行的开源软件的同时，也有相应的公司提供商业支持，使之非常适合企业用户使用。

Elasticsearch 是一个为云构建的分布式 RESTful 搜索引擎，它有如下特点。

- 分布式、高度可用的搜索引擎。
  - 每个索引都使用可配置数量的分片。
  - 每个分片可以有一个或多个副本。
  - 在任何副本分片上执行的读取、搜索操作。
- 多租户与多类型。
  - 支持多个索引。
  - 每个索引支持多个类型。
  - 索引级别配置（分片数、索引存储 ......）。
- 多种 API。
  - HTTP RESTful API。

  - 原生 Java API。
  - 所有 API 执行自动节点操作重新路由。
- 面向文档。
  - 不需要事先定义模式（Schema）。
  - 可以为每个类型定义模式以自定制索引过程。
- 对数据持久化提供可靠、异步的写入。
- （近）实时搜索。
- 建立在 Lucene 之上。
  - 每个分片是一个功能齐全的 Lucene 索引。
  - 容易通过简单的配置 / 插件来发挥 Lucene 所有的功能。
- 保证操作的一致性。
  - 每个文档级的操作都是原子的，具备一致性、隔离性和持久性。
- 遵循 Apache 协议 2（“ALv2”）。

在使用 Elasticsearch 之前，首先来了解一下其中的一些基本概念。

6.2.1 基于 Lucene

Elasticsearch 底层是基于 Lucene 来构建的，无论在开源社区还是专有领域，Lucene 可以被认为是迄今为止最先进、性能最好的、功能最全的 Java 搜索引擎库。

但是，Lucene 是一个相当底层的库，想要使用它，开发人员必须使用 Java 来作为开发语言并将其直接集成到自己的应用中，更糟糕的是，Lucene 在使用上非常复杂，必须要深入了解检索的相关知识来理解它是如何工作的，而这一切都阻碍了开发人员来掌握它。

正是因为考虑到上述难点，Elasticsearch 出现了。Elasticsearch 对底层 Lucene 进行了封装，通过简单易用的 RESTful API 来暴露接口，从而有效减轻了开发人员的学习成本，让全文搜索的实现不再困难。

以下是调用 Elasticsearch 接口的一个例子。该例子没有复杂的实现方式，只需要使用 curl 软件[①]就能实现。

```
curl -XGET 'http://localhost:9200/waylau/_search?pretty=true' -H 'Con-
tent-Type: application/json' -d '
{
    "query" : {
        "match" : { "user": "kimchy" }
    }
```

① curl 是利用 URL 语法在命令行方式下工作的开源文件传输工具，被广泛应用于 RESTful 接口的调试。详见 https://curl.haxx.se/。

```
}'
```

6.2.2 如何查看 Elasticsearch 对应的 Lucene 版本

访问 http://localhost:9200/ 应能看到如下数据返回。

```
{
  "name" : "2RvnJex",
  "cluster_name" : "elasticsearch",
  "cluster_uuid" : "uqcQAMTtTIO6CanROYgveQ",
  "version" : {
    "number" : "5.5.0",
    "build_hash" : "260387d",
    "build_date" : "2017-06-30T23:16:05.735Z",
    "build_snapshot" : false,
    "lucene_version" : "6.6.0"
  },
  "tagline" : "You Know, for Search"
}
```

其中，“lucene_version”一项就是 Lucene 的版本。“number”是当前 Elasticsearch 的版本。

6.2.3 近实时

Elasticsearch 是一个接近实时（Near Realtime，NRT）而非实时的搜索平台，这意味着从索引文档到可搜索的时间有一个轻微的延迟（通常为 1 秒）。之所以会有这个延时，主要考虑查询的性能优化。Elasticsearch 其实是可以做到实时的，因为 Lucene 是可以做到实时的，但是这样做，要么是牺牲索引的效率（每次索引之后刷新），要么就是牺牲查询的效率（每次查询之前都进行刷新），所以 Elasticsearch 采取一种折中的方案，每隔 *n* 秒自动刷新，这样用户创建索引之后，最多在 *n* 秒之内肯定能查到，这就是所谓的近实时查询。

实际上 Elasticsearch 索引新文档后，不会直接写入磁盘，而是首先存入文件系统缓存，之后根据刷新设置，定期同步到磁盘，默认情况下，每个分片每秒自动刷新一次。这就是为什么文档的改动不会立即被搜索，但是会在 1 秒内可见。这个时间可以通过 index.refresh_interval 参数来修改间隔。

6.2.4 集群

集群是一个或多个节点的集合，用来保存应用的全部数据并提供基于全部节点的集成式索引和搜索功能。集群有利于系统性能的水平扩展及保障系统的可用性。

每个 Elasticsearch 集群都需要有一个唯一的名称，默认是“elasticsearch”。此名称很重要，因为有可能节点会设置为通过其名称来加入集群。

当一个集群只有一个节点时，通常不会出现什么问题。然而需要考虑部署到多个独立的集群时，就需要为每个集群配置自己唯一的集群名称了。建议在不同的环境中使用不同的集群名称，这样可以避免由于名称搞混而导致节点加入错误的集群中。例如，可以在针对开发、暂存和生产环境的集群中使用 logging-dev、logging-stage 和 logging-prod 等名称。

6.2.5 节点

节点是一个集群中的单台服务器，用来保存数据并参与整个集群的索引和搜索操作，就像一个集群一样，一个节点有一个名称标识，默认情况下它是一个随机的通用唯一标识符（UUID），在启动时分配给该节点。如果不想使用默认值，可以定义任何所需的节点名称。此名称对于管理集群节点非常重要，开发人员需要确定网络中哪些服务器对应于 Elasticsearch 集群中的哪些节点。

可以将节点配置为通过集群名称加入特定集群。默认情况下，每个节点都设置为加入名为“elasticsearch”的集群，这意味着如果开发人员在网络上启动多个节点，并且假设它们可以发现彼此，则它们将自动形成并加入名为“elasticsearch”的单个集群。

在单个集群中，开发人员可以拥有任意数量的节点。此外，如果网络上当前没有其他 Elasticsearch 节点运行，则启动单个节点将默认形成名为“elasticsearch”的新单节点集群。

6.2.6 索引

索引是相似文档的集合。索引中的内容与应用本身的业务相关，如电子商务应用可以使用索引来保存产品数据、订单数据和客户数据等。每个索引都有一个名称，通过该名称可以对索引中包含的文档进行添加、更新、删除和搜索等操作。

在单个集群中可以根据需要定义任意数量的索引。

6.2.7 类型

类型是对一个索引中包含的文档的进一步细分。一般根据文档的公共属性来进行划分。例如，在电子商务应用的产品数据索引中，可以根据产品的特征划分成不同的类型，如一般产品、虚拟产品、数字产品等。

6.2.8 文档

文档是进行索引的基本单位，与索引中的一个类型相对应。例如，产品数据索引中一般产品类型中的每个具体的产品可以有一个文档与之对应。文档使用 JSON 格式来表示。

在索引 / 类型中，可以存储任意数量的文档。需要注意的是，虽然文档物理上驻留在索引中，但实际上文档必须是索引中的类型。

6.2.9 分片和副本

企业应用需要存储的数据量一般比较巨大，超出单个节点所能处理的范围。Elasticsearch 允许把索引划分成多个分片（Shard）来存储索引的部分数据。Elasticsearch 会负责处理分片的分配和聚合。从可靠性的角度出发，对于一个分片中的数据，应该有至少一个副本（Replica）。Elasticsearch 中每个索引可以划分成多个分片，而且有多个副本。Elasticsearch 会自动管理集群中节点的分片和副本，对开发人员是透明的。

1. 设置分片的原因

设置分片的主要原因如下。

（1）它允许用户水平分割 / 缩放内容卷。

（2）它允许用户跨分片（可能在多个节点上）分布和并行操作，从而提高性能 / 吞吐量。

碎片如何分布、文档如何聚合搜索请求等，这些机制都由 Elasticsearch 完全管理，并且对用户是透明的。

由于在云环境中故障不可避免，因此一定要有故障转移机制，以防某个分片 / 节点故障。为此，Elasticsearch 允许用户将索引的碎片的一个或多个副本转换为所谓的副本碎分片，或简称为副本。

2. 设置副本的原因

设置副本的主要原因如下。

（1）在碎片 / 节点故障的情况下提供高可用性。因此，副本分片不能部署在从其复制的原始 / 主分片相同的节点上。

（2）它允许用户扩展搜索量 / 吞吐量，因为搜索可以并行地在所有副本上执行。

总而言之，每个索引可以拆分成多个分片。索引也可以复制为零（意味着没有副本）或更多次。一旦复制，每个索引将具有主分片和副本分片。可以在创建索引时为每个索引定义分片和副本的数量。创建索引后，用户可以随时动态更改副本数，但不能随后更改分片数。

默认情况下，Elasticsearch 每个索引分配 5 个主分片和 1 个副本，这意味着集群至少有两个节点，索引将有 5 个主要碎片和另外 5 个副本碎片（1 个完整的副本），每个索引共 10 个分片。

注意：每个 Elasticsearch 分片是一个 Lucene 索引。在一个 Lucene 索引中有文档数的最多限制。从 LUCENE-5843 版本（https://issues.apache.org/jira/browse/LUCENE-5843）开始，限制为 2 147 483 519（等于 Integer.MAX_VALUE - 128）个文档。可以使用 _cat/shards api （https://www.elastic.co/guide/en/elasticsearch/reference/current/cat-shards.html）监视分片大小。

6.2.10 使用场景

只要想象力够丰富，Elasticsearch 就能够被广泛地应用到各种场景。以下是真实的使用案例。

（1）维基百科使用 Elasticsearch 来进行全文搜索，并高亮显示关键词，以及提供 search-as-you-type、did-you-mean 等搜索建议功能。

（2）英国卫报使用 Elasticsearch 来处理访客日志，以便能将公众对不同文章的反映实时地反馈给各位编辑。

（3）StackOverflow 将全文搜索与地理位置和相关信息进行结合，以提供 more-like-this 相关问题的展现。

（4）GitHub 使用 Elasticsearch 来检索超过 1300 亿行代码。

（5）每天，Goldman Sachs 使用它来处理 5TB 数据的索引，还有很多投资银行使用它来分析股票市场的变动。

在本书中将使用 Elasticsearch 作为全文搜索引擎，来实现博客文章的关键字搜索、最热文章、最热标签、最热用户等功能。

6.3 Elasticsearch 与 Spring Boot 集成

本节将演示如何在 Spring Boot 中集成 Elasticsearch 的功能。

在 hello-world 项目的基础上创建了一个 elasticsearch-in-action 项目。

6.3.1 所需环境

本例采用的开发环境如下。

- Elasticsearch 5.5.0。
- Spring Data Elasticsearch 3.0.0.M4。

6.3.2 build.gradle

我们需要添加 Spring Data Elasticsearch 的依赖。Spring Boot 已经提供了相关的 Starter 来实现 Spring Data Elasticsearch 开箱即用的功能，所以只需要在 build.gradle 文件中添加 Spring Data Elasticsearch 的 Starter 的库即可。

```
// 依赖关系
dependencies {
    //...

    // 添加Spring Data Elasticsearch的依赖
    compile('org.springframework.boot:spring-boot-starter-data-elastic-
search')
```

```
    //...
}
```

6.3.3 下载并安装 Elasticsearch

下载地址为 https://www.elastic.co/downloads/elasticsearch，解压之后，有一个名为 elasticsearch-5.5.0 的文件夹，将其称为% ES_HOME% 变量。在终端窗口中，切换到% ES_HOME%目录，例如，

```
cd d:\elasticsearch-5.5.0
```

6.3.4 运行与停止服务

命令行执行。

```
.\bin\elasticsearch
```

正常启动后，可以看到控制台输出内容如下。

```
d:\elasticsearch-2.4.4>.\bin\elasticsearch
[2017-07-20T23:12:53,588][INFO ][o.e.n.Node               ] []
initializing ...
[2017-07-20T23:12:53,947][INFO ][o.e.e.NodeEnvironment    ] [2RvnJex]
using [1] data paths, mounts [[(D:)]], net usable_space [218.4gb],
net total_space [291gb], spins? [unknown], types [NTFS]
[2017-07-20T23:12:53,947][INFO ][o.e.e.NodeEnvironment    ] [2RvnJex]
heap size [1.9gb], compressed ordinary object pointers [true]
[2017-07-20T23:12:54,082][INFO ][o.e.n.Node               ] node name
[2RvnJex] derived from node ID [2RvnJexbSxCm04a8P1LR-w]; set [node.name]
to override
[2017-07-20T23:12:54,083][INFO ][o.e.n.Node               ]
version[5.5.0], pid[12860], build[260387d/2017-06-30T23:16:05.735Z],
OS[Windows 10/10.0/amd64], JVM[Oracle Corporation/Java HotSpot(TM)
64-Bit Server VM/1.8.0_112/25.112-b15]
[2017-07-20T23:12:54,084][INFO ][o.e.n.Node               ] JVM
arguments [-Xms2g, -Xmx2g, -XX:+UseConcMarkSweepGC, -XX:
CMSInitiatingOccupancyFraction=75, -XX:+UseCMSInitiatingOccupancyOnly,
-XX:+DisableExplicitGC, -XX:+AlwaysPreTouch, -Xss1m, -Djava.awt.
headless=true, -Dfile.encoding=UTF-8, -Djna.nosys=true, -Djdk.io.
permissionsUseCanonicalPath=true, -Dio.netty.noUnsafe=true, -Dio.netty.
noKeySetOptimization=true, -Dio.netty.recycler.maxCapacityPerThread=0,
-Dlog4j.shutdownHookEnabled=false, -Dlog4j2.disable.jmx=true, -Dlog4j.
skipJansi=true, -XX:+HeapDumpOnOutOfMemoryError, -Delasticsearch, -Des.
path.home=D:\elasticsearch-5.5.0]
[2017-07-20T23:12:56,329][INFO ][o.e.p.PluginsService     ] [2RvnJex]
```

```
loaded module [aggs-matrix-stats]
[2017-07-20T23:12:56,329][INFO ][o.e.p.PluginsService     ] [2RvnJex]
loaded module [ingest-common]
[2017-07-20T23:12:56,338][INFO ][o.e.p.PluginsService     ] [2RvnJex]
loaded module [lang-expression]
[2017-07-20T23:12:56,338][INFO ][o.e.p.PluginsService     ] [2RvnJex]
loaded module [lang-groovy]
[2017-07-20T23:12:56,338][INFO ][o.e.p.PluginsService     ] [2RvnJex]
loaded module [lang-mustache]
[2017-07-20T23:12:56,338][INFO ][o.e.p.PluginsService     ] [2RvnJex]
loaded module [lang-painless]
[2017-07-20T23:12:56,338][INFO ][o.e.p.PluginsService     ] [2RvnJex]
loaded module [parent-join]
[2017-07-20T23:12:56,338][INFO ][o.e.p.PluginsService     ] [2RvnJex]
loaded module [percolator]
[2017-07-20T23:12:56,338][INFO ][o.e.p.PluginsService     ] [2RvnJex]
loaded module [reindex]
[2017-07-20T23:12:56,338][INFO ][o.e.p.PluginsService     ] [2RvnJex]
loaded module [transport-netty3]
[2017-07-20T23:12:56,338][INFO ][o.e.p.PluginsService     ] [2RvnJex]
loaded module [transport-netty4]
[2017-07-20T23:12:56,338][INFO ][o.e.p.PluginsService     ] [2RvnJex]
no plugins loaded
[2017-07-20T23:13:01,218][INFO ][o.e.d.DiscoveryModule    ] [2RvnJex]
using discovery type [zen]
[2017-07-20T23:13:02,070][INFO ][o.e.n.Node               ] initialized
[2017-07-20T23:13:02,070][INFO ][o.e.n.Node               ] [2RvnJex]
starting ...
[2017-07-20T23:13:03,103][INFO ][o.e.t.TransportService   ] [2RvnJex]
publish_address {127.0.0.1:9300}, bound_addresses {127.0.0.1:9300},
{[::1]:9300}
[2017-07-20T23:13:06,241][INFO ][o.e.c.s.ClusterService   ] [2RvnJex]
new_master {2RvnJex}{2RvnJexbSxCm04a8P1LR-w}{JeXig6dMROuEKg91HuPHRA}
{127.0.0.1}{127.0.0.1:9300}, reason: zen-disco-elected-as-master ([0]
nodes joined)
[2017-07-20T23:13:06,534][INFO ][o.e.h.n.Netty4HttpServerTransport]
[2RvnJex] publish_address {127.0.0.1:9200}, bound_addresses
{127.0.0.1:9200}, {[::1]:9200}
[2017-07-20T23:13:06,534][INFO ][o.e.n.Node               ] [2RvnJex]
started
```

可以输入 Ctrl+C 来停止服务。

6.3.5 配置

默认情况下，Elasticsearch 从 %ES_HOME%/config/elasticsearch.yml 文件加载其配置。此配置文件的格式在“配置 Elasticsearch”（https://www.elastic.co/guide/en/elasticsearch/reference/current/

settings.html）中进行了说明，读者可以自行参阅。

还可以在命令行使用 -E 语法指定能在配置文件中指定的任何设置，如下所示。

```
./bin/elasticsearch -Ecluster.name=my_cluster -Enode.name=node_1
```

注意：包含空格的值必须加上双引号，例如，-Epath.logs="C:\My Logs\logs"。

提示：通常，任何集群范围的设置（如 cluster.name）都应该添加到 elasticsearch.yml 配置文件中，而任何特定于节点的设置（如 node.name）都可以在命令行中指定。

6.3.6 确认运行情况

可以通过向 localhost 上的端口 9200 发送 HTTP GET 请求来测试 Elasticsearch 节点是否正在运行。访问 http://localhost:9200/ 应能看到如下数据返回。

```
{
  "name" : "2RvnJex",
  "cluster_name" : "elasticsearch",
  "cluster_uuid" : "uqcQAMTtTIO6CanROYgveQ",
  "version" : {
    "number" : "5.5.0",
    "build_hash" : "260387d",
    "build_date" : "2017-06-30T23:16:05.735Z",
    "build_snapshot" : false,
    "lucene_version" : "6.6.0"
  },
  "tagline" : "You Know, for Search"
}
```

从浏览器来访问 Elasticsearch 的效果，如图 6-1 所示。

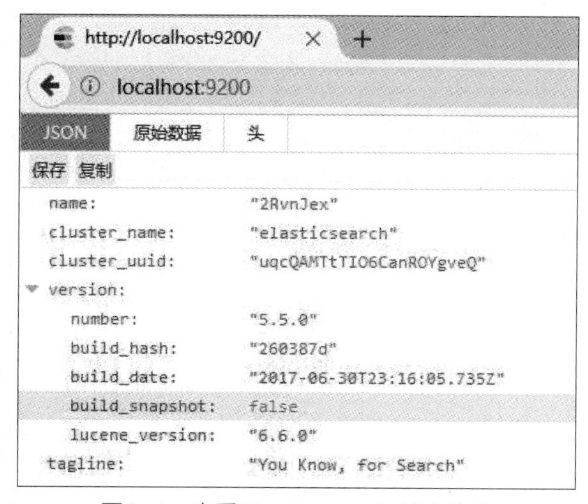

图6-1　查看Elasticsearch 状态信息

6.3.7 作为 Windows 服务

执行 bin 目录下的 elasticsearch-service.bat 即可，允许用户从命令行来安装、删除、管理或配置服务，并能启动和停止服务。

```
D:\>cd D:\elasticsearch-5.5.0\bin

D:\elasticsearch-5.5.0\bin>elasticsearch-service.bat

Usage: elasticsearch-service.bat install|remove|start|stop|manager [SER-
VICE_ID]
```

成功安装为 Windows 服务后，应能看到如下控制台信息。

```
D:\elasticsearch-5.5.0\bin>elasticsearch-service.bat install
Installing service      :  "elasticsearch-service-x64"
Using JAVA_HOME (64-bit):  "C:\Program Files\Java\jdk1.8.0_112"
The service 'elasticsearch-service-x64' has been installed.
```

6.4 Elasticsearch 实战

本节将探索如何使用 Elasticsearch 来存储文本数据，并通过 Spring Data Elasticsearch 快速实现访问 Elasticsearch 数据的能力。6.3 节创建的 elasticsearch-in-action 项目的基础上来实现全文搜索的功能。

6.4.1 启动 Elasticsearch

首先，要确保已经可以正确启动 Elasticsearch 服务器了。在 Linux 下运行 bin/elasticsearch，Windows 平台则运行 bin\elasticsearch.bat。

6.4.2 修改application.properties

在 elasticsearch-in-action 项目中，需要添加 Spring Data Elasticsearch 相关的两项配置。

```
# Elasticsearch 服务地址
spring.data.elasticsearch.cluster-nodes=localhost:9300
# 设置连接超时时间
spring.data.elasticsearch.properties.transport.tcp.connect_timeout=120s
```

其中，一个配置是为了配置 Elasticsearch 服务地址，另一个则是设置连接超时时间。

6.4.3 创建文档类

在 com.waylau.spring.boot.blog.domain.es 包下创建一个注解为 org.springframework.data.elasticsearch.annotations.Document 文档类 EsBlog，专门用于 Elasticsearch 中存储博客的文档。

```
import org.springframework.data.annotation.Id;
import org.springframework.data.elasticsearch.annotations.Document;

@Document(indexName = "blog", type = "blog")
public class EsBlog implements Serializable {

    private static final long serialVersionUID = 1L;

    @Id  // 主键
    private String id;

    private String title;

    private String summary;

    private String content;

    protected EsBlog() {  // JPA的规范要求无参构造函数；设为 protected 防止直
接使用
    }

    public EsBlog(String title, String summary, String content) {
        this.title = title;
        this.summary = summary;
        this.content = content;
    }

    public String getId() {
        return id;
    }

    public void setId(String id) {
        this.id = id;
    }

    public String getTitle() {
        return title;
    }

    public void setTitle(String title) {
        this.title = title;
    }
```

```
    public String getContent() {
        return content;
    }

    public void setContent(String content) {
        this.content = content;
    }

    public String getSummary() {
        return summary;
    }

    public void setSummary(String summary) {
        this.summary = summary;
    }

    @Override
    public String toString() {
        return String.format(
                "User[id=%d, title='%s', summary='%s', content='%s']",
                id, title, summary, content);
    }
}
```

需要注意的是，在 Elasticsearch 中，主键 id 采用 String 类型。

6.4.4 创建资源库

在 com.waylau.spring.boot.blog.repository.es 包下定义资源库的接口 EsBlogRepository，该接口继承自 org.springframework.data.elasticsearch.repository.ElasticsearchRepository。

```
public interface EsBlogRepository extends ElasticsearchRepository<Es-
Blog, String> {

    /**
     * 分页查询博客
     * @param title
     * @param summary
     * @param content
     * @param pageable
     * @return
     */
    Page<EsBlog> findByTitleContainingOrSummaryContainingOrContentCon-
taining(String title, String summary, String content, Pageable pagea-
ble);

}
```

6.4.5 创建资源库测试用例

在 test 目录下建立 com.waylau.spring.boot.blog.repository.es 包，创建资源库测试用例 EsBlogRepositoryTest。

```
@RunWith(SpringRunner.class)
@SpringBootTest
public class EsBlogRepositoryTest {

    @Autowired
    private EsBlogRepository esBlogRepository;

    @Before
    public void initRepositoryData() {

        // 清除所有数据
        esBlogRepository.deleteAll();

        // 初始化数据
        esBlogRepository.save(new EsBlog("Had I not seen the Sun",
                "I could have borne the shade",
                "But Light a newer Wilderness. My Wilderness has
made."));
        esBlogRepository.save(new EsBlog("There is room in the halls of
pleasure",
                "For a long and lordly train",
                "But one by one we must all file on, Through the narrow
aisles of pain."));
        esBlogRepository.save(new EsBlog("When you are old",
                "When you are old and grey and full of sleep",
                "And nodding by the fire, take down this book."));
    }

    @Test
    public void testFindDistinctEsBlogByTitleContainingOrSummaryCon-
tainingOrContentContaining() {
        Pageable pageable = PageRequest.of(0, 20);

        String title = "Sun";
        String summary = "is";
        String content = "down";

        Page<EsBlog> page = esBlogRepository.findByTitleContainingOrSum-
maryContainingOrContentContaining(title, summary, content, pageable);

        System.out.println("---------start 1");
        for (EsBlog blog : page) {
            System.out.println(blog.toString());
        }
```

```
        System.out.println("---------end 1");

        title = "the";
        summary = "the";
        content = "the";

        page = esBlogRepository.findByTitleContainingOrSummaryContainin-
gOrContentContaining(title, summary, content, pageable);

        System.out.println("---------start 2");
        for (EsBlog blog : page) {
            System.out.println(blog.toString());
        }
        System.out.println("---------end 2");
    }
}
```

其中，Pageable pageable = PageRequest.of(0, 20); 是初始化一个分页请求。

在执行测试用例之前，先在 Elasticsearch 的存储库中初始化了 3 首诗，作为测试用的数据。

而后，执行测试，并能看到搜索的结果，并将结果从控制台进行了输出。

```
---------start 1
EsBlog[id='AV1gzYw-yjDMiQfhe8qy',title='When you are old',summary='When
you are old and grey and full of sleep',content='And nodding by the fire,
take down this book.']
EsBlog[id='AV1gzYuhyjDMiQfhe8qw',title='Had I not seen the Sun',summary=
'I could have borne the shade',content='But Light a newer Wilderness.
My Wilderness has made.']
---------end 1
---------start 2
EsBlog[id='AV1gzYvvyjDMiQfhe8qx',title='There is room in the halls of
pleasure',summary='For a long and lordly train',content='But one by one
we must all file on, Through the narrow aisles of pain.']
EsBlog[id='AV1gzYuhyjDMiQfhe8qw',title='Had I not seen the Sun',summary=
'I could have borne the shade',content='But Light a newer Wilderness.
My Wilderness has made.']
EsBlog[id='AV1gzYw-yjDMiQfhe8qy',title='When you are old',summary='When
you are old and grey and full of sleep',content='And nodding by the fire,
take down this book.']
---------end 2
```

第一次搜索使用“Sun”“is”“down”作为了查询参数，其中“Sun”匹配了第一首诗歌，“is”没有匹配任意诗歌，“down”匹配了第三首诗歌，所以一共匹配出来两条数据。

第二次搜索使用“the”作为了 3 个查询参数，由于每首诗歌都包含“the”，因此一共匹配出来 3 条数据。

6.4.6 创建控制器

在 com.waylau.spring.boot.blog.controller 包下创建控制器 BlogController 用于处理博客相关的请求。

```
@RestController
@RequestMapping("/blogs")
public class BlogController {

    @Autowired
    private EsBlogRepository esBlogRepository;

    @GetMapping
    public List<EsBlog> list(@RequestParam(value="title",required=false,
defaultValue="") String title,
            @RequestParam(value="summary",required=false,defaultValue=
"") String summary,
            @RequestParam(value="content",required=false,defaultValue=
"") String content,
            @RequestParam(value="pageIndex",required=false,defaultValue=
"0") int pageIndex,
            @RequestParam(value="pageSize",required=false,defaultValue=
"10") int pageSize) {

        // 数据在Test中先初始化了，这里只负责取数据
        Pageable pageable = PageRequest.of(pageIndex, pageSize);

        Page<EsBlog> page = esBlogRepository.findByTitleContainingOrSum-
maryContainingOrContentContaining(title, summary, content, pageable);

        return page.getContent();
    }
}
```

6.4.7 运行

在启动项目之前，要确保 Elasticsearch 服务器已经启动。在启动项目之后，运行 EsBlogRepositoryTest 测试用例来帮助初始化数据。

而后就能访问控制器所定义的 API，如 http://localhost:8080/blogs?title=i&summary=love&content=you，结果如图 6-2 所示。

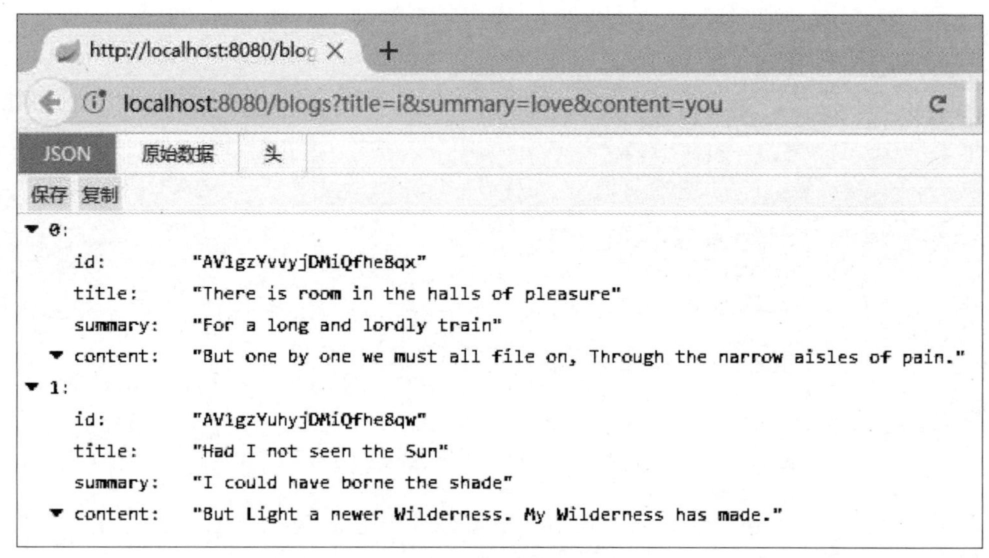

图6-2　执行全文搜索

6.4.8 示例源码

本节示例源码在 elasticsearch-in-action 目录下。

第7章 架构设计与分层

7.1 为什么需要分层

为了便于对程序进行管理，我们倾向于在开发时对应用程序进行分割、分层。在本节将讨论以下话题。

（1）为什么需要将应用程序进行分层？

（2）如果不分层，系统将会出现哪些问题？

（3）常见的分层方式有哪些？

7.1.1 应用的分层

随着面向对象程序设计和设计模式的出现，人们发现，现实生活中的建筑学有很多理论都可以用来指导软件工程（即程序的开发），例如，在开发时，人们会先对要盖的楼房进行评估和核算（软件项目管理）；根据需求设计楼房的图纸（软件设计），而后根据设计把楼房的地基、骨架先搭建出来（搭建框架）；根据不同工种将人员进行分工，有些去砌墙，有些去贴砖（前端编码、后台编码）；最后进行验收测试，交付给用户使用。

软件应用开发与建筑学的分层目的是一致的，旨在根据不同的业务、不同的技术、不同的组织，结合灵活性、可维护性、可扩展性等多种因素，将应用系统划分成不同的部分，并使这些部分彼此之间相互分工、相互协作，从而体现出最大化价值。对于一个良好分层的应用来说，一般具备如下特点。

1. 按业务功能进行分层

分层就是将相关的业务功能的类或组件放置在一起，而将不相关的业务功能的类或组件隔离开。例如，我们会将与用户直接交互的部分分为“表示层”，将实现逻辑计算或者业务处理的部分分为“业务层”，将与数据库打交道的部分分为“数据访问层”。

2. 良好的层次关系

设计良好的架构分层是上层依赖于下层，而下层支撑起上层，但却不能直接访问上层，层与层之间通过协作来共同完成特定的功能。

3. 每一层都能保持独立

层能够被单独构造，也能被单独替换，最终不会影响整体功能。例如，将整个数据持久层的技术从 Hibernate 转成了 EclipseLink，但不能对上层业务逻辑功能造成影响。

7.1.2 不分层的应用架构

为了更好地理解分层的好处，先来看一下不分层的应用架构是如何运作的。

以 Web 应用程序为例。在 Web 应用程序开发的早期，所有的逻辑代码并没有明显的层次区分，因此代码之间的调用是相互交错的，整体代码看上去错综复杂。例如，在早期使用诸如 ASP、JSP 及 PHP 等动态网页技术时，常会将所有的页面逻辑、业务逻辑及数据库访问逻辑放在一起，很多时候就在 JSP 页面中写 SQL 语句了，编码风格完全是过程化的。

以下代码就是一个 JSP 访问 SQL Server 数据库的例子。

```
<%@ page language="java" contentType="text/html; charset=UTF-8" pageEn-
coding="UTF-8"%>
<%@ page import = "java.sql.*"%>
<!DOCTYPE html PUBLIC "-//W3C//DTD XHTML 1.0 Transitional//EN" "http://
www.w3.org/TR/xhtml1/DTD/xhtml1-transitional.dtd">
<html xmlns="http://www.w3.org/1999/xhtml">
<head>
</head>
<body>
<%

// 创建数据库连接
Class.forName("com.microsoft.sqlserver.jdbc.SQLServerDriver");
String url="jdbc:sqlserver://localhost:1433;databaseName=Book;user=sa;
password=";
PreparedStatement pstmt;
String sql = "insert into students (UserName,WebSite) values(?,?)";
int returnValue = 0;
try{
    pstmt = (PreparedStatement) conn.prepareStatement(sql);
    pstmt.setString(1, student.getName());
    pstmt.setString(2, student.getSex());
    returnValue = pstmt.executeUpdate();

    // 判断是添加成功还是失败
    if(returnValue == 1){
        out.print("<li>添加成功！");
        out.print("<li>returnValue = " + returnValue);
    } else {
        out.print("<li>添加失败！");
    }
}catch(Exception ex){
    out.print(ex.getLocalizedMessage());
}finally{
    try{
        if(pstmt != null){
            pstmt.close();
            pstmt = null;
        }
        if(cn != null){
            conn.close();
            conn = null;
        }
```

```
    }catch(Exception e){
        e.printStackTrace();
    }
}
%>
</body>
</html>
```

先且不论这段代码是否正确，从实现功能上来讲，这段代码既处理了数据库的访问操作，还做了页面的表示，其中又夹杂着业务逻辑判断。还好这段代码不长，读下来还能够理解，但如果是更加复杂的功能，这种代码肯定是非常不清晰的，维护起来也相当麻烦。

早期的不分层的架构主要存在如下弊端。

（1）代码不够清晰，难以阅读。

（2）代码职责不明，难以扩展。

（3）代码错综复杂，难以维护。

（4）代码没做分工，难以组织。

7.1.3 应用的三层架构

目前，比较常用的、典型的应用软件倾向于使用三层架构（Three-Tier Architecture），即表示层、业务层和数据访问层。

（1）表示层（Presentation Layer）：提供与用户交互的界面。GUI（图形用户界面）和 Web 页面是表示层的两个典型的例子。

（2）业务层（Business Layer）：也称为业务逻辑层，用于实现各种业务逻辑。如处理数据验证，根据特定的业务规则和任务来响应特定的行为。

（3）数据访问层（Data Access Layer）：也称为数据持久层，负责存放和管理应用的持久性业务数据。

图 7-1 所示为三层架构的架构图。如果仔细看一看这些层，应该能看到，每一个层都需要不同的技能。

（1）表示层需要 HTML、CSS、JavaScript 等之类的前端技能，以及具备 UI 设计能力。

（2）业务层需要编程语言的技能，以便计算机可以处理业务规则。

（3）数据访问层需要具有数据定义语言（DDL）和数据操作语言（DML）及数据库设计形式的 SQL 技能。

虽然一个人有可能拥有所有上述技能，但这样的人是相当罕见的。 在具有大型软件应用程序的大型组织中，将应用程序分割为单独的层，使得每个层都可以由具有相关专业技能的不同团队来开发和维护。

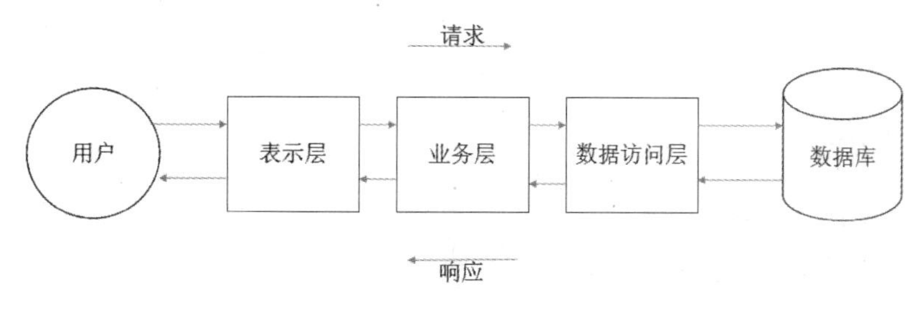

图7-1　三层架构

一个良好分层的架构系统应遵循如下分层原则。

（1）每个层的代码必须包含可以单独维护的单独文件。

（2）每个层只能包含属于该层的代码。因此，业务逻辑只能驻留在业务层，表示逻辑只能在表示层，而数据访问逻辑只驻留在数据访问层中。

（3）表示层只能接收来自外部代理的请求，并向外部代理返回响应。这通常是一个人，但也可能是另一个软件。

（4）表示层只能向业务层发送请求，并从业务层接收响应。它不能直接访问数据库或数据访问层。

（5）业务层只能接收来自表示层的请求，并返回对表示层的响应。

（6）业务层只能向数据访问层发送请求，并从其接收响应。它不能直接访问数据库。

（7）数据访问层只能从业务层接收请求并返回响应。它不能发出请求到除了它支持的数据库管理系统（DBMS）以外的任何地方。

（8）每层应完全不知道其他层的内部工作原理。例如，业务层可以对数据库一无所知，并且可以不知道或不必关心数据访问对象的内部工作原理，它也必须是和表示层无关的，可以不知道或不关心表示层是如何处理它的数据的。表示层可以获取数据并构造 HTML 文档、PDF 文档、CSV 文件或以某种其他方式处理它，但是这应该与业务层完全无关。

（9）每层应当可以用具有类似特征的替代组件来交换这个层，使得整体可以继续工作。

简言之，应用在一开始设计的时候，就要考虑系统的架构设计及如何来将系统进行有效的分层。由于系统架构设计属于比较高级别的话题，本书不会琢磨太多，有这方面兴趣的读者，可以参阅笔者所著的《分布式系统常用技术及案例分析》一书，该书的第 2 章详细介绍了软件系统的常见架构体系。

7.2 系统的架构设计及职责划分

在 7.1 节介绍了系统分层的重要性和必要性。本节按照经典的三层架构来对博客系统进行分层处理。

7.2.1 博客系统的三层架构

我们的博客系统同样也是采用三层架构，通过内聚每一层的工作职责，分为了表示层、业务层和数据访问层。每一层的职责和功能划分如图 7-2 所示。

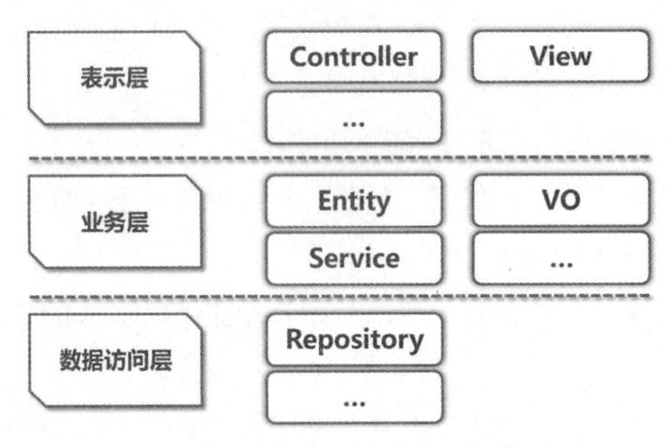

图7-2 博客系统的三层架构

1. 表示层

表示层提供与用户交互的界面，分为 Controller（控制器） 和 View（视图）。Controller 和 View 是 MVC 中重要的两部分，用于处理请求和响应界面。

2. 业务层

业务层也称为业务逻辑层，用于实现各种业务逻辑。例如，处理数据验证，根据特定的业务规则和任务来响应特定的行为。

MVC 中 Model（模型）涵盖的范围比较广，如 Entity（实体）、VO（值对象）、Service（服务）都可以称为领域模型的一种。

3. 数据访问层

数据访问层也称为数据持久层，负责存放和管理应用的持久性业务数据。数据访问层可以用 Dao 或 Repository 来表示。

7.2.2 博客系统的架构设计

博客系统的整体架构如图 7-3 所示。

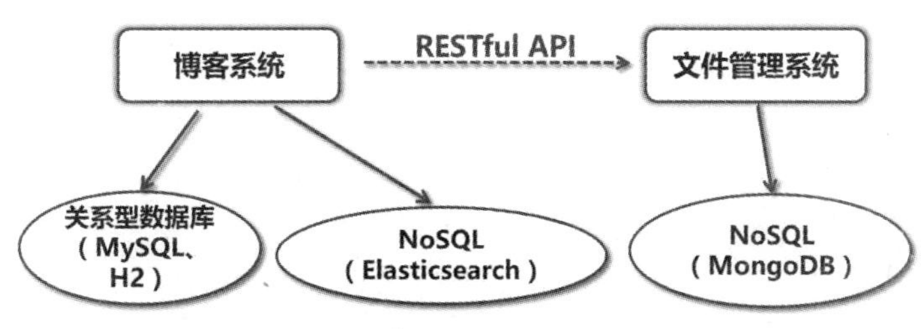

图7-3　博客系统的架构

博客系统主要由“博客系统”和“文件管理系统”两个子系统组成。其中，“博客系统”提供博客系统常见的功能，注册博主、发博客、评论、点赞、全文检索等；“文件管理系统”用于博客系统中所设计的文件的管理，这些文件包括上传的头像，博客正文中的插图等。“文件管理系统”提供了 RESTful API，方便“博客系统”来调用。

数据存储的方式采用了不同的数据库系统。

（1）关系型数据库：作为传统的数据存储方式，系统中的核心数据都是通过关系型数据库来存储的。MySQL 作为开源的大型数据库，是互联网应用的首选。H2 是内存数据库，在 Java 应用中可以方便作为开发阶段数据存储方式。

（2）Elasticsearch：Elasticsearch 自带了 NoSQL 数据的存储，用于全文搜索时的数据本文和索引的存储。

（3）MongoDB：是流行的 NoSQL 数据库。在“文件管理系统”子系统中，MongoDB 承担着存储小型文件的职责。

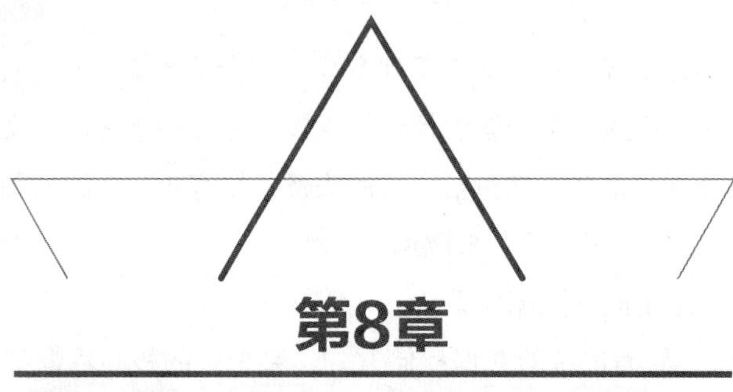

第8章 集成 Bootstrap

8.1 Bootstrap 简介

Bootstrap 是比较受欢迎的 HTML、CSS 和 JS 前端框架，用于开发响应式布局、移动设备优先的 Web 项目。

Bootstrap 让前端开发更快速、更简单。可以说所有开发者、所有应用场景都能适用于 Bootstrap。无论开发者的技能水平是什么层次、无论什么类型的设备、无论项目规模大小，Bootstrap 都能胜任。

Bootstrap 具有以下优点。

（1）支持 Less 和 Sass 预处理器：虽然 Bootstrap 最终提供的是 CSS 文件，但是它的源码是采用 Less（http://getbootstrap.com/css/#less）和 Sass（http://sass-lang.com/）编写的。因此，可以选择直接使用 Bootstrap 的 CSS 文件，也可以直接从源码进行编译。

（2）响应式布局适合所有设备：网站和应用能在 Bootstrap 的帮助下基于同一份代码快速、有效地适配手机、平板电脑、PC 设备，这一切都得益于 CSS3 的媒体查询（Media Query）。

（3）功能组件齐全：Bootstrap 提供了大量的 HTML 和 CSS 组件、jQuery 插件等，并提供全面、美观的文档加以说明，方便学习及应用。

（4）开源：有强大的社区及广泛的开源爱好者为其贡献代码。

8.1.1 HTML5 Doctype

Bootstrap 使用了一些 HTML5 元素和 CSS 属性。为了让这些正常工作，需要使用 HTML5 文档类型（Doctype）。因此，在使用 Bootstrap 项目的开头包含下面的代码段。

```
<!DOCTYPE html>
<html lang="en">
  ...
</html>
```

8.1.2 响应式 meta 标签

Bootstrap 的设计目标是移动设备优先，并能通过 CSS 媒体查询来扩展组件。为了让 Bootstrap 开发的网站对移动设备友好，确保适当的绘制和触屏缩放，需要在网页的 head 中添加 viewport meta 标签，如下所示。

```
<meta name="viewport" content="width=device-width, initial-scale=1,
shrink-to-fit=no">
```

8.1.3 Box-sizing

为了在 CSS 中更简单地调整大小，Bootstrap 将全局框大小值从 content-box 切换到 border-box。

这可以确保填充不会影响元素的最终计算宽度，但它可能会导致某些第三方软件（如 Google 地图和 Google 自定义搜索引擎）出现问题。

在罕见的情况下，需要覆盖它，请使用以下内容。

```
.selector-for-some-widget {
  -webkit-box-sizing: content-box;
     -moz-box-sizing: content-box;
          box-sizing: content-box;
}
```

通过上面的代码片断，嵌套的元素（包括通过 :before 和 :after 生成的内容）都将继承 .selector-for-some-widget 所指定的 box-sizing。

欲了解更多关于盒子模型和尺寸调整 CSS 技巧，请参阅 https://css-tricks.com/box-sizing/。

8.1.4 Normalize.css

Bootstrap 使用 Normalize 来建立跨浏览器的一致性。Normalize.css（<ttps://necolas.github.io/normalize.css/>）是一个很小的 CSS 文件，在 HTML 元素的默认样式中提供了更好的跨浏览器一致性。同时，Bootstrap 也提供了 Reboot（https://v4-alpha.getbootstrap.com/content/reboot/），将所有 HTML 重置样式表整合到 Reboot 中，在用不了 Normalize.css 的地方可以用 Reboot。

8.1.5 模板

开启 Bootstrap，第一个 Bootstrap 页面模板应该是类似于下面这样的。

```
<!DOCTYPE html>
<html lang="en">
  <head>
    <!-- Required meta tags -->
    <meta charset="utf-8">
    <meta name="viewport" content="width=device-width, initial-scale=1, shrink-to-fit=no">

    <!-- Bootstrap CSS -->
    <link rel="stylesheet" href="https://maxcdn.bootstrapcdn.com/bootstrap/4.0.0-alpha.6/css/bootstrap.min.css" integrity="sha384-rwoIResjU2yc3z8GV/NPeZWAv56rSmLldC3R/AZzGRnGxQQKnKkoFVhFQhNUwEyJ" crossorigin="anonymous">
  </head>
  <body>
    <h1>Hello, world!</h1>

    <!-- jQuery first, then Tether, then Bootstrap JS. -->
    <script src="https://code.jquery.com/jquery-3.1.1.slim.min.js"
```

```
integrity="sha384-A7FZj7v+d/sdmMqp/nOQwliLvUsJfDHW+k9Omg/a/EheAdgtzN-
s3hpfag6Ed950n" crossorigin="anonymous"></script>
    <script src="https://cdnjs.cloudflare.com/ajax/libs/tether/1.4.0/js/
tether.min.js" integrity="sha384-DztdAPBWPRXSA/3eYEEUWrWCy7G5KFbe8fFjk
5JAIxUYHKkDx6Qin1DkWx51bBrb" crossorigin="anonymous"></script>
    <script src="https://maxcdn.bootstrapcdn.com/bootstrap/4.0.0-alpha.
6/js/bootstrap.min.js" integrity="sha384-vBWWzlZJ8ea9aCX4pEW3rVHjgjt7zp-
kNpZk+02D9phzyeVkE+jo0ieGizqPLForn" crossorigin="anonymous"></script>
  </body>
</html>
```

8.2 Bootstrap 核心概念

本节将介绍 Bootstrap 的核心概念。

8.2.1 Bootstrap 的网格系统

Bootstrap 提供了一套响应式、移动设备优先的流式网格系统，随着屏幕或视口（Viewport）尺寸的增加，系统会自动分为最多 12 列。

1. 网格（Grid）

在平面设计中，网格是一种由一系列用于组织内容的相交的直线（垂直的、水平的）组成的结构（通常是二维的），它广泛应用于打印设计中的设计布局和内容结构。在网页设计中，它是一种用于快速创建一致的布局和有效地使用 HTML 和 CSS 的方法。 简单地说，网页设计中的网格用于组织内容，让网站易于浏览，并降低用户端的负载。

2. Bootstrap 网格系统（Grid System）

Bootstrap 包含了一个响应式的、移动设备优先的、不固定的网格系统，可以随着设备或视口大小的增加而适当地扩展到 12 列。它包含了用于简单地布局选项的预定义类，也包含了用于生成更多语义布局的功能强大的混合类。Bootstrap 代码从小屏幕设备（如移动设备、平板电脑）开始，然后扩展到大屏幕设备（如笔记本电脑、台式电脑）上的组件和网格。

那么什么是移动设备优先策略呢？概括如下。

（1）内容。决定什么是最重要的。

（2）布局。

①优先设计更小的宽度。

②基础的 CSS 是移动设备优先，媒体查询是针对平板电脑、台式电脑的。

（3）渐进增强。随着屏幕大小的增加而添加元素。

响应式网格系统随着屏幕或视口尺寸的增加，系统会自动分为最多 12 列。图 8-1 所示为 Bootstrap 的网格系统。

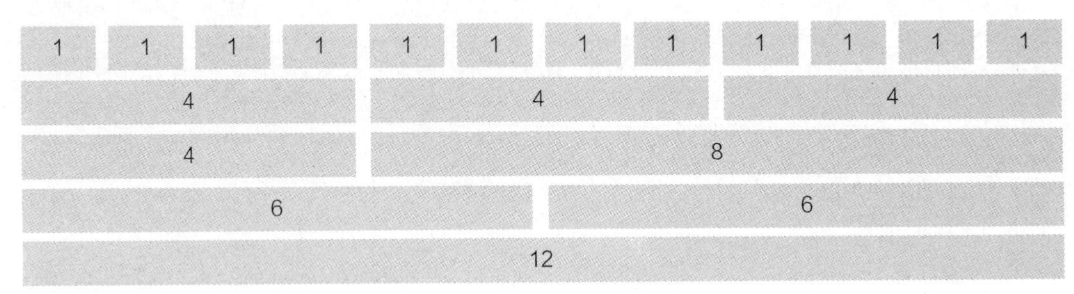

图8-1 Bootstrap的网格系统

3. Bootstrap 网格系统的工作原理

网格系统通过一系列包含内容的行和列来创建页面布局。下面列出了 Bootstrap 网格系统是如何工作的。

（1）行必须放置在 class 名为 .container 或 .container-fluid 内，以便获得适当的对齐（Alignment）和内边距（Padding）。

（2）使用行来创建列的水平组。

（3）内容应该放置在列内，且唯有列可以是行的直接子元素。

（4）预定义的网格类，如 .row 和 .col-xs-4，可用于快速创建网格布局。LESS 混合类可用于更多语义布局。

（5）列通过内边距来创建列内容之间的间隙。该内边距是通过 .row 上的外边距（Margin）取负，表示第一列和最后一列的行偏移。

（6）网格系统是通过指定用户想要横跨的 12 个可用的列来创建的。例如，要创建 3 个相等的列，则使用 3 个 .col-xs-4。

（7）如果在一行内放置超过 12 列，则每组额外的列将作为一个单元包装到新行上。

（8）网格类适用于屏幕宽度大于或等于断点大小的设备，并覆盖针对较小设备的网格类。例如，将任何 .col-md-* 类应用于元素不仅会影响其在中型设备上的样式，而且还会影响没有设置 .col-lg-* 类的大型设备上的样式。

8.2.2 媒体查询

媒体查询是非常别致的“有条件的 CSS 规则”，它只适用于一些基于某些规定条件的 CSS。如果满足那些条件，则应用相应的样式。

Bootstrap 中的媒体查询允许用户基于视口大小移动、显示并隐藏内容。下面的媒体查询在 LESS 文件中使用，用来创建 Bootstrap 网格系统中的关键的断点。

```
/* 超小设备（手机，小于 768px） */
/* Bootstrap 中默认情况下没有媒体查询 */

/* 小型设备（平板电脑，768px 起） */
@media (min-width: @screen-sm-min) { ... }

/* 中型设备（台式电脑，992px 起） */
@media (min-width: @screen-md-min) { ... }

/* 大型设备（大台式电脑，1200px 起） */
@media (min-width: @screen-lg-min) { ... }
```

我们有时也会在媒体查询代码中包含 max-width，从而将 CSS 的影响限制在更小范围的屏幕大小之内。

```
@media (max-width: @screen-xs-max) { ... }
@media (min-width: @screen-sm-min) and (max-width: @screen-sm-max) {
... }
@media (min-width: @screen-md-min) and (max-width: @screen-md-max) {
... }
@media (min-width: @screen-lg-min) { ... }
```

媒体查询有两个部分，先是一个设备规范，然后是一个大小规则。在上面的案例中，设置了下列的规则。

```
@media (min-width: @screen-sm-min) and (max-width: @screen-sm-max) { ... }
```

对于所有带有 min-width: @screen-sm-min 的设备，如果屏幕的宽度小于 @screen-sm-max，则会进行一些处理。

8.2.3 网格选项

表 8-1 总结了 Bootstrap 网格系统如何跨多个设备工作。

表8-1 Bootstrap网格系统的网格选项

	超小设备手机（<768px）	小型设备平板电脑（≥768px）	中型设备台式电脑（≥992px）	大型设备台式电脑（≥1200px）
网格行为	一直是水平的	以折叠开始，断点以上是水平的	以折叠开始，断点以上是水平的	以折叠开始，断点以上是水平的
最大容器宽度	None (auto)	750px	970px	1170px
Class 前缀	.col-xs-	.col-sm-	.col-md-	.col-lg-
列数量和	12	12	12	12
最大列宽	Auto	~61px	~81px	~97px

续表

	超小设备手机（<768px）	小型设备平板电脑（≥768px）	中型设备台式电脑（≥992px）	大型设备台式电脑（≥1200px）
间隙宽度	30px（一个列的每边分别 15px）	30px（一个列的每边分别 15px）	30px（一个列的每边分别 15px）	30px（一个列的每边分别 15px）
可嵌套	Yes	Yes	Yes	Yes
偏移量	Yes	Yes	Yes	Yes
列排序	Yes	Yes	Yes	Yes

8.2.4 实例：移动设备及桌面设备

如何避免列堆叠在较小的设备中？实现方式就是，通过在列中添加 .col-xs-* 、.col-md-* 来使用额外的中小型设备网格类。请参阅下面的示例，以便更好地了解其全部工作原理，实例效果如图 8-2 所示。

```
<!-- 在移动设备上是一个全宽和另一个半宽来堆叠的列 -->
<div class="row">
  <div class="col-xs-12 col-md-8">.col-xs-12 .col-md-8</div>
  <div class="col-xs-6 col-md-4">.col-xs-6 .col-md-4</div>
</div>

<!-- 列宽在移动设备上开始占比是50%，在桌面设备上可达33.3% -->
<div class="row">
  <div class="col-xs-6 col-md-4">.col-xs-6 .col-md-4</div>
  <div class="col-xs-6 col-md-4">.col-xs-6 .col-md-4</div>
  <div class="col-xs-6 col-md-4">.col-xs-6 .col-md-4</div>
</div>

<!-- 不管是移动设备还是桌面设备，列宽占比都是50% -->
<div class="row">
  <div class="col-xs-6">.col-xs-6</div>
  <div class="col-xs-6">.col-xs-6</div>
</div>
```

.col-xs-12 .col-md-8
.col-xs-6 .col-md-4
.col-xs-6 .col-md-4
.col-xs-6 .col-md-4
.col-xs-6 .col-md-4
.col-xs-6
.col-xs-6

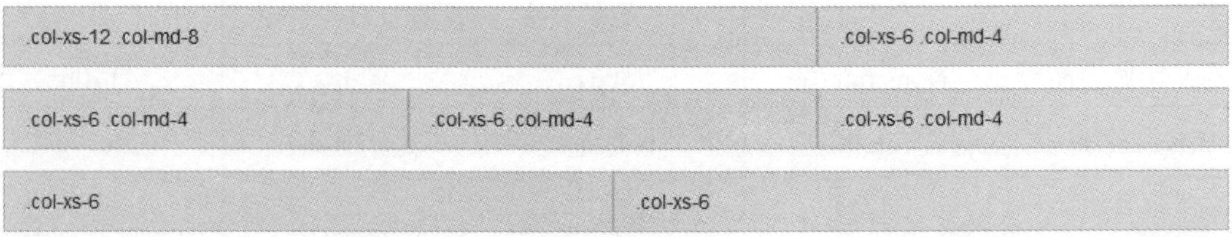

图8-2 实例效果

8.3 Bootstrap 及常用前端框架与 Spring Boot 集成

本节将演示如何在 Spring Boot 中集成 Bootstrap 及常用前端框架。

在 jpa-in-action 项目的基础上，我们构建了一个新的项目 bootstrap-in-action。

8.3.1 所需环境

本例采用的开发环境如下。

- Tether 1.4.0：http://tether.io/。
- Bootstrap v4.0.0-alpha.6：https://v4-alpha.getbootstrap.com/。
- jQuery 3.1.1：http://jquery.com/download/。
- Font Awesome 4.7.0：http://fontawesome.io。
- NProgress 0.2.0：http://ricostacruz.com/nprogress/。
- Thinker-md：http://git.oschina.net/benhail/thinker-md。
- jQuery Tags Input 1.3.6：http://xoxco.com/projects/code/tagsinput/。
- Bootstrap Chosen 1.0.3：https://github.com/haubek/bootstrap4c-chosen。
- toastr 2.1.1：http://www.toastrjs.com/。

注意：上述所用到的前端框架都会在本书的源码中提供。

8.3.2 将 Bootstrap 与 jQuery 集成进项目

在项目的 resources/static 目录下分别建立 css、js、fonts 三个目录，其中，将 Bootstrap、jQuery 发布包中的样式文件、js 脚本文件、字体图标文件放置于上述相应的目录下。整体目录如下。

```
static
│   favicon.ico
│
├─css
│       bootstrap-grid.css
│       bootstrap-grid.css.map
│       bootstrap-grid.min.css
│       bootstrap-grid.min.css.map
│       bootstrap-reboot.css
│       bootstrap-reboot.css.map
│       bootstrap-reboot.min.css
│       bootstrap-reboot.min.css.map
│       bootstrap.css
│       bootstrap.css.map
│       bootstrap.min.css
│       bootstrap.min.css.map
```

```
|
├─fonts
|       glyphicons-halflings-regular.eot
|       glyphicons-halflings-regular.svg
|       glyphicons-halflings-regular.ttf
|       glyphicons-halflings-regular.woff
|       glyphicons-halflings-regular.woff2
|
└─js
        bootstrap.js
        bootstrap.min.js
        jquery-3.1.1.min.js
```

8.3.3 其他前端常用控件的集成

修改 header.html，相关的样式都放置在 <head> 中。

```
<head>
    <meta charset="UTF-8">
    <meta name="viewport" content="width=device-width, initial-scale=1,
shrink-to-fit=no">

    <!-- Tether core CSS -->
    <link href="../../css/tether.min.css" th:href="@{/css/tether.min.
css}" rel="stylesheet">

    <!-- Bootstrap CSS -->
    <link href="../../css/bootstrap.min.css" th:href="@{/css/bootstrap.
min.css}" rel="stylesheet">

    <!-- Font-Awesome CSS -->
    <link href="../../css/font-awesome.min.css" th:href="@{/css/
font-awesome.min.css}" rel="stylesheet">

    <!-- NProgress CSS -->
    <link href="../../css/nprogress.css" th:href="@{/css/nprogress.css}"
rel="stylesheet">

    <!-- thinker-md CSS -->
    <link href="../../css/thinker-md.vendor.css" th:href="@{/css/think-
er-md.vendor.css}" rel="stylesheet">

    <!-- bootstrap tags CSS -->
    <link href="../../css/bootstrap-tagsinput.css" th:href="@{/css/
jquery.tagsinput.min.css}" rel="stylesheet">

    <!-- bootstrap chosen CSS -->
    <link href="../../css/component-chosen.min.css" th:href="@{/css/
```

```
component-chosen.min.css}" rel="stylesheet">

    <!-- toastr CSS -->
    <link href="../../css/toastr.min.css" th:href="@{/css/toastr.min.
css}" rel="stylesheet">

    <!-- jQuery image cropping plugin CSS -->
    <link href="../../css/cropbox.css" th:href="@{/css/cropbox.css}"
rel="stylesheet">

    <!-- Custom styles -->
    <link href="../../css/style.css" th:href="@{/css/style.css}" rel="
stylesheet">
    <link href="../../css/thymeleaf-bootstrap-paginator.css" th:href="
@{/css/thymeleaf-bootstrap-paginator.css}"
          rel="stylesheet">
    <link href="../../css/blog.css" th:href="@{/css/blog.css}" rel="
stylesheet">
</head>
```

同时，th:fragment= “header” 属性移动至 <html> 作为其属性，这样，整个 html 页面都能作为页面片段被引用了。

```
<html xmlns:th="http://www.thymeleaf.org"
      xmlns:layout="http://www.ultraq.net.nz/thymeleaf/layout"
      th:fragment="header">
```

修改 footer.html，相关的 JS 脚本都放置在 <footer> 中。

```
<footer class="blog-footer bg-inverse" th:fragment="footer">

    <a href="https://waylau.com">Welcome to waylau.com</a>

    <!-- JavaScript -->
    <script src="../../js/jquery-3.1.1.min.js" th:src="@{/js/jquery-
3.1.1.min.js}"></script>
    <script src="../../js/jquery.form.min.js" th:src="@{/js/jquery.form.
min.js}"></script>
    <script src="../../js/tether.min.js" th:src="@{/js/tether.min.
js}"></script>
    <script src="../../js/bootstrap.min.js" th:src="@{/js/bootstrap.
min.js}"></script>
    <script src="../../js/nprogress.js" th:src="@{/js/nprogress.js}"></
script>
    <script src="../../js/thinker-md.vendor.min.js" th:src="@{/js/think-
er-md.vendor.min.js}"></script>
    <script src="../../js/bootstrap-tagsinput.js" th:src="@{/js/jquery.
tagsinput.min.js}"></script>
    <script src="../../js/chosen.jquery.js" th:src="@{/js/chosen.jquery.
```

```
js}"></script>
    <script src="../../js/toastr.min.js" th:src="@{/js/toastr.min.
js}"></script>
    <script src="../../js/cropbox.js" th:src="@{/js/cropbox.js}"></
script>
    <script src="../../js/thymeleaf-bootstrap-paginator.js" th:src="@{/
js/thymeleaf-bootstrap-paginator.js}"></script>
    <script src="../../js/catalog-generator.js" th:src="@{/js/catalog-
generator.js}"></script>

</footer>
```

其中，之前的 <div> 已经被更换成 <footer> 标签了。

<title>Thymeleaf in action</title> 可以删除了，因为不会再被引用到。

8.4 Bootstrap 实战

在将 Bootstrap 及常用前端框架集成进 Spring Boot 项目之后，就可以对 bootstrap-in-action 进行改造了。

8.4.1 修改页眉 header.html

下面修改页眉 header.html，主要是使用了 Bootstrap 中的 navbar 和 collapse 组件。

```
<nav class="navbar navbar-inverse bg-inverse navbar-toggleable-md">
    <div class="container">
        <button class="navbar-toggler navbar-toggler-right" type="but-
ton" data-toggle="collapse"
                data-target="#navbarsContainer">
            <span class="navbar-toggler-icon"></span>
        </button>
        <a class="navbar-brand" href="/" th:href="@{~/}">NewStarBlog</
a>

        <div class="collapse navbar-collapse" id="navbarsContainer">

            <ul class="navbar-nav mr-auto">
                <li class="nav-item">
                    <a class="nav-link" th:href="@{~/users}"  >首页
<span class="sr-only">(current)</span></a>
                </li>
            </ul>
```

```
        </div>

    </div>
</nav>
```

8.4.2 修改页脚 footer.html

页脚信息用 .container 样式包裹，修改如下。

```
<div class="container">
    <p class="m-0 text-center text-white">© 2017 <a href="https://
waylau.com">waylau.com</a></p>
</div>
```

8.4.3 修改 list.html

在 <head> 中引用公用的 header 片段。

```
<head th:replace="~{fragments/header :: header}"></head>
```

并将页面的所有内容用 .container 样式包裹。

```
<!-- Page Content -->
<div class="container blog-content-container">
...
</div>
<!-- /.container -->
```

“创建用户”按钮增加了样式 class="btn btn-default"。

<table> 增加了样式 class="table table-striped"。

8.4.4 修改 form.html

在 <head> 中引用公用的 header 片段。

```
<head th:replace="~{fragments/header :: header}"></head>
```

并将页面的所有内容用 .container 样式包裹。

```
<!-- Page Content -->
<div class="container blog-content-container">
...
</div>
<!-- /.container -->
```

form 表单中的 <input> 增加了样式 class="form-control"。

“提交”按钮增加了样式 class="btn btn-default"。

8.4.5 修改 view.html

在 <head> 中引用公用的 header 片段。

```
<head th:replace="~{fragments/header :: header}"></head>
```

并将页面的所有内容用 .container 样式包裹。

```
<!-- Page Content -->
<div class="container blog-content-container">
...
</div>
<!-- /.container -->
```

8.4.6 运行

在完成页面的修改之后，启动 bootstrap-in-action 项目，在浏览器中访问 http://localhost:8080/users 可以看到项目改造后的运行效果。

图 8-3 所示为查看用户列表的界面效果。

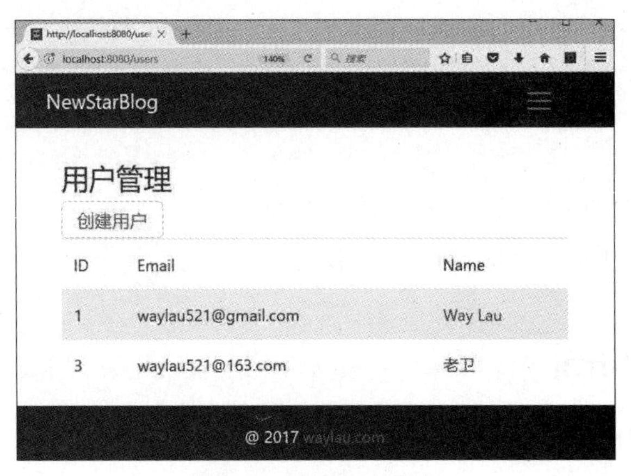

图8-3　查看用户列表

图 8-4 所示为编辑的界面效果。

图8-4　编辑用户

图 8-5 所示为查看用户的界面效果。

图8-5　查看用户

8.4.7 示例源码

本节示例源码在 bootstrap-in-action 目录下。

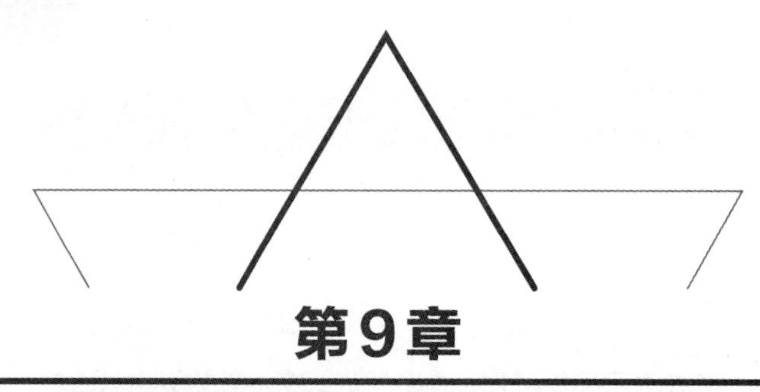

第9章

博客系统的需求分析与设计

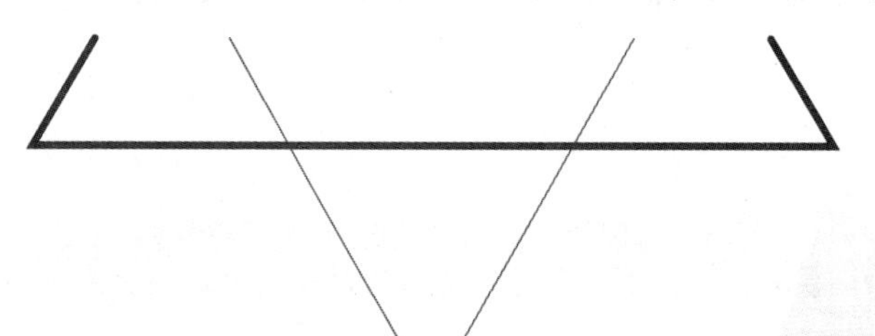

9.1 博客系统的需求分析

在企业级应用中，需求分析是非常重要的一个阶段。一切开发的出发点及前置条件都是要以需求为依据。有了需求，开发工作才能有的放矢。

在目前企业的组织架构中，需求分析的工作往往交给产品经理来处理。产品经理深入客户一线，与客户沟通需求，从而制作出需求文档或者需求说明书。这些需求文档最终交付给开发人员来进行下一步的详细设计及编码实现。

在现实中，用户的需求有时会经常变更。那么，这种变更出现得越早，越容易被实现。反之，如果到项目的后期，才提出变更需求，那么对于开发来说，这种变更所需要的成本将非常高。

9.1.1 如何获得系统的需求

一个项目的需求，往往需要经过调研才能整理出来。也就是说需求的获得并不是用大脑想出来的。以博客系统为例，市面上本身也有很多现成的博客系统，作为技术人员来说，很多人也喜欢逛论坛，去看博客，写博客，所以博客的需求可以通过现有的系统去参考，将现有的一些博客功能作为需求的参考。

好的需求又不要拘泥于现状。既然市面上有了成熟的、类似的功能的产品，那么就不能“重复造轮子”，需要有一些创新思维，才能赢得用户的认可。创新并不一定要大刀阔斧，可以是“微创新”，只要能突出自己的特色即可。

9.1.2 博客系统的整体功能需求

经过调研之后，博客系统的整体功能就整理出来了，如图 9-1 所示。

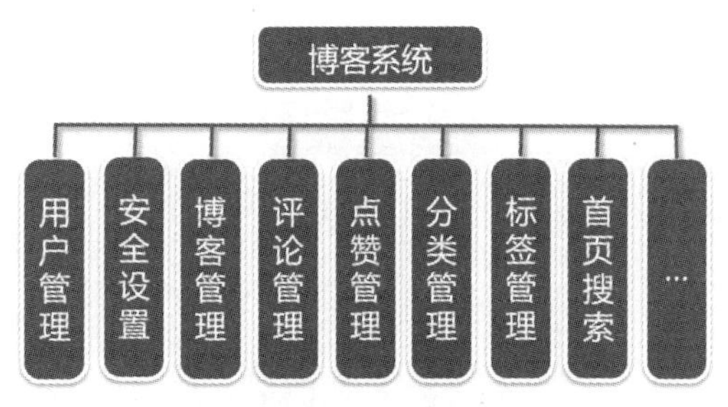

图9-1　博客系统整体功能

博客系统主要由用户管理、安全设置、博客管理、评论管理、点赞管理、分类管理、标签管理、首页搜索等 8 个核心功能模块组成。下面将一一介绍这些功能。

9.1.3 用户管理

任何一个信息化系统都少不了用户管理。一方面，用户要使用系统，必然要先注册成为系统的用户；另一方面，用户在使用系统的过程中，也经常会涉及对自己的信息进行修改，例如，用户可以更换头像、重置密码等。

对于博客系统的管理员来说，管理员可以通过用户管理的功能来实现对普通用户（博主）进行管理的目的。图 9-2 所示为用户管理的核心功能。对于普通用户来说，可以注册用户、登录系统；对于管理员来说，可以增加用户、修改用户、删除用户、搜索用户。

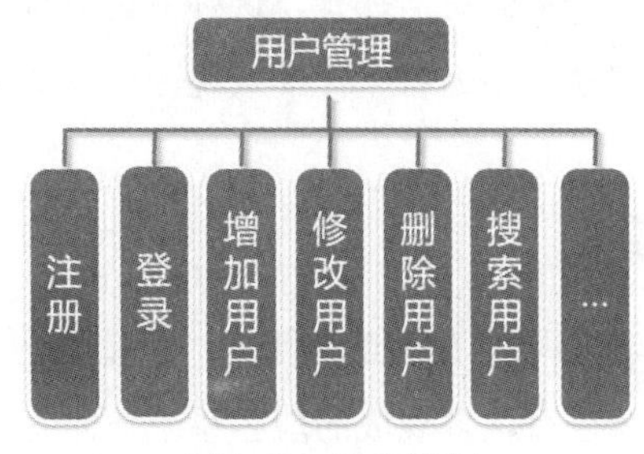

图9-2　用户管理

9.1.4 安全设置

图 9-3 所示为安全设置的核心功能。对于安全设置来说，一方面，需要对非法的用户请求进行拦截；另一方面，对于合法的用户来说，也要设置角色权限。也就是说，某些角色允许可以做某些事情。以用户管理为例，对于用户管理这个模块来说，只有管理员这个角色才能访问，其他的一般用户是没有权限访问的。再举一个例子，用户要修改自己的个人信息，那只能是用户自己来修改，其他人是没有修改他人的用户信息的权限的。

系统除了对非法的请求进行拦截外，还要做更多的安全防御工作，如 CSRF 攻击、Session Fixation 攻击等。

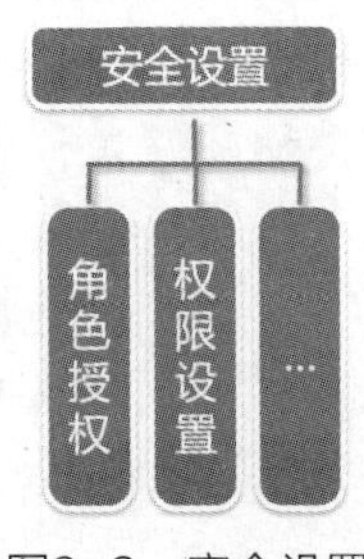

图9-3　安全设置

9.1.5 博客管理

既然是博客系统，博客管理自然是整个系统的核心功能，如图 9-4 所示。用户能够通过博客管理来发表博客、编辑博客、删除博客，还能对博客进行分类管理。

在设计的博客系统中，是采用 Markdown 编辑器来撰写博客的。同时，在博客正文中，也支持插入图片。

博客可以通过关键字被模糊搜索到。同时博客也能按照最新、最热等方式进行排序。其中，阅读量是衡量一篇文章热度的一个重要指标。

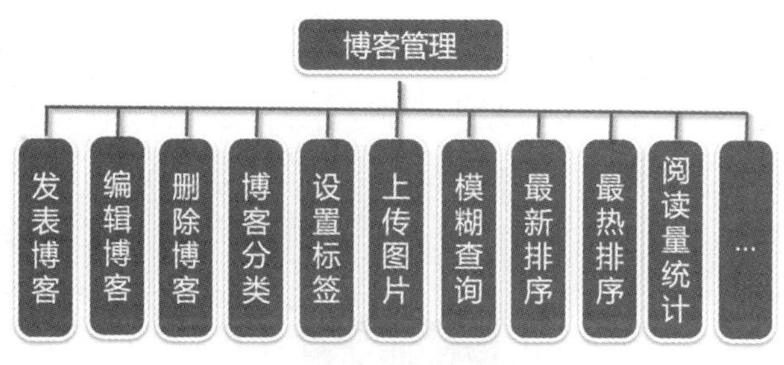

图9-4　博客管理

9.1.6 评论管理

评论管理是社交化软件的重要组成部分，如图 9-5 所示。评论可以让网友之间建立起交互和管理。

评论的内容可以被删除。其中，评论量是衡量一篇文章热度的一个重要指标。

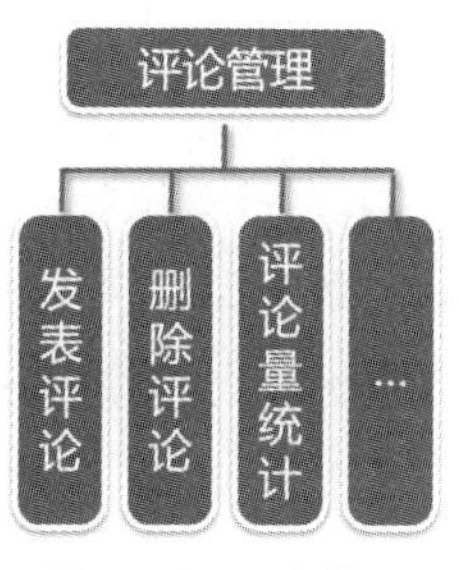

图9-5　评论管理

9.1.7 点赞管理

点赞管理是另外一种社交化的方式，如图 9-6 所示。有时候，如果对某篇博客很赞赏，但又没有那么多时间来撰写评论的内容，这时采用点赞的方式就非常方便。

点赞同样也可以被取消。其中，点赞量是衡量一篇文章热度的一个重要指标。

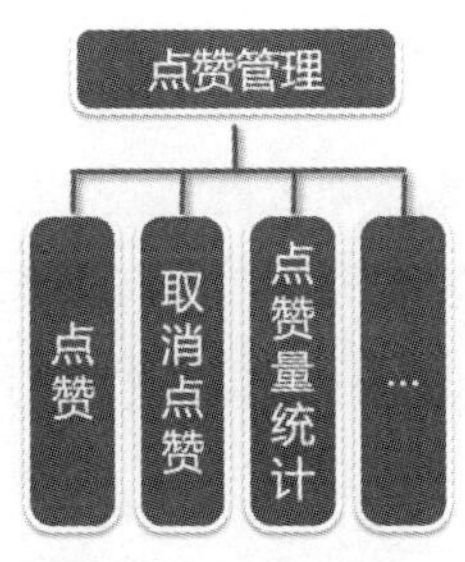

图9-6　点赞管理

9.1.8 分类管理

分类管理可以方便地将自己的个人博客进行分门别类。分类后的博客更加容易管理和查找。

对于分类管理来说，支持创建分类、修改分类、删除分类的功能，同时，也支持按照分类查询的功能，如图 9-7 所示。

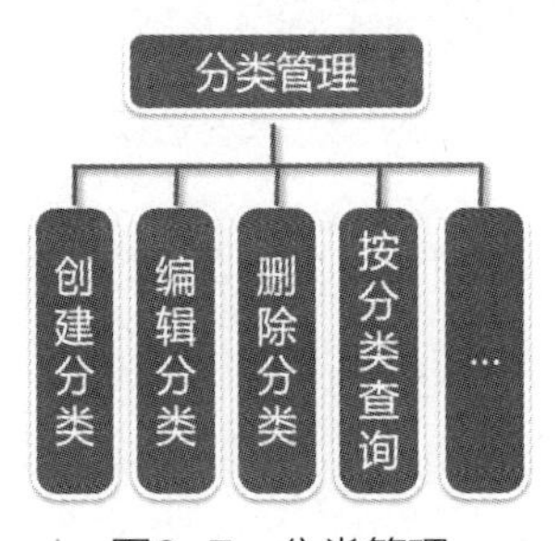

图9-7　分类管理

9.1.9 标签管理

在社交化流行的今天，如何能让自己的博客跟其他人的博客关联起来呢？答案就是标签。标签可以让具有相同的标签的博客建立起关联关系。标签同时也是一个可以被快速搜索的关键字。

对于标签管理这个功能来说，支持创建标签、删除标签的功能，同时，也支持按照标签查询的功能，如图 9-8 所示。

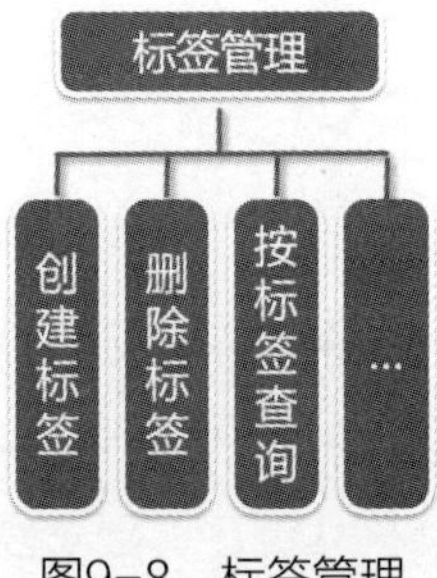

图9-8　标签管理

9.1.10 首页搜索

对于博客系统来说，首页是整个博客系统的门面。首页所展示的博客内容都与搜索分不开。由于博客数量众多，搜索往往会涉及全文搜索的需求，所以需要采用大数据搜索的方式来实现。在本书中，采用 Elasticsearch 来实现首页搜索的功能。图 9-9 所示为首页搜索的核心功能。

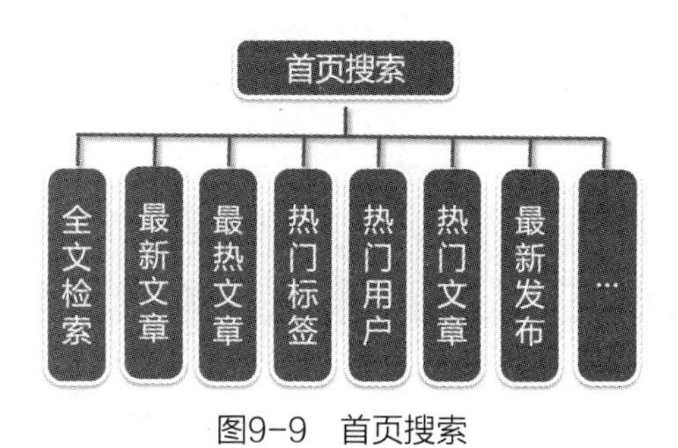

图9-9 首页搜索

9.2 博客系统的原型设计

在 9.1 节讨论了博客系统的需求分析。既然有了需求文档，那么为什么还需要原型设计呢？

9.2.1 原型设计的重要性

原型设计在敏捷开发中往往占有非常重要的地位。

1. 原型设计可以将最终的界面效果展示出来

产品经理可以在项目早期通过原型设计，来向客户解释系统的界面，表达需求。原型往往可以通过工具来快速实现，而不必等到最终功能开发完成才能看到最终的效果。这样，通过原型设计，产品经理可以更早地向客户演示系统的原型效果，也就能更早地得到用户的反馈，及时纠正问题。

2. 某些需求无法通过文字来表达

很多需求都是无法用文字很好地表达出来的，特别是涉及界面、用户体验等方面的内容，文字的表达能力就远没有原型设计强。

3. 原型可以指导开发工作

有了原型作为参考，开发工作更加可以有的放矢。参考原型来做实现，开发人员能更好地把握住产品经理的需求，也就能更好地完成需求。

4. 原型可以提升开发效率

特别是采用 Thymeleaf 模板的系统，通过“原型即界面”的理念，原型通过极少量的代码工作，就能直接转化为最终的界面，从而提高了开发效率。

另外，由于原型设计是经过用户确认的，这样实现的界面，也就能更好地被用户所接受。

9.2.2 博客系统的原型设计

1. 首页

图 9-10 和图 9-11 展示了博客系统首页的原型设计效果。首页的上方是头部（页首），其中“NewStarBlog”是博客系统的名称。博客可以按照最新、最热来排序。博客可以通过关键字来搜索。在头部的右侧是登录、注册的通道。

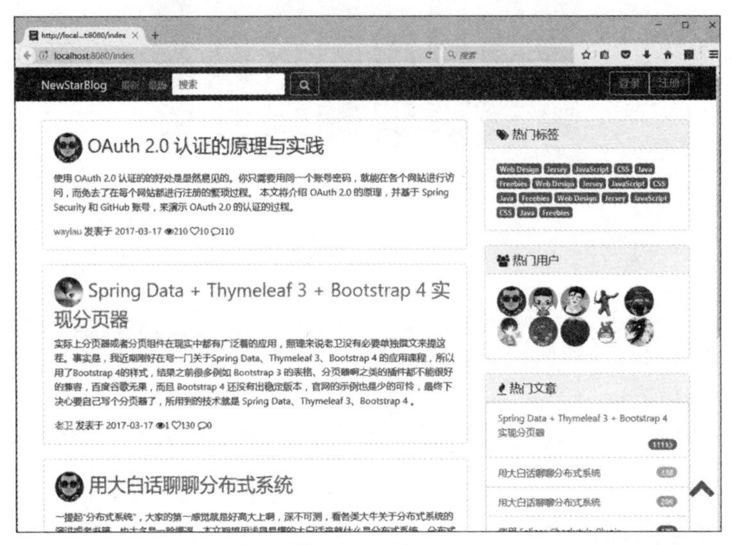

图9-10　首页1

图9-11　首页2

系统中的博客显示了博客的标题、作者、发表时间、阅读量、评论量、点赞量、摘要等信息。博客采用分页的形式来展示。

博客还能按照热门标签、热门用户、热门文章、最新发布等来排序。

博客系统的底部是系统的版权信息。

2. 博主空间

博主空间是用户的个人空间，如图 9-12 所示。博主空间展示了博主的头像、博主名称、联系方式等。同时，也展示了博主发表的博客和博客的分类，这些博客同样也可以按照最新、最热来排序。

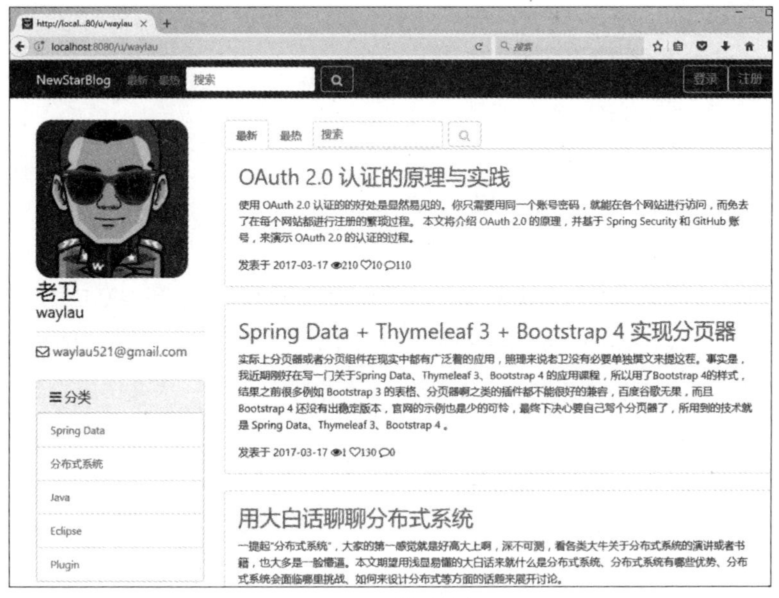

图9-12　博主空间

3. 博客展示

图 9-13 和图 9-14 是博客正文的展示界面。除了显示博客的标题、作者、发表时间、阅读量、评论量、点赞量、摘要、正文等信息外，在博客展示的右侧还包含了文章目录的功能，这样用户可以点击文章的目录，来快速定位到文章的位置。

图9-13　博客展示1

在博客的下方，显示了博客的分类和标签，并包含了博客的评论和点赞功能。其中，评论的展示方式是按照“楼层”来排序的。第一个评论称为“1 楼”，第二个评论称为“2 楼”，其他的以此类推。当然，某些评论系统还采用了“沙发”“板凳”等术语，读者可以自行了解，这里不再赘述。

图9-14　博客展示2

4. 博客编辑

图 9-15 所示的是博客的编辑界面，在博客编辑界面中采用了流行的 Markdown 编辑器来撰写博客。Markdown 是一种可以使用普通文本编辑器编写的标记语言，通过简单的标记语法，可以使普通文本内容具有一定的格式。

同时，在博客正文中也支持插入图片。正文中的插图是由专门的文件管理器来实现管理的。在发表博客之前，可以对博客进行分类，并打上标签。

图9-15 博客编辑

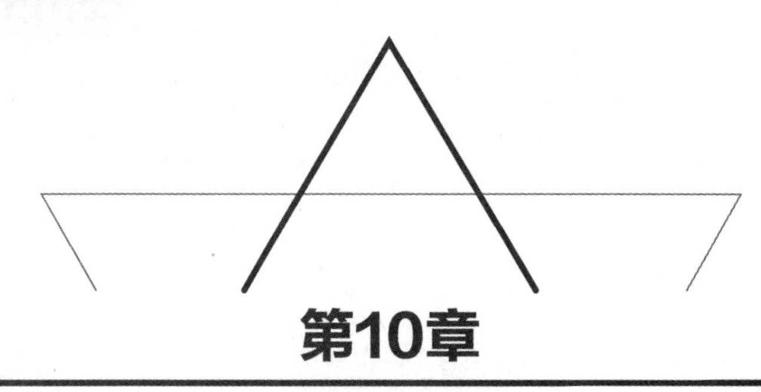

第10章

集成 Spring Security

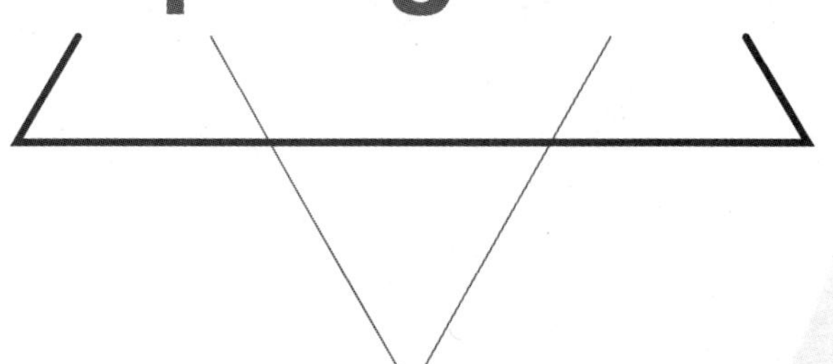

10.1 基于角色的权限管理

本节将讨论在权限管理中角色的概念，以及基于角色的机制来进行权限管理。

10.1.1 角色的概念

当说到程序的权限管理时，人们往往想到“角色”这一概念。角色是代表一系列行为或责任的实体，用于限定在软件系统中能做什么、不能做什么。用户账号往往与角色相关联，因此，一个用户在软件系统中能“做”什么取决于与之关联的具有什么样的角色。

例如，一个用户以关联了“项目管理员”角色的账号登录系统，那这个用户就可以做项目管理员能做的所有事情，如列出项目中的应用、管理项目组成员、产生项目报表等。

从这个意义上来说，角色更多的是一种行为的概念，表示用户能在系统中进行的操作。

10.1.2 基于角色的访问控制

既然角色代表了可执行的操作这一概念，一个合乎逻辑的做法是在软件开发中使用角色来控制对软件功能和数据的访问。这种权限控制方法就称为基于角色的访问控制（Role Based Access Control），或简称为 RBAC。

有两种正在实践中使用的 RBAC 访问控制方式：隐式的方式和显式的方式。

1. 隐式访问控制（ImplicitAccessControl）

前面提到，角色代表一系列的可执行的操作。但如何知道一个角色到底关联了哪些可执行的操作呢？

答案是，目前大多数的应用，用户并不能明确地知道一个角色到底关联了哪些可执行操作。可能用户心里是清楚的（你知道一个有“管理员”角色的用户可以锁定用户账号、进行系统配置；一个关联了“消费者”这一角色的用户可在网站上进行商品选购），但这些系统并没有明确定义一个角色到底包含了哪些可执行的行为。

以“项目管理员”来说，系统中并没有对“项目管理员”能进行什么样的操作进行明确定义，它仅是一个字符串名词。开发人员通常将这个名词写在程序里以进行访问控制。例如，判断一个用户是否能查看项目报表，程序员可能会进行如下编码。

```
if (user.hasRole("Project Manager") ) {

    // 显示报表按钮
} else {

    // 不显示按钮
}
```

在上面的示例代码中，开发人员判断用户是否有“项目管理员”角色来决定是否显示查看项目报表按钮。请注意上面的代码，它并没有明确语句来定义“项目管理员”这一角色到底包含哪些可执行的行为，它只是假设一个关联了项目管理员角色的用户可查看项目报表，而开发人员也是基于这一假设来写 if/else 语句。这种方式就是基于角色的隐式访问控制。

2. 脆弱的权限策略

像上面的权限访问控制是非常脆弱的。一个极小的权限方面的需求的变动都可能导致上面的代码需要重新修改。

举例来说，假如某一天这个开发团队被告知，需要一个“部门管理员”角色，他们也可以查看项目报表。那么，之前的隐式访问控制的代码被修改成了如下的样子。

```
if (user.hasRole("Project Manager") || user.hasRole("Department Manag-
er") ) {

    // 显示报表按钮
} else {

    // 不显示按钮
}
```

随后，开发人员需要更新测试用例、重新编译系统，还可能需要重走软件质量控制（QA）流程，然后再重新部署上线。这一切仅仅是因为一个微小的权限方面的需求变动。

后面如果需求方说又有另一个角色可查看报表，或是前面关于“部门管理员可查看报表”的需求不再需要了，怎么办?

如果需求方要求动态地创建、删除角色，以便他们自己配置角色，又该如何应对呢?

像上面的情况，这种隐式的（静态字符串）形式的基于角色的访问控制方式难以满足需求。理想的情况是如果权限需求变动，不需要修改任何代码。怎样才能做到这一点呢?

3. 显式访问控制（ExplicitAccessControl）

从上面的例子可以看到，当权限需求发生变动时，隐式的权限访问控制方式会给程序开发带来沉重的负担。如果能有一种方式在权限需求发生变化时不需要去修改代码，就能满足需求，那就好了。理想的情况是，即使是正在运行的系统，也可以修改权限策略却又不影响最终用户的使用。当你发现某些错误的或危险的安全策略时，可以迅速地修改策略配置，同时系统还能正常使用，而不需要重构代码重新部署系统。

怎样才能达到上面的理想效果呢? 实际上，可以通过显式的（明确的）界定在应用中能做的操作来进行。

回顾上面隐式的权限控制的例子，思考一下这些代码最终的目的及最终是要做什么样的控制。从根本上说，这些代码最终是在保护资源（项目报表），是要界定一个用户能对这些资源进行什么样的操作（查看 / 修改）。当将权限访问控制分解到这种最原始的层次，就可以用一种更细粒度、更富有弹性的方式来表达权限控制策略。

可以修改上面的代码块，以基于资源的语义来更有效地进行权限访问控制。

```
if (user.isPermitted("projectReport:view:12345")) {

    // 显示报表按钮
} else {

    // 不显示按钮
}
```

上面的例子中，可明确地看到是在控制什么。不要太在意冒号分隔的语法，这仅是一个例子，重点是上面的语句明确地表示了“如果当前用户允许查看编号为 12345 的项目报表，则显示项目报表按钮”。也就是说，明确地说明了一个用户账号可对一个资源实例进行的具体的操作。

10.1.3 哪种方式更好

上面最后的示例代码块与前面的代码的主要区别在于，最后的代码块是基于什么是受保护的，而不是谁可能有能力做什么。看似简单的区别，但后者对系统开发及部署有着深刻的影响。显式访问控制方式与隐式访问控制方式相比，具有以下优势。

（1）更少的代码重构：我们是基于系统的功能（系统的资源及对资源的操作）来进行权限控制，而相应来说，系统的功能需求一旦确定下来后，一段时间内对它的改动相应还是比较少的。只是当系统的功能需求改变时，才会涉及权限代码的改变。例如，上面提到的查看项目报表的功能，显式的权限控制方式不会像传统隐式的 RBAC 权限控制那样，因不同的用户 / 角色要进行这个操作就需要重构代码；只要这个功能存在，显式方式的权限控制代码是不需要改变的。

（2）资源和操作更直观：保护资源对象、控制对资源对象的操作，这样的权限控制方式更符合人们的思想习惯。正因为符合这种直观的思维方式，面向对象的编辑思想及 REST 通信模型变得非常成功。

（3）安全模型更有弹性：上面的示例代码中没有确定哪些用户、组或角色可对资源进行什么操作。这意味着它可支持任何安全模型的设计。例如，可以将操作（权限）直接分配给用户 ，或者他们可以被分配到一个角色，然后再将角色与用户关联，或者将多个角色关联到组（Group）上，等等。完全可以根据应用的特点定制权限模型。

（4）外部安全策略管理：由于源代码只反映资源和行为，而不是用户、组和角色，这样资源 / 行为与用户、组、角色的关联可以通过外部的模块或专用工具或管理控制台来完成。这意味着在权限需求变化时，开发人员并不需要花费时间来修改代码，业务分析师甚至最终用户就可以通过相应的管理工具修改权限策略配置。

（5）运行时做修改：因为基于资源的权限控制代码并不依赖于行为的主体（如组、角色、用户），开发者并没有将行为的主体的字符名词写在代码中，所以开发者甚至可以在程序运行的时候通过修改主体能对资源进行的操作这样一些方式，通过配置的方式就可应对权限方面需求的变动，再也不需要像隐式的 RBAC 方式那样需要重构代码。

显式访问控制方式更适合于当前的软件应用。

10.1.4 真实的案例

不管是隐式访问控制，还是显式访问控制，它们都有其合适的场景。庆幸的是，在 Java 平台，都有很多现成的现代权限管理框架可供选择，有 Apache Shiro（http://shiro.apache.org/）和 Spring Security（http://projects.spring.io/spring-security/）。一个是以简洁好用而被业界广泛应用，而另一个则以功能强大而著称。

关于这两个框架的用法，读者也可以参考笔者的另外两本开源书《Apache Shiro 1.2.x 参考手册》（https://github.com/waylau/apache-shiro-1.2.x-reference）及《Spring Security 教程》（https://github.com/waylau/spring-security-tutorial）。以下主要介绍 Spring Security 框架的应用。

10.2 Spring Security 概述

Spring Security 为基于 Java EE 的企业软件应用程序提供全面的安全服务。特别是使用 Spring 框架构建的项目，可以更好地使用 Spring Security 来加快构建的速度。

Spring Security 的出现有很多原因，但主要是基于 Java EE 的 Servlet 规范或 EJB 规范的缺乏对企业应用的安全性方面的支持。而使用 Spring Security 就能克服这些问题，并带来了数十个其他有用的可自定义的安全功能。

在 Java 领域，另一个值得关注的安全框架是 Apache Shiro。但与 Apache Shiro 相比，Spring Security 的功能更加强大，与 Spring 的兼容性也更加好。

10.2.1 Spring Security 的认证模型

在应用程序安全性的两个主要领域是认证（Authentication）与授权（Authorization）。

（1）认证：认证是建立主体（Principal）的过程。主体通常是指可以在应用程序中执行操作的用户、设备或其他系统。

（2）授权：或称为访问控制（Access-Control），授权是指决定是否允许主体在应用程序中执行操作。为了到达需要授权决定的点，认证过程已经建立了主体的身份。这些概念是常见的，并不是特定于 Spring Security。

在认证级别，Spring Security 支持各种各样的认证模型。这些认证模型中的大多数由第三方提供，或者由诸如互联网工程任务组的相关标准机构开发。此外，Spring Security 提供了自己的一组认证功能。具体来说，Spring Security 目前支持以下所有这些技术的身份验证集成。

- HTTP BASIC 认证头（基于 IETF RFC 的标准）。
- HTTP Digest 认证头（基于 IETF RFC 的标准）。
- HTTP X.509 客户端证书交换（基于 IETF RFC 的标准）。
- LDAP（一种常见的跨平台身份验证需求，特别是在大型环境中）。
- 基于表单的身份验证（用于简单的用户界面需求）。
- OpenID 身份验证。
- 基于预先建立的请求头的验证（如 Computer Associates Siteminder）。
- Jasig Central Authentication Service，也称为 CAS，这是一个流行的开源单点登录系统。
- 远程方法调用（RMI）和 HttpInvoker（Spring 远程协议）的透明认证上下文传播。
- 自动“remember-me”身份验证（所以可以选中一个框，以避免在预定时间段内重新验证）。
- 匿名身份验证（允许每个未经身份验证的调用来自动承担特定的安全身份）。
- Run-as 身份验证（一个调用应使用不同的安全身份继续运行，这是有用的）。
- Java 认证和授权服务（Java Authentication and Authorization Service，JAAS）。
- Java EE 容器认证（因此，如果需要，仍然可以使用容器管理身份验证）。
- Kerberos。
- Java Open Source Single Sign-On（JOSSO）*。
- OpenNMS Network Management Platform *。
- AppFuse *。
- AndroMDA *。
- Mule ESB *。
- Direct Web Request （DWR）*。
- Grails *。
- Tapestry *。
- JTrac *。
- Jasypt *。
- Roller *。
- Elastic Path *。
- Atlassian Crowd *。
- 自己创建的认证系统。

其中加“*”是指由第三方提供，由 Spring Security 来集成。

许多独立软件供应商（ISV）选择采用 Spring Security，都是出于这种灵活的认证模型。这样，可以快速地将解决方案与最终客户需求进行组合，从而避免了进行大量的工作或者要求变更。如果上述认证机制都不符合需求，Spring Security 作为一个开发平台，也可以基于它很容易就实现自己的认证机制。

如果不考虑上述认证机制，Spring Security 还提供了一组深层次的授权功能。有三个主要领域。

- 对 Web 请求进行授权。
- 授权某个方法是否可以被调用。
- 授权访问单个领域对象实例。

10.2.2 Spring Security 的安装

Spring Security 的安装非常简单，以下展示了采用 Maven 和 Gradle 两种方式来安装。

1. 使用Maven

使用 Maven 的最少依赖如下所示。

```
<dependencies>
    <!-- ... -->
    <dependency>
        <groupId>org.springframework.security</groupId>
        <artifactId>spring-security-web</artifactId>
        <version>5.0.0.M2</version>
    </dependency>
    <dependency>
        <groupId>org.springframework.security</groupId>
        <artifactId>spring-security-config</artifactId>
        <version>5.0.0.M2</version>
    </dependency>
    <!-- ... -->
</dependencies>
```

2. 使用Gradle

使用 Gradle 的最少依赖如下所示。

```
dependencies {
    //...
    compile 'org.springframework.security:spring-security-web:5.0.0.M2'
    compile 'org.springframework.security:spring-security-config:
5.0.0.M2'
    //...
}
```

其中 Gradle 是本书所推荐的方式。使用 Gralde 可以获得更简洁的配置方式，以及更快的构建速度。

10.2.3 模块

自 Spring 3 开始，Spring Security 将代码划分到不同的 jar 中，这使不同的功能模块和第三方依赖显得更加清晰。Spring Security 主要包括以下几个核心模块。

1. Core-spring-security-core.jar

包含核心的 authentication 和 authorization 的类和接口、远程支持和基础配置 API。任何使用 Spring Security 的应用都需要引入这个 jar。支持本地应用、远程客户端、方法级别的安全和 JDBC 用户配置。主要包含的顶级包如下。

（1）org.springframework.security.core：核心。

（2）org.springframework.security.access：访问，即 authorization 的作用。

（3）org.springframework.security.authentication：认证。

（4）org.springframework.security.provisioning：配置。

2. Remoting-spring-security-remoting.jar

提供与 Spring Remoting 整合的支持，开发者并不需要这个，除非开发者需要使用 Spring Remoting 写一个远程客户端。主包为 org.springframework.security.remoting。

3. Web-spring-security-web.jar

包含 filter 和相关 Web 安全的基础代码。如果需要使用 Spring Security 进行 Web 安全认证和基于 URL 的访问控制。主包为 org.springframework.security.web。

4. Config-spring-security-config.jar

包含安全命名空间解析代码和 Java 配置代码。 如果使用 Spring Security XML 命名空间进行配置或 Spring Security 的 Java 配置支持，则需要它。主包为 org.springframework.security.config。不应该在代码中直接使用这个 jar 中的类。

5. LDAP-spring-security-ldap.jar

LDAP 认证和配置代码。如果需要进行 LDAP 认证或管理 LDAP 用户实体。顶级包为 org.springframework.security.ldap。

6. ACL-spring-security-acl.jar

特定领域对象的 ACL（访问控制列表）实现。使用它可以对特定对象的实例进行一些安全配置。顶级包为 org.springframework.security.acls。

7. CAS-spring-security-cas.jar

Spring Security CAS 客户端集成。如果需要使用一个单点登录服务器进行 Spring Security Web 安全认证，需要引入。顶级包为 org.springframework.security.cas。

8. OpenID-spring-security-openid.jar

OpenID Web 认证支持。基于一个外部 OpenID 服务器对用户进行验证。顶级包为 org.spring

framework.security.openid，需要使用 OpenID4Java。

一般情况下，spring-security-core 和 spring-security-config 都会引入，在 Web 开发中，通常还会引入 spring-security-web。

9. Test-spring-security-test.jar

用于测试 Spring Security。在开发环境中通常需要添加该包。

10.2.4 Spring Security 5 的新特性及高级功能

本书案例采用 Spring Security 5 来进行编写。对于 Spring Security 5 版本来说，相对于之前的版本，主要提供了如下特性。

（1）为 OAuth 2.0 登录添加的支持：详见 https://github.com/spring-projects/spring-security/issues/3907。

（2）支持初始响应式编程：详见 https://github.com/spring-projects/spring-security/issues/4128。

其中，基于 OAuth 2.0 登录的编程案例可以参阅笔者所著的《Spring Security 教程》[①]中的“OAuth 2.0 认证的原理与实践”相关内容。

针对 Web 方面的开发，Spring Security 提供了如下高级功能。

1. Remember-Me认证

Remember-Me 身份验证是指网站能够记住身份之间的会话，这通常是通过发送 cookie 到浏览器，cookie 在未来会话中被检测到，并导致自动登录发生。Spring Security 为这些操作提供了必要的钩子，并且有两个具体的实现。

（1）使用散列来保存基于 cookie 的令牌的安全性。

（2）使用数据库或其他持久存储机制来存储生成的令牌。

在本书的案例中，将会通过散列的方式来实现 Remember-Me 认证。

2. 使用HTTPS

可以使用 <intercept-url> 上的 requires-channel 属性直接支持这些 URL 采用 HTTPS 协议。

```
<http>
<intercept-url pattern="/secure/**" access="ROLE_USER" requires-channel=
"https"/>
<intercept-url pattern="/**" access="ROLE_USER" requires-channel="any"/>

...
</http>
```

注意：如果用户尝试使用 HTTP 访问与“/secure/**”模式匹配的任何内容时，都会首先将其重定向到 HTTPS 的 URL 上。

---

① 教程地址参见 https://github.com/waylau/spring-security-tutorial。

如果开发者的应用程序想使用 HTTP/HTTPS 的非标准端口，则可以指定端口映射列表，如下所示。

```
<http>
...
<port-mappings>
    <port-mapping http="9080" https="9443"/>
</port-mappings>
</http>
```

注意：为了安全，应用程序应该始终采用 HTTPS 在整个过程中使用安全连接，以避免中间人发生攻击的可能性。

3. 会话管理

在会话管理方面，Spring Security 提供了诸如检测超时、控制并发会话、防御会话固定攻击等方面的内容。

（1）检测超时。开发者可以配置 Spring Security，以检测提交的无效会话 ID，并将用户重定向到适当的 URL。这是通过 session-management 元素实现的。

```
<http>
...
<session-management invalid-session-url="/invalidSession.htm" />
</http>
```

注意：使用此机制来检测会话超时，如果用户注销，然后在不关闭浏览器的情况下重新登录，则可能会报告错误。这是因为当会话 cookie 无效时，会话 cookie 不会被清除，即使用户已经注销也将被重新提交。开发者可能需要在注销时显式删除 JSESSIONID cookie，例如，在注销处理程序中使用以下语法。

```
<http>
<logout delete-cookies="JSESSIONID" />
</http>
```

但这种用法并不是每个 servlet 容器都支持，所以开发者需要在自己的环境中测试它。

（2）并发会话控制。如果开发者希望对单个用户登录应用程序的能力施加限制，Spring Security 将支持以下简单添加功能。首先，需要将以下监听器添加到 web.xml 文件中，以使 Spring Security 更新有关会话生命周期事件。

```
<listener>
<listener-class>
    org.springframework.security.web.session.HttpSessionEventPublisher
</listener-class>
</listener>
```

然后将以下内容添加到应用程序上下文中。

```
<http>
...
<session-management>
    <concurrency-control max-sessions="1" />
</session-management>
</http>
```

这将阻止用户多次登录（第二次登录将导致第一个无效）。通常，开发者更希望防止第二次登录，在这种情况下可以使用：

```
<http>
...
<session-management>
    <concurrency-control max-sessions="1" error-if-maximum-exceeded=
"true" />
</session-management>
</http>
```

第二次登录将被拒绝。通过“被拒绝”，我们的意思是，如果正在使用基于表单的登录，用户将被发送到 authentication-failure-url。如果第二次认证是通过另一个非交互机制进行的，如“remember-me”，则会向客户端发送“unauthorized”（401）错误。如果想要使用错误页面，则可以将 session-authentication-error-url 属性添加到 session-management 元素中。

（3）会话固定攻击防护。会话固定攻击（Session Fixation Attacks）是潜在的风险，恶意攻击者可能通过访问站点来创建会话而后通过这个会话进行攻击（拥有了会话，一定程度上表明了攻击者通过了认证）。而 Spring Security 通过创建新会话或在用户登录时以其他方式更改会话 ID 来自动防范此问题。如果不需要此保护或与其他要求冲突，则可以在 <session-management> 元素中使用 session-fixation-protection 属性来进行设置。它有 4 个选项。

① none：不要做任何事情，原始会话将被保留。

② newSession：创建一个新的“干净”会话，而不复制现有的会话数据（Spring Security 相关属性仍将被复制）。

③ migrateSession：创建新会话并将所有现有会话属性复制到新会话。这是 Servlet 3.0 或更旧容器中的默认值。

④ changeSessionId：不创建新会话，而是使用 Servlet 容器提供的会话固定保护（HttpServletRequest#changeSessionId()）。此选项仅适用于 Servlet 3.1（Java EE 7）和较新的容器。在较旧的容器中肯定会导致异常。这是 Servlet 3.1 和更新容器中的默认值。

当会话固定保护发生时，会导致 SessionFixationProtectionEvent 事件在应用程序上下文中发布。如果使用 changeSessionId，则此保护也将导致任何 javax.servlet.http.HttpSessionIdListener 被通知，因此，如果所设计的代码侦听这两个事件，请谨慎使用。

4. 支持OpenID

Spring Security 命名空间支持 OpenID 登录，例如，

```
<http>
<intercept-url pattern="/**" access="ROLE_USER" />
<openid-login />
</http>
```

通过向 OpenID 提供商进行注册，并将用户信息添加到内存中的 <user-service>。

```
<user name="http://jimi.hendrix.myopenid.com/" authorities="ROLE_USER"
/>
```

开发者应该可以使用 myopenid.com 网站登录进行身份验证，还可以通过在 openid-login 元素上设置 user-service-ref 属性来选择特定的 UserDetailsService bean 来使用 OpenID。

Spring Security 也支持 OpenID 属性交换。 例如，以下配置将尝试从 OpenID 提供程序中检索电子邮件和全名，供应用程序使用。

```
<openid-login>
<attribute-exchange>
    <openid-attribute name="email" type="http://axschema.org/contact/
email" required="true"/>
    <openid-attribute name="name" type="http://axschema.org/namePerson"
/>
</attribute-exchange>
</openid-login>
```

5. 自定义过滤器

如果开发者以前使用过 Spring Security，将知道该框架维护一连串的过滤器。开发者可能希望在特定位置将自己的过滤器添加到堆栈中，或使用当前没有命名空间配置选项（如 CAS）的 Spring Security 过滤器，或者开发者可能希望使用自定义版本的标准命名空间过滤器，例如，由 <form-login> 元素创建的 UsernamePasswordAuthenticationFilter，利用可以通过明确使用 bean 的一些额外配置选项。

使用命名空间时，始终严格执行过滤器的顺序。当创建应用程序上下文时，过滤器 bean 将通过命名空间处理代码进行排序，并且标准 Spring Security 过滤器在命名空间中具有别名，并且是众所周知的位置。表 10-1 展示了 Spring Security 标准过滤器的别名和排序情况。

表10-1 Spring Security 标准过滤器的别名和排序情况

别名	过滤器类	命名空间元素或属性
CHANNEL_FILTER	ChannelProcessingFilter	http/intercept-url@ requires-channel
SECURITY_CONTEXT_ FILTER	SecurityContextPersistenceFilter	http

续表

别名	过滤器类	命名空间元素或属性
CONCURRENT_SESSION_FILTER	ConcurrentSessionFilter	session-management/concurrency-control
HEADERS_FILTER	HeaderWriterFilter	http/headers
CSRF_FILTER	CsrfFilter	http/csrf
LOGOUT_FILTER	LogoutFilter	http/logout
X509_FILTER	X509AuthenticationFilter	http/x509
PRE_AUTH_FILTER	AbstractPreAuthenticatedProcessingFilter 子类	N/A
CAS_FILTER	CasAuthenticationFilter	N/A
FORM_LOGIN_FILTER	UsernamePasswordAuthenticationFilter	http/form-login
BASIC_AUTH_FILTER	BasicAuthenticationFilter	http/http-basic
SERVLET_API_SUPPORT_FILTER	SecurityContextHolderAw areRequestFilter	http/@servlet-api-provision
JAAS_API_SUPPORT_FILTER	JaasApiIntegrationFilter	http/@jaas-api-provision
REMEMBER_ME_FILTER	RememberMeAuthenticationFilter	http/remember-me
ANONYMOUS_FILTER	AnonymousAuthenticationFilter	http/anonymous
SESSION_MANAGEMENT_FILTER	SessionManagementFilter	session-management
EXCEPTION_TRANSLATION_FILTER	ExceptionTranslationFilter	http
FILTER_SECURITY_INTERCEPTOR	FilterSecurityInterceptor	http
SWITCH_USER_FILTER	SwitchUserFilter	N/A

10.3 Spring Security 与 Spring Boot 集成

本节将演示如何在 Spring Boot 中集成 Spring Security 框架的功能。在 bootstrap-in-action 项目

的基础上构建了一个新的项目 security-in-action。

1. 所需环境

本例采用的开发环境如下。

- Spring Security 5.0.0.M2。
- Thymeleaf Spring Security 3.0.2.RELEASE。

2. build.gradle

我们需要添加 Spring Security 的依赖。Spring Boot 已经提供了相关的 Starter 来实现 Spring Security 开箱即用的功能，所以只需要在 build.gradle 文件中添加 Spring Security 的 Starter 的库即可。

Thymeleaf 社区还提供了 Thymeleaf 与 Spring Security 集成模块 thymeleaf-extras-springsecurity 4，需要手动添加这个依赖。

```
// 依赖关系
dependencies {
    //...

    // 添加Spring Security依赖
    compile('org.springframework.boot:spring-boot-starter-security')

    // 添加Thymeleaf Spring Security依赖
    compile('org.thymeleaf.extras:thymeleaf-extras-springsecurity4:
3.0.2.RELEASE')

    //...
}
```

注意：截至目前，Thymeleaf 还没有推出关于 Spring Security 5 版本的集成包，但仍可以使用 Spring Security 4 版本的集成包来代替。

10.4 Spring Security 实战

本节将用 Spring Security 来实现对系统的安全管理。为了更好地演示，在 10.3 节创建的 security-in-action 项目的基础上进行修改。

10.4.1 后台代码

对于后台的代码来说，主要实现安全配置类和控制器即可。

1. 安全配置类

增加 com.waylau.spring.boot.blog.config 包，用于放置项目的配置类。在该包下创建 SecurityConfig.java。

```
import org.springframework.beans.factory.annotation.Autowired;
import org.springframework.security.config.annotation.authentication.
builders.AuthenticationManagerBuilder;
import org.springframework.security.config.annotation.web.builders.
HttpSecurity;
import org.springframework.security.config.annotation.web.configuration.
EnableWebSecurity;
import org.springframework.security.config.annotation.web.configuration.
WebSecurityConfigurerAdapter;

@EnableWebSecurity
public class SecurityConfig extends WebSecurityConfigurerAdapter {

    /**
     * 自定义配置
     */
    @Override
    protected void configure(HttpSecurity http) throws Exception {
        http.authorizeRequests()
            .antMatchers("/css/**", "/js/**", "/fonts/**", "/index").
permitAll() // 都可以访问
            .antMatchers("/users/**").hasRole("ADMIN")    // 需要相应的角色
才能访问
            .and()
            .formLogin()    //基于 Form 表单登录验证
            .loginPage("/login").failureUrl("/login-error"); // 自定义登
录界面
    }

    /**
     * 认证信息管理
     * @param auth
     * @throws Exception
     */
    @Autowired
    public void configureGlobal(AuthenticationManagerBuilder auth) throws
Exception {
        auth
            .inMemoryAuthentication()   // 认证信息存储于内存中
                .withUser("waylau").password("123456").roles("ADMIN");
    }
}
```

安全配置类 SecurityConfig 继承自 WebSecurityConfigurerAdapter。WebSecurityConfigurerAdapter 提供用于创建一个 Websecurityconfigurer 实例方便的基类，允许自定义重写其方法。这里重写了

configure 方法。

（1）permitAll() 是指允许任何人访问的方法，包括匹配 css、js、fonts 路径的 URL 及 index 页面。

（2）所有匹配 users 的 URL 请求需要用户进行身份验证。在本例中，用户必须并且具备 ADMIN 角色，才有权限访问 users 路径下的资源。

（3）formLogin() 表明这是个基于表单的身份验证，指明了登录的 URL 路径及登录失败的 URL。

configureGlobal 方法创建了基于内存的身份认证管理器。在本例，存储了用户名为 waylau、密码为 123456、角色为 ADMIN 的身份信息。configureGlobal 方法可以是任意名称，但在类中必须要有 @EnableWebSecurity、@EnableGlobalMethodSecurity 或 @EnableGlobalAuthentication 注解。

3. 控制器

在 com.waylau.spring.boot.blog.controller 包下创建了 MainController 控制器。

```
@Controller
public class MainController {

    @GetMapping("/")
    public String root() {
        return "redirect:/index";
    }

    @GetMapping("/index")
    public String index() {
        return "index";
    }

    @GetMapping("/login")
    public String login() {
        return "login";
    }

    @GetMapping("/login-error")
    public String loginError(Model model) {
        model.addAttribute("loginError", true);
        model.addAttribute("errorMsg", "登录失败，用户名或者密码错误！");
        return "login";
    }

}
```

该控制器说明如下。

（1）当访问根路径或 /index 路径时，将会跳转到 index.html 页面。

（2）访问 /login 路径时，将会跳转到 login.html 页面。

（3）登录失败，将会重定向到 /login-error 路径，最终会跳转到 login.html 页面。其中，在页面中绑定了错误提示信息。

10.4.2 前端代码

下面将演示前端功能的实现过程。

1. index.html

在前端创建了 index.html 作为主页（可以直接复制 list.html 页面来修改）。

这里需要在 <html> 中增加 Thymeleaf Spring Security 的命名空间。

```
<html xmlns:th="http://www.thymeleaf.org"
      xmlns:layout="http://www.ultraq.net.nz/thymeleaf/layout"
      xmlns:sec="http://www.thymeleaf.org/thymeleaf-extras-springsecu
rity4">
```

页面内容如下。

```
...
<!-- Page content -->
<div class="container blog-content-container">
    <div sec:authorize="isAuthenticated()">
        <p>已有用户登录</p>
        <p>登录的用户为: <span sec:authentication="name"></span></p>
        <p>用户角色为: <span sec:authentication="principal.authorities">
</span></p>
    </div>
    <div sec:authorize="isAnonymous()">
        <p>未有用户登录</p>
    </div>
</div><!-- /.container -->
...
```

其中，sec:authorize 和 sec:authentication 属性是由 Thymeleaf Spring Security 库提供的扩展支持，可以方便地用 sec 标签来获取认证、授权方法的信息。

（1）sec:authorize="isAnonymous()"：判断用户是不是匿名的（未认证）。

（2）sec:authorize="isAuthenticated()"：判断用户是否经过认证。

（3）sec:authentication="name"：获取到用户的名称。

（4）sec:authentication="principal.authorities"：获取到用户的角色。

想了解该库的更多信息，可以参阅 https://github.com/thymeleaf/thymeleaf-extras-springsecurity。

2. header.html

在前端修改了 header.html。

```
...
<div class="collapse navbar-collapse" id="navbarsContainer">
```

```
    <ul class="navbar-nav mr-auto">
        <li class="nav-item">
            <a class="nav-link" th:href="@{~/users}"  >首页 <span
class="sr-only">(current)</span></a>
        </li>
    </ul>

    <!-- 登录判断 -->
    <div sec:authorize="isAuthenticated()"  class="row" >
        <ul class="navbar-nav mr-auto">
            <li class="nav-item">
                <span class="nav-link"  sec:authentication="name"></
span>
             </li>
        </ul>
        <form action="/logout" th:action="@{/logout}" method="post">
            <input class="btn btn-outline-success " type="submit"
value="退出">
        </form>
    </div>

    <div sec:authorize="isAnonymous()">
        <a href="/login" th:href="@{~/login}" class="btn btn-outline-
success my-2 my-sm-0" type="submit">登录</a>
    </div>
</div>
...
```

菜单栏会根据用户是否认证来显示不同的信息。

（1）未认证：显示“登录”按钮。

（2）已认证：显示用户的名称。

（3）要退出时发送 logout 表单请求到后台。

3. login.html

在前端增加了 login.html。

```
...
<!-- Page content -->
<div class="container blog-content-container">
    <form  th:action="@{/login}" method="post">
        <h2 >请登录</h2>

        <div class="form-group col-md-5">
            <label for="username" class="col-form-label">账号</label>
            <input type="text" class="form-control" id="username"
name="username" maxlength="50" placeholder="请输入账号">

        </div>
```

```
        <div class="form-group col-md-5">
            <label for="password" class="col-form-label">密码</label>
            <input type="password" class="form-control" id="password"
name="password" maxlength="30" placeholder="请输入密码" >
        </div>
        <div class="form-group col-md-5">
            <button type="submit" class="btn btn-primary">登录</button>
        </div>
            <div class=" col-md-5" th:if="${loginError}">
            <p class="blog-label-error" th:text="${errorMsg}"></p>
        </div>
    </form>
</div><!-- /.container -->
...
```

10.4.3 运行

启动 security-in-action 项目后，在浏览器访问 http://localhost:8080，可以看到项目的运行效果。图 10-1 所示为用户未认证访问主页时的效果。

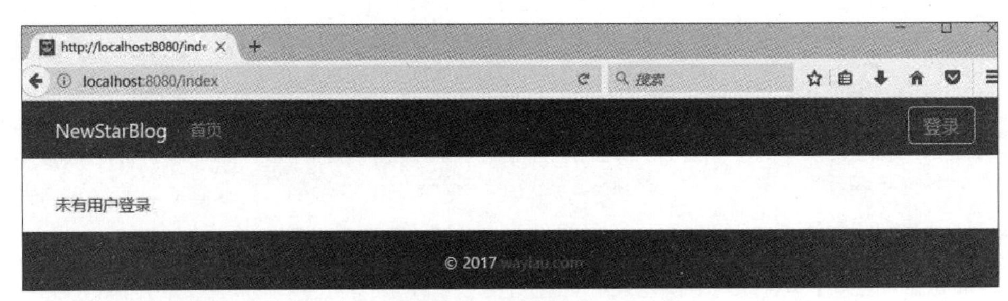

图10-1　未认证

当试图访问“用户管理”时，被重定向到了登录页面，如图 10-2 所示。

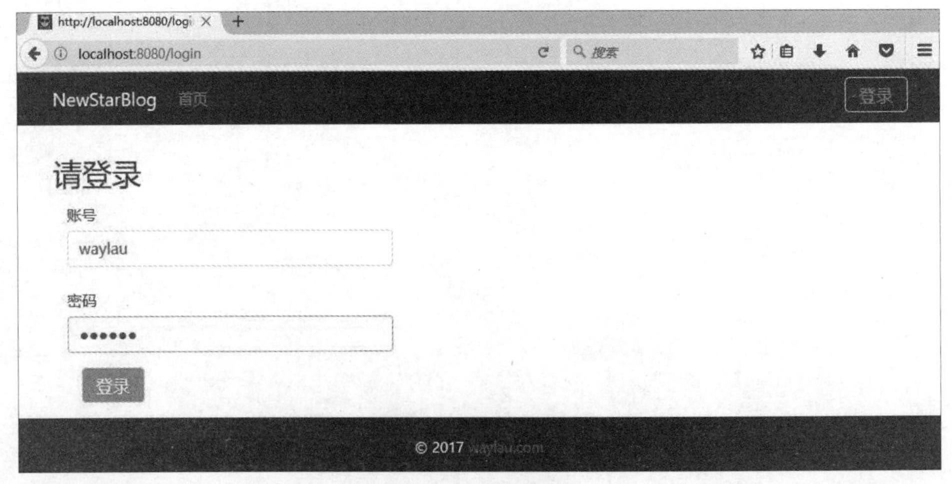

图10-2　登录页面

用默认的 waylau 用户进行登录后可以访问“用户管理”，如图 10-3 所示。

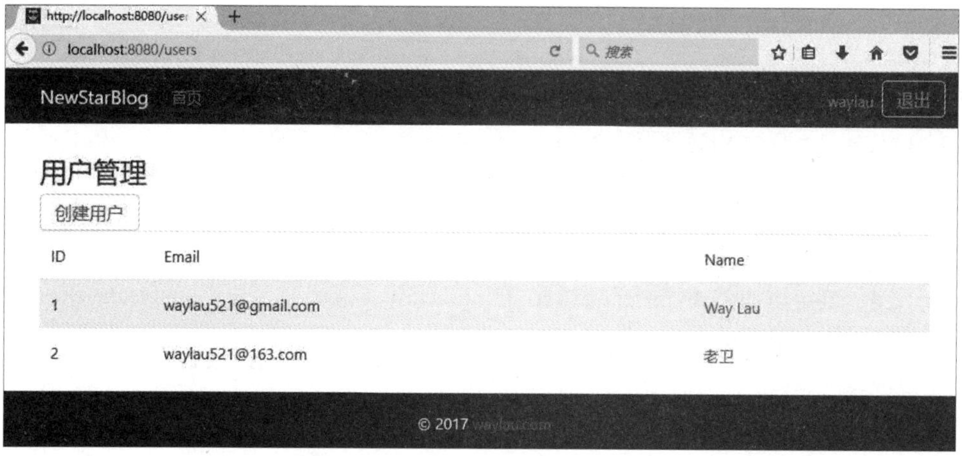

图10-3 已认证

用户经过认证后，再次访问主页时的效果如图 10-4 所示。

图10-4 已认证后访问主页

10.4.4 相关问题解决

1. 问题1

```
Invalid CSRF Token 'null' was found on the request parameter '_csrf' or
header 'X-CSRF-TOKEN'.
```

自 Spring Security 3.2 起，启用了 CSRF 保护机制，所以 Form 表单提交必须满足以下条件。

（1）HTTP 方法必须是 POST。

（2）CSRF token 必须添加到请求。由于使用了 @EnableWebSecurity 和 Thymeleaf，CSRF token 将自动添加到一个隐藏的 <input> 中（查看源码看到）；类似于 <input type="hidden" name="_csrf "value="f912aef3-f9a2-4c22-852e-db8cecf4175a"/>。

解决方法是加上 Thymeleaf 标签，例如，将

```
<form action="/users" method="post">
    <!-- ... -->
</form>
```

改为

```
<form th:action="@{/users}" method="post">    <!-- ... --></form>
```

2. 问题2

sec:authorize 和 sec:authentication 属性不起作用。

解决方法 1：确定添加了 thymeleaf-extras-springsecurity4 依赖，且与 Thymeleaf 版本一致。

解决方法 2：在 Spring 中注入相关的方言。

```
<bean id="templateEngine" class="org.thymeleaf.spring4.SpringTemplate
Engine">
  ...
  <property name="additionalDialects">
    <set>
      <!-- 如果使用的是Spring Security 3版本，则修改成'springsecurity3'-->
      <bean class="org.thymeleaf.extras.springsecurity4.dialect.
SpringSecurityDialect"/>
    </set>
  </property>
  ...
</bean>
```

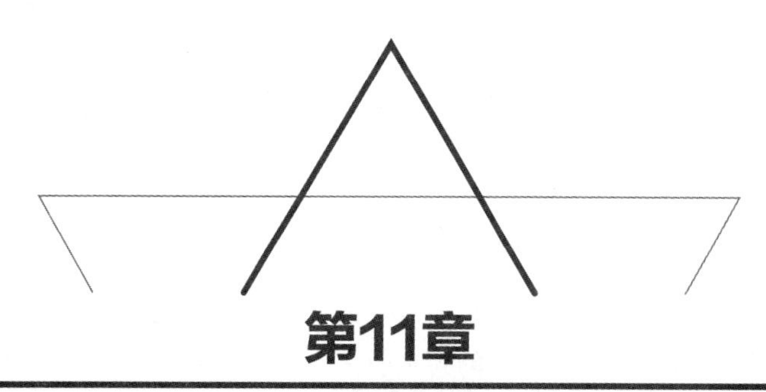

第11章

博客系统的整体框架实现

11.1 如何设计 API

接下来，搭建整个博客系统的框架。在企业级应用的开发流程中，需求定义完成之后，下面一个流程就是要进行架构设计。所谓架构设计，就是厘清整个系统的需求，并合理划分模块，以及定义模块之间的通信方式。

在本节来设计整个系统的 API。API 的设计遵循 REST 的风格。

11.1.1 REST的含义

要设计 RESTful API, 首要先要了解什么是 REST。

REST（Representational State Transfer，表述性状态转移）是由 Roy Fielding 在他的博士论文《Architectural Styles and the Design of Network-based Software Architectures》[①] 中提出的。REST 并非标准，而是一种开发 Web 应用的架构风格，可以将其理解为一种设计模式。REST 基于 HTTP、URI 及 XML 这些现有的广泛流行的协议和标准。伴随着 REST 的流行，HTTP 协议得到了更加正确的使用。

相较于基于 SOAP 和 WSDL 的 Web 服务，REST 模式提供了更为简洁的实现方案。REST Web 服务（RESTful Web Services）是松耦合的，这特别适用于为客户创建在互联网传播的轻量级的 Web 服务 API。REST 应用是围绕“资源表述的转移”（the transfer of representations of resources）为中心来做请求和响应。数据和功能均被视为资源，并使用统一资源标识符（URI）来访问资源。网页中的链接就是典型的 URI。该资源由文档表述，并通过使用一组简单的、定义明确的操作来执行。

例如，一个 REST 资源可能是一个城市当前的天气情况，该资源的表述可能是一个 XML 文档、图像文件 或 HTML 页面。客户端可以检索特定表述，通过更新其数据修改的资源，或者完全删除该资源。

目前，越来越多的 Web 服务开始采用 REST 风格设计和实现，真实世界中比较著名的 REST 服务包括 Google AJAX 搜索 API、Amazon Simple Storage Service（Amazon S3）等。

11.1.2 REST 设计原则

基于 REST 的 Web 服务遵循一些基本的设计原则，使 RESTful 应用更加简单、轻量，开发速度也更快。

（1）通过 URI 来标识资源：系统中的每一个对象或资源都可以通过一个唯一的 URI 来进行寻址，URI 的结构应该简单、可预测且易于理解，如定义目录结构式的 URI。

① 论文地址可参见 http://www.ics.uci.edu/~fielding/pubs/dissertation/top.htm。

（2）统一接口：以遵循 RFC-2616 所定义的协议的方式显式地使用 HTTP 方法，建立创建、检索、更新和删除（CRUD：Create, Retrieve, Update and Delete）操作与 HTTP 方法之间的一对一映射。

①若要在服务器上创建资源，应该使用 POST 方法。

②若要检索某个资源，应该使用 GET 方法。

③若要更新或添加资源，应该使用 PUT 方法。

④若要删除某个资源，应该使用 DELETE 方法。

（3）资源多重表述：URI 所访问的每个资源都可以使用不同的形式加以表示（如 XML 或 JSON），具体的表现形式取决于访问资源的客户端，客户端与服务提供者使用一种内容协商的机制（请求头与 MIME 类型）来选择合适的数据格式，最小化彼此之间的数据耦合。

（4）无状态：对服务器端的请求应该是无状态的，完整、独立的请求不要求服务器在处理请求时检索任何类型的应用程序上下文或状态。无状态约束使服务器的变化对客户端是不可见的，因为在两次连续的请求中，客户端并不依赖于同一台服务器。一个客户端从某台服务器上收到一份包含链接的文档，当它要做一些处理时，这台服务器宕机，可能是硬盘坏掉而被拿去修理，可能是软件需要升级重启——如果这个客户端访问了从这台服务器接收的链接，它不会察觉到后台的服务器已经改变了。通过超链接实现有状态交互，即请求消息是自包含的（每次交互都包含完整的信息），有多种技术实现了不同请求间状态信息的传输，如 URI 重定向、cookies 和隐藏表单字段等，状态可以嵌入应答消息中，这样一来状态在接下来的交互中仍然有效。REST 风格应用可以实现交互，它却天然地具有服务器无状态的特征。在状态迁移的过程中，服务器不需要记录任何 Session，所有的状态都通过 URI 的形式记录在了客户端。更准确地说，这里的无状态服务器是指服务器不保存会话状态（Session）；而资源本身则是天然的状态，通常是需要被保存的；这里所指的无状态服务器均指无会话状态服务器。

表 11-1 所示为 HTTP 请求方法在 RESTful Web 服务中的典型应用。

表11-1　请求方法在RESTful Web服务中的典型应用

资源	GET	PUT	POST	DELETE
一组资源的URI，例如，https://waylau.com/resources/	列出URI及该资源组中每个资源的详细信息（后者可选）	使用给定的一组资源替换当前整组资源	在本组资源中创建/追加一个新的资源。该操作往往返回新资源的URL	删除整组资源
单个资源的URI，例如，https://waylau.com/resources/142	获取指定的资源的详细信息，格式可以自选一个合适的网络媒体类型（如XML、JSON等）	替换/创建指定的资源，并将其追加到相应的资源组中。	把指定的资源当作一个资源组，并在其下创建/追加一个新的元素，使其隶属于当前资源	删除指定的元素

11.1.3 Java REST

针对 REST 在 Java 中的规范，主要是 JAX-RS（Java API for RESTful Web Services），该规范使得 Java 程序员可以使用一套固定的接口来开发 REST 应用，避免了依赖于第三方框架。同时，JAX-RS 使用 POJO 编程模型和基于标注的配置，并集成了 JAXB，从而可以有效缩短 REST 应用的开发周期。Java EE 6 引入了对 JSR-311 的支持，Java EE 7 支持 JSR-339 规范。

JAX-RS 定义的 API 位于 javax.ws.rs 包中。

伴随着 JSR 311 规范的发布，Sun 同步发布该规范的参考实现 Jersey。JAX-RS 的具体实现第三方还包括 Apache 的 CXF 及 JBoss 的 RESTEasy 等。未实现该规范的其他 REST 框架还包括 Spring-MVC 等。

截至目前，JAX-RS 最新的版本是 2.0（JSR-339）。

11.1.4 常用 REST 框架

在 Java 平台中，很多框架都提供了开发 REST 风格 API 的能力，如 Spring MVC、Jersey、Apache CXF，其中 Jersey、Apache CXF 遵循了 JAX-RS 规范，而 Spring MVC 并没有参考这个规范。所以在选用 REST 框架时，也结合自己的实际需要，例如，在项目中，我们需要完全符合规范来做事，那么毫无疑问，就会选型 Jersey 或 Apache CXF。如果项目是基于 Spring 框架来构建的，需要很好地与 Spring 相关技术栈做兼容，那么 Spring MVC 自然就是首选。

本书是介绍 Spring Boot 的书籍，所以理所当然，在项目中采用了 Spring MVC 来实现 RESTful API 。

读者如果对 Jersey 框架感兴趣，也可以参阅笔者所著的开源书《Jersey 2.x 用户指南》（https://github.com/waylau/Jersey-2.x-User-Guide）和《REST 实战》（https://github.com/waylau/rest-in-action）。

11.1.5 博客系统的 API

下面是整个博客系统的 API。在后面将介绍如何来一一实现这些 API 。

- index : 主页，含最新、最热文章，最热标签，最热用户等。
  - /blogs：GET。
    - order：排序类型， new/hot，默认是 new.。
    - keyword：搜索关键字。博客的标签，即为关键字。
    - async：是否异步请求页面。
    - pageIndex。
    - pageSize。
- user space：用户主页空间。

  - /u/{username}：GET 具体某个用户的主页。
    - username：用户账号。
  - /u/{username}/profile：GET 获取个人设置页面。
    - username：用户账号。
  - /u/{username}/profile：POST 保存个人设置。
    - username：用户账号。
    - User：待保存的对象。
  - /u/{username}/avatar ：GET 获取个人头像。
    - username：用户账号。
  - /u/{username}/avatar：POST 保存个人头像。
    - username：用户账号。
    - User：待保存的对象。
  - /u/{username}/blogs：GET 查询用户博客，以下三个条件任选一个。
    - order: 排序类型， new/hot ， 默认是 new。
    - catalog：博客分类 Id，默认是空。
    - keyword：搜索关键字。博客的标签，即为关键字。
      - async：是否异步请求页面。
      - pageIndex。
      - pageSize。
  - /u/{username}/blogs/edit：GET 获取新增博客的界面。
    - username：用户账号。
  - /u/{username}/blogs/edit：POST 新增、编辑博客。
    - username：用户账号。
    - Blog ：待保存的博客对象。
  - /u/{username}/blogs/edit/{id}：GET 获取编辑博客的界面。
    - username：用户账号。
    - id：博客的 id。
  - /u/{username}/blogs/edit/{id}：DELETE 删除博客。
    - username：用户账号。
    - id：博客的 id。
- login：登录。
  - /login :GET 获取登录的界面。
  - /login :POST 登录。

    - username：用户账号。
    - password：用户密码。
    - remember-me：是否记住我。
- register：注册。
  - /register :GET 获取注册的界面。
  - /register :POST 注册， 注册成功跳转至 登录界面。
    - User：待保存的用户对象。
- admins：后台管理。
  - /admins：获取后台管理的界面。
- users：用户管理。
  - /users：GET 用户列表。
    - async：是否异步请求页面。
    - pageIndex。
    - pageSize。
    - name：用户名称的关键字。
  - /users/add：GET 获取用户新增页面。
  - /users：POST 保存用户。
    - User：待保存用户对象。
    - authorityId：角色 ID。
  - /users/{id}：DELETE 删除用户。
    - id：用户 id。
  - /users/edit/{id} :GET 获取具体某个用户编辑界面。
    - id：某个用户的 id。
- comments：评论管理。
  - /comments：GET 获取评论列表。
    - blogId：博客 id。
  - /comments：POST 保存评论。
    - blogId：博客 id。
    - commentContent：评论内容。
  - /comments/{id}：DELETE 删除评论。
    - id：评论 id。
    - blogId：博客 id。
- votes：点赞管理。

    - /votes：POST 保存点赞。
        - blogId：博客 id。
    - /votes/{id}：DELETE 删除点赞。
        - id：点赞 id。
        - blogId：博客 id。
- catalogs：分类管理。
    - /catalogs：GET 获取用户分类列表。
        - username：用户账号。
    - /catalogs：POST 保存用户分类。
        - username：用户账号。
        - CatalogVO：含 username、Catalog。
    - /catalogs/edit：GET 获取编辑分类的界面。
    - /catalogs/edit/{id}：GET 获取某 ID 分类的编辑界面。
    - /catalogs/{id}：DELETE 删除分类。
        - id：分类 id。
        - username：用户账号。

11.2 实现后台整体控制层

在 bootstrap-in-action 项目的基础上新建了 blog-prototype 项目，该项目有以下几个作用。

（1）让读者重新回顾博客系统的整体需求。

（2）让读者了解到整体界面的原型设计。

（3）固化了整个系统的架构。

（4）定义了 API。

大家知道，在 Spring MVC 中，所有的客户端请求首先会被控制层处理，所以，API 的定义也是属于控制层的职责。

下面先来实现整个博客系统的控制层及对 API 的定义。控制层主要由 MainController.java、BlogController.java、UserspaceController.java、AdminController.java、UserController.java 等组成。

11.2.1 MainController

新增了 MainController 用于处理整个系统相关的控制，包括登录、退出、注册等功能。详细的实现方式如下。

```
@Controller
public class MainController {

    @GetMapping("/")
    public String root() {
        return "redirect:/index";
    }

    @GetMapping("/index")
    public String index() {
        return "index";
    }

    @GetMapping("/login")
    public String login() {
        return "login";
    }

    @GetMapping("/login-error")
    public String loginError(Model model) {
        model.addAttribute("loginError", true);
        model.addAttribute("errorMsg", "登录失败，用户名或者密码错误！");
        return "login";
    }

    @GetMapping("/register")
    public String register() {
        return "register";
    }

}
```

11.2.2 BlogController

新增了 BlogController 用于处理博客相关的控制，这里主要涉及查询博客列表。详细的实现方式如下。

```
@Controller
@RequestMapping("/blogs")
public class BlogController {

    @GetMapping
    public String listBlogs(@RequestParam(value="order",required=false,-
defaultValue="new") String order,
            @RequestParam(value="keyword",required=false,defaultValue=
"" ) String keyword){
```

```
        System.out.println("order:" +order + ";keyword:" +keyword );
        return "redirect:/index?order="+order+"&keyword="+keyword;
    }

}
```

11.2.3 UserspaceController

新增了 UserspaceController 用于处理用户空间相关的控制，这里主要涉及展示用户空间主页、查询用户博客列表、显示用户某篇博客等功能。详细的实现方式如下。

```
@Controller
@RequestMapping("/u")
public class UserspaceController {

    @GetMapping("/{username}")
    public String userSpace(@PathVariable("username") String username)
{
        System.out.println("username" + username);
        return "/userspace/u";
    }

    @GetMapping("/{username}/blogs")
    public String listBlogsByOrder(@PathVariable("username") String
username,
            @RequestParam(value="order",required=false,defaultValue=
"new") String order,
            @RequestParam(value="category",required=false ) Long
category,
            @RequestParam(value="keyword",required=false ) String key-
word) {

        if (category != null) {

            System.out.print("category:" +category );
            System.out.print("selflink:" + "redirect:/u/"+ username +"/
blogs?category="+category);
            return "/userspace/u";

        } else if (keyword != null && keyword.isEmpty() == false) {

            System.out.print("keyword:" +keyword );
            System.out.print("selflink:" + "redirect:/u/"+ username +"/
blogs?keyword="+keyword);
            return "/userspace/u";
        }
```

```
        System.out.print("order:" +order);
        System.out.print("selflink:" + "redirect:/u/"+ username +"/
blogs?order="+order);
        return "/userspace/u";
    }

    @GetMapping("/{username}/blogs/{id}")
    public String listBlogsByOrder(@PathVariable("id") Long id) {

        System.out.print("blogId:" + id);
        return "/userspace/blog";
    }


    @GetMapping("/{username}/blogs/edit")
    public String editBlog() {

        return "/userspace/blogedit";
    }

}
```

11.2.4 AdminController

新增了 AdminController 用于处理后台管理相关的控制，这里主要涉及返回后台管理页面的功能。详细的实现方式如下。

```
@Controller
@RequestMapping("/admins")
public class AdminController {

    /**
     * 获取后台管理主页面
     * @return
     */
    @GetMapping
    public ModelAndView listUsers(Model model) {
        return new ModelAndView("admins/index", "menuList", model);
    }

}
```

11.2.5 UserController.java

UserController 实现暂时与 bootstrap-in-action 项目中的一致，不做任何修改。在后面“用户管理实现”部分再按照 RESTful API 的定义来修改。

11.3 实现前台整体布局

前台整体布局在之前的“博客系统的原型设计”部分也做了展示。由于前台的代码量比较大，因此本节只对比较核心的功能做讲解。其他部分读者可以自行参阅源码。

11.3.1 登录界面

登录界面采用了 Bootstrap 中的 form-group 类，该 form 表单通过 Bootstrap 的 container 和 blog-content-container 类来定义。

```
...
<div class="container blog-content-container">

    <form>
        <h2>请登录</h2>

        <div class="form-group col-md-5">
            <label for="username" class="col-form-label">账号</label>
<input
                type="text" class="form-control" id="username" name="us-
ername"
                maxlength="50" placeholder="请输入账号">

        </div>
        <div class="form-group col-md-5">
            <label for="password" class="col-form-label">密码</label>
<input
                type="password" class="form-control" id="password"
name="password"
                maxlength="30" placeholder="请输入密码">
        </div>
        <div class="form-group col-md-5">
            <input type="checkbox" value="remember-me"> 记住我
        </div>
        <div class="form-group col-md-5">
            <button type="submit" class="btn btn-primary">登录</button>
        </div>

    </form>
</div><!-- /container -->
...
```

其中 blog-content-container 在 blog.css 文件中定义如下。

```
...
.blog-content-container {
    margin-top: 2.0em;
    background-color: #fff;
}
...
```

11.3.2 注册界面

注册界面的整体布局与登录界面基本类似，也是使用了 form 表单。

```
...
<div class="container blog-content-container">

    <form>
        <h2 class="form-signin-heading">注册成为博主</h2>

        <div class="form-group col-md-5">
            <label for="username" class="col-form-label">账号</label>
<input
                type="text" class="form-control" id="username" name="us-
ername"
                maxlength="50" placeholder="请输入账号，至少3个字符，至多20
个">

        </div>
        <div class="form-group col-md-5">
            <label for="email" class="col-form-label">邮箱</label>
<input
                type="email" class="form-control" id="email" name=
"email"
                maxlength="50" placeholder="请输入邮箱">
        </div>
        <div class="form-group col-md-5">
            <label for="username" class="col-form-label">姓名</label>
<input
                type="text" class="form-control" id="username" name="us-
ername"
                maxlength="20" placeholder="请输入姓名，至少2个字符，至多20
个">
        </div>
        <div class="form-group col-md-5">
            <label for="password" class="col-form-label">密码</label>
<input
                type="password" class="form-control" id="password"
name="password"
```

```
                maxlength="30" placeholder="请输入密码，字母或特殊符号和数字
结合">
        </div>
        <div class="form-group col-md-5">
            <button type="submit" class="btn btn-primary">提交</button>
        </div>

    </form>

</div><!-- /container -->
...
```

11.3.3 页首

首页的上方是头部（页首），其中“NewStarBlog”是博客系统的名称。博客可以按照最新、最热来排序，可以通过关键字来搜索。在头部的右侧是登录、注册的通道。

页首主要由 Bootstrap 中的 navbar、navbar-inverse、collapse 类来实现菜单的折叠效果。

```
<nav class="navbar navbar-inverse bg-inverse navbar-toggleable-md">
    <div class="container">
        <button class="navbar-toggler navbar-toggler-right" type=
"button"
            data-toggle="collapse" data-target="#navbarsContainer"
            aria-controls="navbarsExampleContainer" aria-expanded=
"false"
            aria-label="Toggle navigation">
            <span class="navbar-toggler-icon"></span>
        </button>
        <a class="navbar-brand" href="/" th:href="@{/}">NewStarBlog</a>

        <div class="collapse navbar-collapse" id="navbarsContainer">

            <ul class="navbar-nav mr-auto">
                <li class="nav-item"><a class="nav-link"
                    href="/blogs?order=new">最新 <span class="sr-only">
(current)</span></a>
                </li>
                <li class="nav-item"><a class="nav-link"
                    href="/blogs?order=hot">最热</a></li>
                <form class="form-inline mt-2 mt-md-0">
                    <input class="form-control mr-sm-2" type="text"
placeholder="搜索">
                    <a href="/blogs?keyword=ww"
                        class="btn btn-outline-secondary my-2 my-sm-0">
<i
                        class="fa fa-search" aria-hidden="true"></i></
a>
```

```
                </form>

            </ul>

            <a href="/login" class="btn btn-outline-success my-2 my-
sm-0"
                type="submit">登录</a> <a href="/register"
                class="btn btn-outline-success my-2 my-sm-0" type="sub-
mit">注册</a>
        </div>

    </div>
</nav>
```

11.3.4 分页组件

分页组件使用了 Bootstrap 中的 pagination 类，每个页码按钮都是一个 page-item 类。

```
...
<!-- Pagination -->
<nav>
    <ul class="pagination justify-content-center mb-4">
        <li class="page-item">
            <a href="#" class="page-link" aria-label="Previous">
                <span aria-hidden="true">&laquo;</span>
            </a>
        </li>
        <li class="page-item active"><a href="#" class="page-link">1</
a></li>
        <li class="page-item"><a href="#" class="page-link">2</a></li>
        <li class="page-item"><a href="#" class="page-link">3</a></li>
        <li class="page-item"><a href="#" class="page-link">4</a></li>
        <li class="page-item"><a href="#" class="page-link">5</a></li>
        <li class="page-item"><a href="#" class="page-link">6</a></li>
        <li class="page-item"><a href="#" class="page-link">7</a></li>
        <li class="page-item">
            <a href="#" class="page-link" aria-label="Next">
                <span aria-hidden="true">&raquo;</span>
            </a>
        </li>
    </ul>
</nav>
...
```

11.3.5 文章列表

文章列表的每一个列都是固定格式的，包含了博客的标题、作者、发表时间、阅读量、评论量、点赞量、摘要等信息，格式如下所示。

```
...
<!-- Blog Post -->
<div class="card mb-4">
    <div class="card-block">
        <h2 class="card-title">

            <a href="/u/waylau/blogs/1" class="card-link" title="way-
lau">
                OAuth 2.0 认证的原理与实践 </a>
        </h2>
        <p class="card-text">使用 OAuth 2.0认证的好处是显而易见的。只需要用同
一个账和密码，就能在各个网站进行访问，而免去了在每个网站都进行注册的烦琐过程。本文将介
绍 OAuth 2.0 的原理，并基于Spring Security和GitHub账号，来演示OAuth 2.0的认证
过程。
        </p>
        <div class="card-text">
            发表于 2017-03-17 <i class="fa fa-eye" aria-hidden=
"true">210</i> <i
                class="fa fa-heart-o" aria-hidden="true">10</i> <i
                class="fa fa-comment-o" aria-hidden="true">110</i>
        </div>
    </div>
</div>
...
```

card 类是 Bootstrap 4 版本中新添加的内容。一个 card 可以由很多个卡片块 card-block 组成，配合 card-title 和 card-text 可以很好地实现卡片式布局。

11.3.6 示例源码

本节原型设计的源码在 blog-prototype 目录下。

第12章

用户管理实现

12.1 用户管理的需求回顾

自本节开始将正式开始实现博客系统的具体功能。

本节首先要实现的是用户管理功能。

在blog-prototype项目基础上构建了一个新的项目blog-user，用于演示整个用户管理实现的过程。

12.1.1 所需环境

本例采用的开发环境如下。

- Apache Commons Lang 3.6。

Apache Commons Lang 在本节中主要用于支持字符串的常用处理。

12.1.2 build.gradle

修改 build.gradle 文件，需要添加 Apache Commons Lang 依赖。

```
// 依赖关系
dependencies {
    //...

    // 添加Apache Commons Lang依赖
    compile('org.apache.commons:commons-lang3:3.6')

    //...
}
```

12.1.3 用户管理的需求

首先来回顾一下用户管理所具有的功能需求。

任何一个信息化系统，都少不了用户管理。一方面，用户要使用系统，必然要先注册成为系统的用户；另一方面，用户在使用系统的过程中，也经常会涉及对自己的信息进行修改，如用户可以更换头像、重置密码等。

博客系统主要分为两类用户：普通用户（博主）和管理员。对于普通用户来说，可以注册用户、登录系统；对于管理员来说，可以增加、删除、修改、搜索用户。

概括起来，在本节将实现的功能如下。

- 访问后台管理界面。
- 新增用户、删除用户、编辑用户。
- 分页组件、分页查询。

- 批量删除。
- 用户名模糊搜索。
- 注册。

12.1.4 API

在了解完需求后，就要开始设计需求功能的 API 了。

1. 注册

注册功能的接口设计如下。

- register：注册。
  - /register：GET 获取注册的界面。
  - /register：POST 注册， 注册成功跳转至登录界面。
    - User：待保存的用户对象。

2. 后台管理

用户管理是通过后台管理界面来实现的，后台管理功能的接口设计如下。

- admins：后台管理
  - /admins：获取后台管理的界面

3. 用户管理

用户管理功能的接口设计如下。

- users：用户管理。
  - /users：GET 用户列表。
    - async：是否异步请求页面。
    - pageIndex。
    - pageSize。
    - name：用户名称的关键字。
  - /users/add：GET 获取用户新增页面。
  - /users：POST 保存用户。
    - User：待保存用户对象。
    - authorityId：角色 ID。
  - /users/{id}：DELETE 删除用户。
    - id：用户 ID。
  - /users/edit/{id} :GET 获取具体某个用户编辑界面。
    - id：某个用户的 ID。

12.2 用户管理的后台实现

本节将进行用户管理的后台编码实现。

12.2.1 修改 User

在原有的 User 基础上增加了 username、password、avatar 三个字段，分别用于存放用户账号、登录密码和头像地址，并且所有的字段都加上 Bean Validation。

```
import javax.persistence.Column;
import javax.persistence.Entity;
import javax.persistence.GeneratedValue;
import javax.persistence.GenerationType;
import javax.persistence.Id;
import javax.validation.constraints.Size;

import org.hibernate.validator.constraints.Email;
import org.hibernate.validator.constraints.NotEmpty;

@Entity // 实体
public class User {

    @Id // 主键
    @GeneratedValue(strategy=GenerationType.IDENTITY) // 自增策略
    private Long id; // 实体一个唯一标识

    @NotEmpty(message = "姓名不能为空")
    @Size(min=2, max=20)
    @Column(nullable = false, length = 20) // 映射为字段，值不能为空
    private String name;

    @NotEmpty(message = "邮箱不能为空")
    @Size(max=50)
    @Email(message= "邮箱格式不对" )
    @Column(nullable = false, length = 50, unique = true)
    private String email;

    @NotEmpty(message = "账号不能为空")
    @Size(min=3, max=20)
    @Column(nullable = false, length = 20, unique = true)
    private String username; // 用户账号，用户登录时的唯一标识

    @NotEmpty(message = "密码不能为空")
    @Size(max=100)
    @Column(length = 100)
    private String password; // 登录时密码
```

```
    @Column(length = 200)
    private String avatar; // 头像图片地址

    protected User() { // 无参构造函数;设为protected防止直接使用
    }

    public User(String name, String email,String username,String pass-
word) {
        this.name = name;
        this.email = email;
        this.username = username;
        this.password = password;
    }

    // 省略getter/setter方法

    @Override
    public String toString() {
        return String.format("User[id=%d,name='%s',username='%s',email=
'%s']", id, name, username, email);
    }
}
```

其中，注解 @NotEmpty、@Email 是 Hibernate 框架提供的特性。

12.2.2 修改 UserRepository

修改 UserRepository 继承自 JpaRepository，而非 CrudRepository，这样 UserRepository 就有了分页的功能，并新增了两个接口。

```
import org.springframework.data.domain.Page;
import org.springframework.data.domain.Pageable;
import org.springframework.data.jpa.repository.JpaRepository;
...

public interface UserRepository extends JpaRepository<User, Long>{

    /**
     * 根据用户姓名分页查询用户列表
     * @param name
     * @param pageable
     * @return
     */
    Page<User> findByNameLike(String name, Pageable pageable);

    /**
     * 根据用户账号查询用户
     * @param username
```

```
     * @return
     */
    User findByUsername(String username);
}
```

12.2.3 新增 UserService 接口及实现

创建 com.waylau.spring.boot.blog.service 包，并新增 UserService 接口，以提供用户服务。

```
import java.util.Optional;
...

public interface UserService {
     /**
     * 新增、编辑、保存用户
     * @param user
     * @return
     */
    User saveOrUpateUser(User user);

    /**
     * 注册用户
     * @param user
     * @return
     */
    User registerUser(User user);

    /**
     * 删除用户
     * @param id
     */
    void removeUser(Long id);

    /**
     * 根据id获取用户
     * @param id
     * @return
     */
    Optional<User> getUserById(Long id);

    /**
     * 根据用户名进行分页模糊查询
     * @param name
     * @param pageable
     * @return
     */
    Page<User> listUsersByNameLike(String name, Pageable pageable);
}
```

再新增 UserService 的实现 UserServiceImpl。

```
import java.util.Optional;
import javax.transaction.Transactional;
import org.springframework.beans.factory.annotation.Autowired;
import org.springframework.stereotype.Service;
...

@Service
public class UserServiceImpl implements UserService {

    @Autowired
    private UserRepository userRepository;

    @Transactional
    @Override
    public User saveOrUpateUser(User user) {
        return userRepository.save(user);
    }

    @Transactional
    @Override
    public User registerUser(User user) {
        return userRepository.save(user);
    }

    @Transactional
    @Override
    public void removeUser(Long id) {
        userRepository.deleteById(id);
    }

    @Override
    public Optional<User> getUserById(Long id) {
        return userRepository.findById(id);
    }

    @Override
    public Page<User> listUsersByNameLike(String name, Pageable pagea-
ble) {

        // 模糊查询
        name = "%" + name + "%";
        Page<User> users = userRepository.findByNameLike(name, pageable);

        return users;
    }

}
```

12.2.4 增加 Response

创建 com.waylau.spring.boot.blog.vo 包，并新增 Response 类。Response 是 REST 接口中统一返回的值对象。

```
public class Response {

    private boolean success;
    private String message;
    private Object body;

    /** 响应处理是否成功 */
    public boolean isSuccess() {
        return success;
    }
    public void setSuccess(boolean success) {
        this.success = success;
    }

    /** 响应处理的消息 */
    public String getMessage() {
        return message;
    }
    public void setMessage(String message) {
        this.message = message;
    }

    /** 响应处理的返回内容 */
    public Object getBody() {
        return body;
    }
    public void setBody(Object body) {
        this.body = body;
    }

    public Response(boolean success, String message) {
        this.success = success;
        this.message = message;
    }

    public Response(boolean success, String message, Object body) {
        this.success = success;
        this.message = message;
        this.body = body;
    }
}
```

12.2.5 增加 ConstraintViolationExceptionHandler

创建 com.waylau.spring.boot.blog.util 包，并新增 ConstraintViolationExceptionHandler 类，用于处理 ConstraintViolationException 异常的处理器。

```
import javax.validation.ConstraintViolation;
import javax.validation.ConstraintViolationException;
import org.apache.commons.lang3.StringUtils;
...

public class ConstraintViolationExceptionHandler {

    /**
     * 获取批量异常信息
     * @param e
     * @return
     */
    public static String getMessage(ConstraintViolationException e) {
        List<String> msgList = new ArrayList<>();
        for (ConstraintViolation<?> constraintViolation : e.getCon-
straintViolations()) {
            msgList.add(constraintViolation.getMessage());
        }
        String messages = StringUtils.join(msgList.toArray(), ";");
        return messages;
    }
}
```

12.2.6 修改 UserController

修改 UserController，用于处理对 API 的不同请求。

```
@RestController
@RequestMapping("/users")
public class UserController {

    @Autowired
    private UserService userService;

    /**
     * 查询所有用户
     *
     * @param async
     * @param pageIndex
     * @param pageSize
     * @param name
     * @param model
```

```
     * @return
     */
    @GetMapping
    public ModelAndView list(@RequestParam(value = "async", required =
false) boolean async,
            @RequestParam(value = "pageIndex", required = false, default
Value = "0") int pageIndex,
            @RequestParam(value = "pageSize", required = false, default-
Value = "10") int pageSize,
            @RequestParam(value = "name", required = false, defaultValue =
"") String name, Model model) {

        Pageable pageable = PageRequest.of(pageIndex, pageSize);
        Page<User> page = userService.listUsersByNameLike(name, page
able);
        List<User> list = page.getContent(); // 当前所在页面数据列表

        model.addAttribute("page", page);
        model.addAttribute("userList", list);
        return new ModelAndView(async == true ? "users/list :: #main
ContainerRepleace" : "users/list", "userModel",
                model);
    }

    /**
     * 获取创建表单页面
     *
     * @param model
     * @return
     */
    @GetMapping("/add")
    public ModelAndView createForm(Model model) {
        model.addAttribute("user", new User(null, null, null, null));
        return new ModelAndView("users/add", "userModel", model);
    }

    /**
     * 保存或修改用户
     *
     * @param user
     * @return
     */
    @PostMapping
    public ResponseEntity<Response> saveOrUpateUser(User user) {

        try {
            userService.saveOrUpateUser(user);
        } catch (ConstraintViolationException e) {
            return ResponseEntity.ok().body(new Response(false, Con-
```

```
straintViolationExceptionHandler.getMessage(e)));
        }

        return ResponseEntity.ok().body(new Response(true, "处理成功",
user));
    }

    /**
     * 删除用户
     *
     * @param id
     * @param model
     * @return
     */
    @DeleteMapping(value = "/{id}")
    public ResponseEntity<Response> delete(@PathVariable("id") Long id,
Model model) {
        try {
            userService.removeUser(id);
        } catch (Exception e) {
            return ResponseEntity.ok().body(new Response(false, e.get
Message()));
        }
        return ResponseEntity.ok().body(new Response(true, "处理成功"));
    }

    /**
     * 获取修改用户的界面
     *
     * @param id
     * @param model
     * @return
     */
    @GetMapping(value = "edit/{id}")
    public ModelAndView modifyForm(@PathVariable("id") Long id, Model
model) {
        Optional<User> user = userService.getUserById(id);
        model.addAttribute("user", user.get());
        return new ModelAndView("users/edit", "userModel", model);
    }

}
```

12.2.7 增加 Menu

在 com.waylau.spring.boot.blog.vo 包新增 Menu 类，用于放置用户管理的菜单项。

```
public class Menu {
```

```
    private String name; // 菜单名称
    private String url; // 菜单URL

    public Menu(String name, String url) {
        this.name = name;
        this.url = url;
    }

    // 省略getter/setter方法
}
```

12.2.8 修改 AdminController

修改 AdminController，当请求进入后台管理界面时，就返回菜单列表。当然，由于目前菜单项只有一个，因此，暂时是编写在代码中。如果业务需要扩展，也可以将该菜单项存储在数据库中。

```
@RestController
@RequestMapping("/admins")
public class AdminController {

    /**
     * 获取后台管理主页面
     * @param model
     * @return
     */
    @GetMapping
    public ModelAndView listUsers(Model model) {
        List<Menu> list = new ArrayList<>();
        list.add(new Menu("用户管理", "/users"));
        model.addAttribute("list", list);
        return new ModelAndView("/admins/index", "model", model);
    }
}
```

12.2.9 修改 MainController

在原有 MainController 的基础上，增加注册用户 registerUser 方法。

```
@Controller
public class MainController {
    @Autowired
    private UserService userService;

    ...
```

```
    /**
     * 注册用户
     * @param user
     * @return
     */
    @PostMapping("/register")
    public String registerUser(User user) {
        userService.registerUser(user);
        return "redirect:/login";
    }

}
```

用户注册成功之后，就会重定向到 login 接口上，从而进入登录界面。

12.2.10 问题解决

在编写项目过程中，遇到了如下问题。

```
org.springframework.beans.factory.UnsatisfiedDependencyException: Error
creating bean with name 'org.springframework.boot.autoconfigure.orm.jpa.
HibernateJpaAutoConfiguration': Unsatisfied dependency expressed through
constructor parameter 0; nested exception is org.springframework.beans.
factory.BeanCreationException: Error creating bean with name 'dataSource':
Invocation of init method failed; nested exception is java.lang.Ille-
galStateException: Failed to replace DataSource with an embedded data-
base for tests. If you want an embedded database please put a supported
one on the classpath or tune the replace attribute of @AutoconfigureTest
Database.
```

解决方法：在测试类上添加 @AutoconfigureTestDatabase 即可。

12.3 用户管理的前台实现

在完成后台接口之后，本节将进行用户管理的前端编码。

12.3.1 用户注册

在原型的基础上修改 register.html，增加了 POST 的接口地址。

```
...
<form th:action="@{~/register}"  method="post">
...
```

12.3.2 后台管理

1. 修改admins/index.html

修改 admins/index.html，增加了对菜单的处理。

```
<body>
    <!-- Page Content -->
    <div class="container blog-content-container">

        <div class="row">

            <!-- 左侧栏目 -->
            <div class="col-md-4 col-xl-3">


                <!-- 分类菜单 -->
                <div class="card">
                    <h5 class="card-header">
                        <i class="fa fa-bars" aria-hidden="true"></i>
菜单
                    </h5>

                    <ul class="list-group blog-menu" th:each="menu :
${model.list}">
                        <a href="javascript:void(0)" class="list-group-
item "
                            th:title="${menu.name}" th:text="${menu.
name}"
                            th:attr="url=${menu.url}"> 用户管理 </a>
                    </ul>
                </div>

            </div>

            <!-- 右侧栏目 -->
            <div class="col-md-8 col-xl-9">
                <div class="card" id="rightContainer"></div>
            </div>
        </div>
        <!-- /.row -->

    </div>
    <!-- /.container -->

    <div th:replace="~{fragments/footer :: footer}">...</div>

    <!-- JavaScript -->
    <script src="../../js/admins/main.js" th:src="@{/js/admins/main.
```

```
js}"></script>

</body>
```

通过 th:each 将菜单项遍历动态生成菜单按钮。

2. 修改admin/main.js

修改 admin/main.js，增加了菜单单击事件。

```
$(function() {

    // 菜单事件
    $(".blog-menu .list-group-item").click(function() {

        var url = $(this).attr("url");

        // 先移除其他的单击样式，再添加当前的单击样式
        $(".blog-menu .list-group-item").removeClass("active");
        $(this).addClass("active");

        // 加载其他模块的页面到右侧工作区
         $.ajax({
             url: url,
             success: function(data){
                 $("#rightContainer").html(data);
         },
         error : function() {
             alert("error");
             }
         });
    });

    // 选中菜单第一项
     $(".blog-menu .list-group-item:first").trigger("click");
});
```

菜单初始化的时候，会默认选中第一个菜单。

12.3.3 用户管理

1. 修改users/list.html

修改 users/list.html。

```
<div class="card-header bg-dark font-white">

    <div class="input-group col-md-7 col-xl-6">

        <input type="text" class="form-control" id="searchName"
```

```
            placeholder="输入用户名称进行搜索"> <span class="input-group-
btn">
            <button class="btn btn-secondary" type="button" id="
searchNameBtn">
                <i class="fa fa-search" aria-hidden="true"></i>
            </button>
        </span> <a class="btn btn-primary" data-toggle="modal"
            data-target="#flipFlop" role="button" id="addUser"><i
            class="fa fa-plus" aria-hidden="true"></i></a>

    </div>
</div>

<div id="mainContainer" class="container">
    <div id="mainContainerRepleace" class="row">
        <table class="table table-striped">
            <thead>
                <tr>
                    <th data-field="id">ID</th>
                    <th data-field="username">账号</th>
                    <th data-field="name">姓名</th>
                    <th data-field="email">邮箱</th>
                    <th data-field="operation">操作</th>

                </tr>
            </thead>
            <tbody>

                <tr th:each="user : ${userModel.userList}">
                    <td th:text="${user.id}">1</td>
                    <td th:text="${user.username}">1</td>
                    <td th:text="${user.name}">waylau</td>
                    <td th:text="${user.email}">waylau</td>
                    <td>
                        <div>
                            <a class="blog-edit-user" data-toggle="mod
al"
                                data-target="#flipFlop" role="button"
                                data-th-attr="userId=${user.id}"> <i
                                class="fa fa-pencil-square-o" aria-hid
den="true"></i>
                            </a> <a class="blog-delete-user" role="but
ton"
                                data-th-attr="userId=${user.id}"> <i
class="fa fa-times"
                                aria-hidden="true"></i>
                            </a>
                        </div>
                    </td>
```

```
                </tr>

            </tbody>

        </table>
        <div th:replace="~{fragments/page :: page}">...</div>

    </div>
</div>

<!-- The modal -->
<div class="modal fade" id="flipFlop" tabindex="-1" role="dialog"
    aria-labelledby="modalLabel" aria-hidden="true">
    <div class="modal-dialog" role="document">
        <div class="modal-content">
            <div class="modal-header">
                <h4 class="modal-title" id="modalLabel">新增/编辑</h4>
                <button type="button" class="close" data-dismiss="mod
al"
                    aria-label="Close">
                    <span aria-hidden="true">&times;</span>
                </button>

            </div>
            <div class="modal-body" id="userFormContainer"></div>
            <div class="modal-footer">
                <button class="btn btn-primary" data-dismiss="modal"
id="submitEdit">提交</button>
                <button type="button" class="btn btn-secondary" data-
dismiss="modal">Close</button>
            </div>
        </div>
    </div>
</div>

<!-- JavaScript -->
<script src="../../js/users/main.js" th:src="@{/js/users/main.js}"></
script>
```

2. 修改users/main.js

修改 users/main.js。

```
// DOM加载完再执行
$(function() {

    var _pageSize; // 存储用于搜索

    // 根据用户名、页面索引、页面大小获取用户列表
    function getUersByName(pageIndex, pageSize) {
```

```
        $.ajax({
            url: "/users",
            contentType : 'application/json',
            data:{
                "async":true,
                "pageIndex":pageIndex,
                "pageSize":pageSize,
                "name":$("#searchName").val()
            },
            success: function(data){
                $("#mainContainer").html(data);
            },
            error : function() {
                toastr.error("error!");
            }
        });
    }

    // 分页
    $.tbpage("#mainContainer", function (pageIndex, pageSize) {
        getUersByName(pageIndex, pageSize);
        _pageSize = pageSize;
    });

    // 搜索
    $("#searchNameBtn").click(function() {
        getUersByName(0, _pageSize);
    });

    // 获取添加用户的界面
    $("#addUser").click(function() {
        $.ajax({
            url: "/users/add",
            success: function(data){
                $("#userFormContainer").html(data);
            },
            error : function(data) {
                toastr.error("error!");
            }
        });
    });

    // 获取编辑用户的界面
    $("#rightContainer").on("click",".blog-edit-user", function () {
        $.ajax({
            url: "/users/edit/" + $(this).attr("userId"),
            success: function(data){
                $("#userFormContainer").html(data);
            },
```

```
            error : function() {
                toastr.error("error!");
            }
        });
    });

    // 提交变更后，清空表单
    $("#submitEdit").click(function() {
        $.ajax({
            url: "/users",
            type: 'POST',
            data:$('#userForm').serialize(),
            success: function(data){
                $('#userForm')[0].reset();

                if (data.success) {
                    // 重新刷新主界面
                    getUersByName(0, _pageSize);
                } else {
                    toastr.error(data.message);
                }

            },
            error : function() {
                toastr.error("error!");
            }
        });
    });

    // 删除用户
    $("#rightContainer").on("click",".blog-delete-user", function () {

        $.ajax({
            url: "/users/" + $(this).attr("userId") ,
            type: 'DELETE',
            success: function(data){
                if (data.success) {
                    // 重新刷新主界面
                    getUersByName(0, _pageSize);
                } else {
                    toastr.error(data.message);
                }
            },
            error : function() {
                toastr.error("error!");
            }
        });
    });
});
```

3. 删除form.html、view.html

在前面提到了，使用 form.html 和 view.html 是为了简化 Thymeleaf 的使用，而没有遵循 RESTful API 的规划。现在需要采用 REST 风格的架构，所以这些页面已经没有使用价值，可以将其删除。取而代之的是 add.html 和 edit.html。

4. 增加add.html

增加 add.html，用于添加用户的信息。

```
<div class="container">

    <form th:action="@{/users}" method="post" th:object="${userModel.
user}"
        id="userForm">
        <input type="hidden" name="id" th:value="*{id}">

        <div class="form-group ">
            <label for="username" class="col-form-label">账号</label>
<input
                type="text" class="form-control" id="username" name="us-
ername"
                th:value="*{username}" maxlength="50"
                placeholder="请输入账号，至少3个字符，至多20个">

        </div>
        <div class="form-group">
            <label for="email" class="col-form-label">邮箱</label> <in
put
                type="email" class="form-control" id="email" name=
"email"
                th:value="*{email}" maxlength="50" placeholder="请输入邮
箱">
        </div>
        <div class="form-group">
            <label for="name" class="col-form-label">姓名</label> <input
                type="text" class="form-control" id="name" name="name"
                th:value="*{name}" maxlength="20" placeholder="请输入姓名，
至少2个字符，至多20个">
        </div>
        <div class="form-group">
            <label for="password" class="col-form-label">密码</label>
<input
                type="password" class="form-control" id="password"
name="password"
                th:value="*{password}" maxlength="30"
                placeholder="请输入密码，字母或特殊符号和数字结合">
        </div>

    </form>
```

```
</div><!-- /.container -->
```

5. 增加edit.html

增加 edit.html，用于用户的信息编辑。

```
<div class="container">

    <form th:action="@{/users}" method="post" th:object="${userModel.
user}"
        id="userForm">
        <input type="hidden" name="id" th:value="*{id}">

        <div class="form-group ">
            <label for="username" class="col-form-label">账号</label>
<input
                type="text" class="form-control" id="username" name="us-
ername"
                th:value="*{username}" maxlength="50"
                placeholder="请输入账号，至少3个字符，至多20个">

        </div>
        <div class="form-group">
            <label for="email" class="col-form-label">邮箱</label> <in
put
                type="email" class="form-control" id="email" name=
"email"
                th:value="*{email}" maxlength="50" placeholder="请输入邮
箱">
        </div>
        <div class="form-group">
            <label for="name" class="col-form-label">姓名</label> <input
                type="text" class="form-control" id="name" name="name"
                th:value="*{name}" maxlength="20" placeholder="请输入姓名，
至少2个字符，至多20个">
        </div>
        <div class="form-group">
            <label for="password" class="col-form-label">密码</label>
<input
                type="password" class="form-control" id="password"
name="password"
                th:value="*{password}" maxlength="30"
                placeholder="请输入密码，字母或特殊符号和数字结合">
        </div>

    </form>

</div><!-- /.container -->
```

12.3.4 分页组件

分页组件是笔者写的一个插件，名字叫作 thymeleaf-bootstrap-paginator.js，适用于 Thymeleaf 与 Bootstrap 4 整合开发的项目。其实现方式也可以在项目的源码中找到。该插件需要配合 page.html 页面片段来结合使用。

其原理和实现方式在此就不再赘述，可以参阅笔者的博文《Spring Data + Thymeleaf 3 + Bootstrap 4 实现分页器》（https://waylau.com/spring-data-thymeleaf-bootstrap-paginator/）。

12.3.5 问题解决

在编写本项目的过程中，遇到了 Thymeleaf 取值 ${page.isLast} 报错。

```
Caused by: org.attoparser.ParseException: Exception evaluating SpringEL
expression: "page.isLast" (template: "fragments/page" - line 28, col53)
    at org.attoparser.MarkupParser.parseDocument(MarkupParser.java:393) ~
[attoparser-2.0.2.RELEASE.jar:2.0.2.RELEASE]
    at org.attoparser.MarkupParser.parse(MarkupParser.java:257) ~[atto-
parser-2.0.2.RELEASE.jar:2.0.2.RELEASE]
    at org.thymeleaf.templateparser.markup.AbstractMarkupTemplatePars
er.parse(AbstractMarkupTemplateParser.java:230) ~[thymeleaf-3.0.3.RE-
LEASE.jar:3.0.3.RELEASE]
    ... 87 common frames omitted
```

原因：返回的 org.springframework.data.domain.Page 对象中找不到 isLast 属性。

解决方法：这个是序列化转化的问题。布尔值 isLast 映射属性为 last，所以改为 ${page.last} 即可。

12.3.6 示例源码

本节示例源码在 blog-prototype 目录下。

第13章 角色管理实现

13.1 角色管理的需求回顾

本章将实现博客系统的角色管理。

在前面讲到了博客系统的用户主要分为两大类：普通用户（博主）和管理员。相应地，系统应该提供普通用户（博主）和管理员相应的角色 USER 和 ADMIN。

在 blog-user 项目基础上构建了一个新的项目 blog-role，用于演示整个角色管理实现的过程。

13.1.1 所需环境

本例采用的开发环境与“Spring Security 与 Spring Boot 集成”的环境一致，具体配置如下。

- Spring Security 5.0.0.M2。
- Thymeleaf Spring Security 3.0.2.RELEASE。

13.1.2 build.gradle

同样，在 build.gradle 中需要添加 Spring Security 的依赖。Spring Boot 已经提供了相关的 Starter 来实现 Spring Security 开箱即用的功能，所以只需要在 build.gradle 文件中添加 Spring Security 的 Starter 库即可。

Thymeleaf 社区还提供了 Thymeleaf 与 Spring Security 集成模块 thymeleaf-extras-springsecurity4，开发者需要手动添加这个依赖。

```
// 依赖关系
dependencies {
    //...

    // 添加Spring Security依赖
    compile('org.springframework.boot:spring-boot-starter-security')

    // 添加Thymeleaf Spring Security依赖
    compile('org.thymeleaf.extras:thymeleaf-extras-springsecurity4:
3.0.2.RELEASE')

    //...
}
```

13.1.3 角色管理的需求

首先来回顾一下角色管理所具有的功能需求。

角色管理是安全设置的重要方面。角色定义了用户可以执行的操作的集合。角色是代表一系列行为或责任的实体，用于限定在软件系统中能做什么、不能做什么。用户账号往往与角色相关联，因此，一个用户在软件系统中能“做”什么取决于与之关联的具有什么样的角色。

概括起来，在本章将实现的功能如下。

- 建立权限（角色）的实体。
- 建立用户与权限（角色）的关系。
- 创建用户时关联角色。
- 修改用户的角色。
- 初始化权限（角色）、用户的数据。

13.2 角色管理的后台实现

本节将进行用户管理的后台编码实现。

13.2.1 创建 Authority

建立 Authority 实体，该实体代表权限，也等同于角色。

```
@Entity
public class Authority implements GrantedAuthority {

    private static final long serialVersionUID = 1L;

    @Id // 主键
    @GeneratedValue(strategy = GenerationType.IDENTITY) // 自增长策略
    private Long id; // 用户的唯一标识

    @Column(nullable = false) // 映射为字段，值不能为空
    private String name;

    public Long getId() {
        return id;
    }

    public void setId(Long id) {
        this.id = id;
    }

    public void setName(String name) {
        this.name = name;
    }
```

```
    @Override
    public String getAuthority() {
        return name;
    }

}
```

这里需要实现 org.springframework.security.core.GrantedAuthority 接口，并重写 getAuthority() 方法。

13.2.2 修改 User

修改 User，将其实现 org.springframework.security.core.userdetails.UserDetails 接口，并建立了与 Authority 的关系。

```
import javax.persistence.CascadeType;
import javax.persistence.FetchType;
import javax.persistence.GeneratedValue;
import javax.persistence.GenerationType;
import javax.persistence.JoinColumn;
import javax.persistence.JoinTable;
import javax.persistence.ManyToMany;
import org.springframework.security.core.GrantedAuthority;
import org.springframework.security.core.authority.SimpleGrantedAuthor-
ity;
import org.springframework.security.core.userdetails.UserDetails;

...

@Entity // 实体
public class User implements UserDetails {

    private static final long serialVersionUID = 1L;

    ...

    @ManyToMany(cascade = CascadeType.DETACH, fetch = FetchType.EAGER)
    @JoinTable(name = "user_authority", joinColumns = @JoinColumn(name
= "user_id", referencedColumnName = "id"),
        inverseJoinColumns = @JoinColumn(name = "authority_id", refer-
encedColumnName = "id"))
    private List<Authority> authorities;

    ...

    @Override
    public Collection<? extends GrantedAuthority> getAuthorities() {
```

```
        //需将List<Authority>转成List<SimpleGrantedAuthority>，否则前端拿不
到角色列表名称
        List<SimpleGrantedAuthority> simpleAuthorities = new Array-
List<>();
        for(GrantedAuthority authority : this.authorities){
            simpleAuthorities.add(new SimpleGrantedAuthority(authority.
getAuthority()));
        }
        return simpleAuthorities;
    }

    public void setAuthorities(List<Authority> authorities) {
        this.authorities = authorities;
    }

    @Override
    public boolean isAccountNonExpired() {
        return true;
    }

    @Override
    public boolean isAccountNonLocked() {
        return true;
    }

    @Override
    public boolean isCredentialsNonExpired() {
        return true;
    }

    @Override
    public boolean isEnabled() {
        return true;
    }

    ...
}
```

authorities 就是用户的角色列表。

13.2.3 新增 AuthorityRepository

在 com.waylau.spring.boot.blog.repository 包下创建 AuthorityRepository。AuthorityRepository 是 Authority 的资源库接口，继承自 JpaRepository。

```
import org.springframework.data.jpa.repository.JpaRepository;
import com.waylau.spring.boot.blog.domain.Authority;
```

```
public interface AuthorityRepository extends JpaRepository<Authority,
Long>{

}
```

13.2.4 新增 AuthorityService 接口及实现

AuthorityService 是 Authority 的服务接口。在 com.waylau.spring.boot.blog.service 包下增加 AuthorityService。

```
public interface AuthorityService {
    /**
     * 根据ID查询Authority
     * @param id
     * @return
     */
    Optional<Authority> getAuthorityById(Long id);
}
```

AuthorityServiceImpl 是 AuthorityService 服务接口的实现。

```
@Service
public class AuthorityServiceImpl implements AuthorityService {

    @Autowired
    private AuthorityRepository authorityRepository;

    @Override
    public Optional<Authority> getAuthorityById(Long id) {
        return authorityRepository.findById(id);
    }

}
```

13.2.5 修改 UserController

主要修改了 saveOrUpdateUser 方法，增加了 权限 ID 作为参数。

```
@RestController
@RequestMapping("/users")
public class UserController {

    @Autowired
    private AuthorityService  authorityService;

    ...
```

```
    /**
     * 保存或修改用户
     * @param user
     * @param authorityId
     * @return
     */
    @PostMapping
    public ResponseEntity<Response> saveOrUpdateUser(User user, Long
authorityId) {

        List<Authority> authorities = new ArrayList<>();
        authorities.add(authorityService.getAuthorityById(authorityId));
        user.setAuthorities(authorities);

        try {
            userService.saveOrUpateUser(user);
        }  catch (ConstraintViolationException e)   {
            return ResponseEntity.ok().body(new Response(false, Constra
intViolationExceptionHandler.getMessage(e)));
        }

        return ResponseEntity.ok().body(new Response(true, "处理成功",
user));
    }

    ...

}
```

13.2.6 修改 MainController

对注册用户的接口进行调整，增加了用户的权限。

```
@Controller
public class MainController {
    private static final Long ROLE_USER_AUTHORITY_ID = 2L; // 用户权限（博
主）

    @Autowired
    private AuthorityService  authorityService;

    ...

    /**
     * 注册用户
     * @param user
     * @return
```

```
     */
    @PostMapping("/register")
    public String registerUser(User user) {
        List<Authority> authorities = new ArrayList<>();
        authorities.add(authorityService.getAuthorityById(ROLE_USER_AU-
THORITY_ID));
        user.setAuthorities(authorities);
        userService.registerUser(user);
        return "redirect:/login";
    }

}
```

这里设置了用户权限的 ID 为 2，这是因为在系统初始化的时候，会先初始化角色的相关信息（类似于码表的概念），而这些信息是一个系统能够正常运行的必要条件。

初始化系统数据的过程会在后面的章节中讲述。

13.2.7 安全配置类

增加 com.waylau.spring.boot.blog.config 包，用于放置项目的配置类。在该包下创建 Security-Config.java。

```
import org.springframework.security.config.annotation.web.builders.
HttpSecurity;
import org.springframework.security.config.annotation.web.configuration.
EnableWebSecurity;
import org.springframework.security.config.annotation.web.configuration.
WebSecurityConfigurerAdapter;

@EnableWebSecurity
public class SecurityConfig extends WebSecurityConfigurerAdapter {

    /**
     * 自定义配置
     */
    @Override
    protected void configure(HttpSecurity http) throws Exception {
        http.authorizeRequests()
            .antMatchers("/css/**", "/js/**", "/fonts/**", "/index").
permitAll(); // 都可以访问
    }
}
```

目前还不涉及具体的权限拦截。权限管理的内容会在第 14 章介绍。所以，这里设置允许任何角色访问任何资源。

13.3 角色管理的前台实现

在完成后台接口之后，本节将进行角色管理的前端编码。

13.3.1 用户管理

增加了用户角色（权限）的显示。

1. 修改users/list.html

修改 users/list.html 的内容，主要修改了列表，增加了显示角色这一列。

```
...
<div id="mainContainer" class="container">
    <div id="mainContainerRepleace" class="row">
        <table class="table table-striped">
            <thead>
                <tr>
                    <th data-field="id">ID</th>
                    <th data-field="username">账号</th>
                    <th data-field="name">姓名</th>
                    <th data-field="email">邮箱</th>
                    <th data-field="authorities">角色</th>
                    <th data-field="operation">操作</th>

                </tr>
            </thead>
            <tbody>

                <tr th:each="user : ${userModel.userList}">
                    <td th:text="${user.id}">1</td>
                    <td th:text="${user.username}">1</td>
                    <td th:text="${user.name}">waylau</td>
                    <td th:text="${user.email}">waylau</td>
                    <td th:text="${user.authorities}">waylau</td>
                    <td>
                        <div>
                            <a class="blog-edit-user" data-toggle="mod-
al"
                                data-target="#flipFlop" role="button"
                                data-th-attr="userId=${user.id}"> <i
                                class="fa fa-pencil-square-o" aria-hid-
den="true"></i>
                            </a> <a class="blog-delete-user" role="but-
ton"
                                data-th-attr="userId=${user.id}"> <i
class="fa fa-times"
                                aria-hidden="true"></i>
```

```
                                </a>
                            </div>
                        </td>
                    </tr>

                </tbody>

            </table>
            <div th:replace="~{fragments/page :: page}">...</div>

        </div>
    </div>
...
```

2. 修改add.html

在 add.html 中增加了一列表单的选项，用于选择角色。

```
...
<div class="form-group">
    <label for="authorities" class="col-form-label">角色</label>
    <select id="authorities" name="authorityId" class="form-control
form-control-chosen" data-placeholder="请选择">
        <option value="1">管理员</option>
        <option value="2">博主</option>
    </select>
</div>
...
```

3. 修改edit.html

在 edit.html 中也增加了一列表单的选项，用于选择角色。

```
...
<div class="form-group">
    <label for="authorities" class="col-form-label">角色</label>
    <select id="authorities" name="authorityId" class="form-control
form-control-chosen" data-placeholder="请选择">
        <option value="1" data-th-selected="${#strings.contains(userMod-
el.user.authorities[0],'ROLE_ADMIN')}">管理员</option>
        <option value="2" data-th-selected="${#strings.contains(userMod-
el.user.authorities[0],'ROLE_USER')}">博主</option>
    </select>
</div>
...
```

需要注意的是，修改时要选中用户已经有的角色，这里是通过 data-th-selected="${#strings.contains(userModel.user.authorities[0],'ROLE_ADMIN')}" 来判断实现的。

13.3.2 初始化数据

考虑到角色信息是基础数据，并且在项目运行中基本不会发生变更，所以，在项目启动时，就需要自动把相应的基础数据导入数据库。

Spring Boot 就提供了这样的功能，在 src/main/resources 下新建一个 import.sql 文件，把相应的 SQL 脚本写入其中，这样启动项目时，就会自动去执行这个 SQL 脚本，从而完成数据的导入。

相应的 SQL 脚本如下。

```
INSERT INTO user (id, username, password, name, email) VALUES (1, 'ad-
min', '123456', '老卫', 'i@waylau.com');
INSERT INTO user (id, username, password, name, email)  VALUES (2, 'way-
lau', '123456', 'Way Lau', 'waylau@waylau.com');

INSERT INTO authority (id, name) VALUES (1, 'ROLE_ADMIN');
INSERT INTO authority (id, name) VALUES (2, 'ROLE_USER');

INSERT INTO user_authority (user_id, authority_id) VALUES (1, 1);
INSERT INTO user_authority (user_id, authority_id) VALUES (2, 2);
```

除了初始化管理员和博主角色信息外，还初始化了相应的两个用户账号。

13.3.3 运行

启动 blog-role 项目后，观察控制台，可以看到初始化 import.sql 文件的过程。

```
2017-07-29 00:46:00.318  INFO 11012 --- [           main] o.h.t.schema.
internal.SchemaCreatorImpl   : HHH000476: Executing import script
'ScriptSourceInputFromUrl(file:/D:/workspaceGitosc/spring-boot-enter-
prise-application-development/samples/blog-role/bin/import.sql)'
Hibernate: INSERT INTO user (id, username, password, name, email) VAL-
UES (1, 'admin', '123456', '老卫', 'i@waylau.com')
Hibernate: INSERT INTO user (id, username, password, name, email)  VAL-
UES (2, 'waylau', '123456', 'Way Lau', 'waylau@waylau.com')
Hibernate: INSERT INTO authority (id, name) VALUES (1, 'ROLE_ADMIN')
Hibernate: INSERT INTO authority (id, name) VALUES (2, 'ROLE_USER')
Hibernate: INSERT INTO user_authority (user_id, authority_id) VALUES
(1, 1)
Hibernate: INSERT INTO user_authority (user_id, authority_id) VALUES
(2, 2)
```

通过浏览器访问 http://localhost:8080/admins，可以看到项目的运行效果。

图 13-1 所示为查看用户列表的界面。

图13-1　用户列表

图 13-2 所示为编辑用户的界面。

新增/编辑

账号

waylau

邮箱

waylau@waylau.com

姓名

Way Lau

密码

••••••

角色

博主

提交　Close

图13-2　编辑用户

需要注意的是，当试图单击“删除”按钮时，发现会报 403 的错误提示。这是由于 Spring Security 自动启用了 CSRF 防御，对 DELETE 方法进行了拦截。

那么，如何才能正常删除用户呢？ CSRF 防护处理又要如何设置呢？

这些将在第 14 章进行解答。

13.3.4 示例源码

本节示例源码在 blog-role 目录下。

第14章 权限管理实现

14.1 权限管理的需求回顾

本节将实现博客系统的权限管理。权限管理和角色管理是系统安全设置中重要的两个方面。对于权限管理来说，除了考虑正常用户是否对某些资源具有访问权限外，还需要考虑非法用户的请求、应对网络攻击、密码加密等非功能性需求。

在 blog-role 项目基础上构建了一个新的项目 blog-auth，用于演示整个权限管理实现的过程。

14.1.1 权限管理的需求

首先来回顾一下权限管理所具有的功能需求。

概括起来，在本节将实现如下功能。

（1）角色与资源的关系。建立角色与资源的关系。一个角色对应多个资源，或者说一个角色拥有操作某些资源的权限。通过匹配 URL 路径规则，可以实现哪些 URL 能够被哪些角色所访问。

（2）CSRF 问题的解决。回顾第 13 章中提到，用户在发出 DELETE 请求试图删除用户时，系统提示了 403 的错误。这个错误是 Spring Security 默认启用了 CSRF 防护引起的。也就是说，当时的这个 DELETE 请求被认为具有 CSRF 攻击风险，所以被系统挡住了。本节介绍如何来处理 CSRF 的问题。

（3）启用方法级别的安全设置。方法级别的安全设置相比于 URL 路径规则而言，具有更加灵活的资源授权。

（4）使用 BCrypt 加密密码。一些敏感数据，如密码等，在数据库中如果是以明文进行存储的，那么就会具有极大的安全风险。本节将介绍如何使用 BCrypt 加密方式来对博客系统的密码进行加密。

（5）用户登录。实现用户的登录功能。

（6）记住我。用户在首次登录系统时，应提供“记住我”功能。这样，在一段时间内，用户无须再次输入密码就能自动登录系统。

14.1.2 API 设计

权限管理的整体 API 设计如下。

- login：登录
  - /login：GET 获取登录的界面。
  - /login：POST 登录。
    - username：用户账号。
    - password：用户密码。
    - remember-me：是否记住我。

14.2 权限管理的后台实现

本节将进行权限管理的后台编码实现。

14.2.1 修改 SecurityConfig

修改 SecurityConfig，已设置如下规则。

（1）配置角色与资源的关系，规定了哪些角色可以访问哪些资源。

（2）启用了 CSRF 防护，并设置部分资源无须 CSRF 防护。

（3）启用了"记住我"功能。

```
private static final String KEY = "waylau.com";

/**
 * 自定义配置
 */
@Override
protected void configure(HttpSecurity http) throws Exception {
    http.authorizeRequests().antMatchers("/css/**", "/js/**", "/
fonts/**", "/index").permitAll() //都可以访问
            .antMatchers("/h2-console/**").permitAll() //都可以访问
            .antMatchers("/admins/**").hasRole("ADMIN")
                                        //需要相应的角色才能访问
            .and()
            .formLogin()    //基于Form 表单登录验证
            .loginPage("/login").failureUrl("/login-error")//自定义登录界面
            .and().rememberMe().key(KEY) //启用remember me
            .and().exceptionHandling().accessDeniedPage("/403");
//处理异常，拒绝访问就重定向到403页面
    http.csrf().ignoringAntMatchers("/h2-console/**");
//禁用H2控制台的CSRF防护
    http.headers().frameOptions().sameOrigin();
//允许来自同一来源的H2控制台的请求
}
...
```

上述代码的含义，代码中的注释已经非常详细了，在此不再赘述。读者也可以回顾"集成 Spring Security"章节的内容。需要注意以下几个问题。

（1）由于管理员需要访问 H2 数据库的控制台，因此 /h2-console 路径下的资源被设置为 permitAll（允许所有人访问）。同时，为了不对 H2 资源进行 CSRF 防护，采用 ttp.csrf().ignoringAntMatchers("/h2-console/**") 的方式来禁用。

（2）常量 KEY，是设置用于识别"记住我"身份验证而创建的令牌的键。

（3）注入 org.springframework.security.core.userdetails.UserDetailsService 接口的实现，意味着

认证信息是从数据库中来获取的。

（4）使用 BCrypt 加密方式来加密密码。

接着添加如下代码。

```
...
@Autowired
private UserDetailsService userDetailsService;

@Autowired
private PasswordEncoder passwordEncoder;

@Bean
public PasswordEncoder passwordEncoder() {
    return new BCryptPasswordEncoder();    //使用BCrypt加密
}

@Bean
public AuthenticationProvider authenticationProvider() {
    DaoAuthenticationProvider authenticationProvider = new DaoAuthenti-
cationProvider();
    authenticationProvider.setUserDetailsService(userDetailsService);
    authenticationProvider.setPasswordEncoder(passwordEncoder);
// 设置密码加密方式
    return authenticationProvider;
}

...

/**
 * 认证信息管理
 * @param auth
 * @throws Exception
 */
@Autowired
public void configureGlobal(AuthenticationManagerBuilder auth) throws
Exception {
    auth.userDetailsService(userDetailsService);
    auth.authenticationProvider(authenticationProvider());
}
```

为了启用方法级别的安全设置，需要在配置类上加上 @EnableGlobalMethodSecurity(prePostEn-abled = true) 注解。

```
import org.springframework.security.authentication.AuthenticationPro-
vider;
import org.springframework.security.authentication.dao.DaoAuthentication
Provider;
import org.springframework.security.config.annotation.authentication.
```

```
builders.AuthenticationManagerBuilder;
import org.springframework.security.config.annotation.web.builders.
HttpSecurity;
import org.springframework.security.config.annotation.web.configuration.
EnableWebSecurity;
import org.springframework.security.config.annotation.web.configuration.
WebSecurityConfigurerAdapter;
import org.springframework.security.core.userdetails.UserDetailsService;

import org.springframework.security.crypto.bcrypt.BCryptPasswordEncod-
er;
import org.springframework.security.crypto.password.PasswordEncoder;
...

@EnableGlobalMethodSecurity(prePostEnabled = true) // 启用方法级别安全设置
public class SecurityConfig extends WebSecurityConfigurerAdapter {
    ...
}
```

14.2.2 UserDetailsService 接口的实现

修改 UserServiceImpl 并实现 UserDetailsService 接口。该接口只需要重新使用 loadUserByUsername 方法即可。

```
import org.springframework.security.core.userdetails.UserDetails;
import org.springframework.security.core.userdetails.UserDetailsService;

import org.springframework.security.core.userdetails.UsernameNotFound-
Exception;

...

@Service
public class UserServiceImpl implements UserService, UserDetailsService
{

    ...

    @Override
    public UserDetails loadUserByUsername(String username) throws User
nameNotFoundException {
        return userRepository.findByUsername(username);
    }
}
```

14.3 CSRF 防护处理

Thymeleaf 在 POST 请求中会自动添加 CSRF 的 token 码，但其他操作（如 PUT、DELETE 等）则需要自行处理。为此修改 header.html，添加 _csrf 和 _csrf_header 两个 meta 标签。

```
<head>
    ...
    <!-- CSRF -->
    <meta name="_csrf" th:content="${_csrf.token}"/>
    <!-- default header name is X-CSRF-TOKEN -->
    <meta name="_csrf_header" th:content="${_csrf.headerName}"/>
    ...
</head>
```

在发送 PUT、DELETE 等 ajax 请求前，对请求头进行处理，添加上述 _csrf 和 _csrf_header 两个 meta 标签的值。

下面是用户管理（users/main.js）删除用户时，进行 DELETE 操作时的处理。

```
// 删除用户
$("#rightContainer").on("click",".blog-delete-user", function () {
    // 获取CSRF Token
    var csrfToken = $("meta[name='_csrf']").attr("content");
    var csrfHeader = $("meta[name='_csrf_header']").attr("content");


    $.ajax({
         url: "/users/" + $(this).attr("userId") ,
         type: 'DELETE',
         beforeSend: function(request) {
             request.setRequestHeader(csrfHeader, csrfToken); // 添加
CSRF Token
         },
         success: function(data){
             if (data.success) {
                 // 重新刷新主界面
                 getUersByName(0, _pageSize);
             } else {
                 toastr.error(data.message);
             }
         },
         error : function() {
             toastr.error("error!");
         }
     });
});
```

在发送请求前，通过 request.setRequestHeader(csrfHeader, csrfToken); 方式，在请求头中添加了 _csrf 和 _csrf_header 两个 meta 标签的值。这样请求就能被 Spring Security 的 CSRF 防护所处理。

14.4 权限管理的前台实现

在完成后台接口和配置之后，本节将进行权限管理的前端编码。

14.4.1 用户登录

修改 login.html，在原型的基础上添加 action。

```
...
<form th:action="@{~/login}" method="post">
    <h2>请登录</h2>

    <div class="form-group col-md-5">
        <label for="username" class="col-form-label">账号</label> <input
            type="text" class="form-control" id="username" name="user
name"
            maxlength="50" placeholder="请输入账号">

    </div>
    <div class="form-group col-md-5">
        <label for="password" class="col-form-label">密码</label> <input
            type="password" class="form-control" id="password" name=
"password"
            maxlength="30" placeholder="请输入密码">
    </div>
    <div class="form-group col-md-5">
        <input type="checkbox" value="remember-me"> 记住我
    </div>
    <div class="form-group col-md-5">
        <button type="submit" class="btn btn-primary">登录</button>
    </div>

</form>
...
```

为了登录后能显示用户相关信息，在 header.html 页面中，将：

```
...
<a href="/login" class="btn btn-outline-success my-2 my-sm-0" type=
"submit">登录</a>
<a href="/register" class="btn btn-outline-success my-2 my-sm-0" type=
"submit">注册</a>
...
```

修改为：

```
...
<div sec:authorize="isAuthenticated()" class="row">
    <div class="dropdown">
```

```
        <a class=" dropdown-toggle" href="/u/waylau"
            th:href="@{'/u/' + ${#authentication.principal.username}}"
            data-toggle="dropdown"><span sec:authentication="name"></
span></a>
        <div class="dropdown-menu">
            <a class=" dropdown-item" href="/u/waylau"
                th:href="@{'/u/' + ${#authentication.principal.user
name}}">个人主页</a> <a
                class="dropdown-item" href="/u/waylau/profile"
                th:href="@{'/u/' + ${#authentication.principal.user
name}} + '/profile'">个人设置</a>
        </div>
    </div>
    <div>
        <a href="/u/waylau/blogs/edit"
            th:href="'/u/' + ${#authentication.principal.username} + '/
blogs/edit'"
            class="btn btn-outline-success my-2 my-sm-0">写博客</a>
    </div>

    <form action="/logout" th:action="@{/logout}" method="post">
        <input class="btn btn-outline-success " type="submit" value="退
出">
    </form>
</div>

<div sec:authorize="isAnonymous()">
    <a href="/login" class="btn btn-outline-success my-2 my-sm-0"
        type="submit">登录</a> <a href="/register"
        class="btn btn-outline-success my-2 my-sm-0" type="submit">注册
</a>
</div>
...
```

14.4.2 初始化数据

由于已经启用了密码加密，因此初始化的用户信息中的密码，应该提前加密好，再写入数据库。在 src/main/resources 下修改 import.sql 文件，相应的用户信息修改为：

```
...
INSERT INTO user (id, username, password, name, email) VALUES (1, 'ad-
min', '$2a$10$N.zmdr9k7uOCQb376NoUnuTJ8iAt6Z5EHsM8lE9lBOsl7iKTVKIUi',
'老卫', 'i@waylau.com');
INSERT INTO user (id, username, password, name, email)  VALUES (2, 'way-
lau', '$2a$10$N.zmdr9k7uOCQb376NoUnuTJ8iAt6Z5EHsM8lE9lBOsl7iKTVKIUi',
'Way Lau', 'waylau@waylau.com');
...
```

原来的密码“123456”最终被加密成了“$2aN.zmdr9k7uOCQb376NoUnuTJ8iAt6Z5EHsM8lE9l-

BOsl7iKTVKIUi”。

14.4.3 运行

启动项目后，当试图访问受保护的资源时，如 http://localhost:8080/admins，如果当前用户没有权限，会被重定向到登录界面。

图 14-1 所示为登录界面的效果。

图14-1　登录界面

当用具有“ADMIN”权限的管理员账号 admin（密码 123456）登录成功后，可以访问保护的资源。如果登录时，选择了“记住我”选项，下次登录时，就能记住用户的信息，而无须再次进行登录。图 14-2 所示为当登录成功后，进入后台管理界面的效果。

图14-2　后台管理界面

单击“退出”按钮，用户就能退出账号。

14.4.4 示例源码

本节示例源码在 blog-auth 目录下。

第15章 文件服务器实现

15.1 文件服务器的需求分析

博客系统往往有文件的存储需求，如用户的头像、博客中的插图等。这些文件往往都有一个共同点，那就是所占的存储空间很小。这就意味着网络负载比较小，对存储的性能要求也不是很高。在这种情况下，非常适合使用支持小型文件存储的文件服务器。

本节将介绍如何基于 MongoDB 技术来存储二进制文件，从而实现一个文件服务器 MongoDB File Server，以提供给博客系统使用。

15.1.1 文件服务器的需求

本文件服务器致力于小型文件的存储，如博客中的图片、普通文档等。由于 MongoDB 支持多种数据格式的存储，当然也包括二进制的存储，因此可以很方便地用于存储文件。由于 MongoDB 的 BSON 文档对于数据量大小的限制（每个文档不超过 16MB），因此本文件服务器主要针对的是小型文件的存储。对于大型文件的存储（如超过 16MB），MongoDB 官方已经提供了成熟的产品 GridFS，读者可以自行了解一下。

文件服务器应能够提供与平台无关的 RESTful API，以便博客系统调用。

15.1.2 API

文件服务器整体的 API 设计如下。

（1）GET /files/{pageIndex}/{pageSize}：分页查询已经上传了的文件。

（2）GET /files/{id}：下载某个文件。

（3）GET /view/{id}：在线预览某个文件，如显示图片。

（4）POST /upload：上传文件。

（5）DELETE /{id}：删除文件。

15.2 MongoDB 简介

MongoDB 是一个介于关系数据库和非关系数据库之间的产品，是非关系数据库中功能较丰富、较像关系数据库的，旨在为 Web 应用提供可扩展的高性能数据存储解决方案。它支持的数据结构非常松散，是类似 JSON 的 BSON 格式，因此可以存储比较复杂的数据类型。MongoDB 最大的特点是它支持的查询语言非常强大，其语法有点类似于面向对象的查询语言，几乎可以实现类似关系数据库单表查询的绝大部分功能，而且还支持对数据建立索引。

本节不会对 MongoDB 的概念、基本用法做过多的介绍，有兴趣的读者可自行查阅其他文献。例如，笔者所著的《分布式系统常用技术及案例分析》一书，对 MongoDB 方面也有所涉及。

15.2.1 MongoDB 特点

MongoDB Server 是用 C++ 编写的、开源的、面向文档的数据库（Document Database），它的特点是高性能、高可用性，以及可以实现自动化扩展，存储数据非常方便。其主要功能特性如下。

MongoDB 将数据存储为一个文档，数据结构由 field-value（字段 - 值）对组成，如图 15-1 所示。MongoDB 文档类似于 JSON 对象。字段的值可以包含其他文档、数组及文档数组。

```
{
   name: "sue",                          ←—— field: value
   age: 26,                              ←—— field: value
   status: "A",                          ←—— field: value
   groups: [ "news", "sports" ]          ←—— field: value
}
```

图15-1　MongoDB 文档结构

使用文档的优点如下。

①文档（即对象），在许多编程语言里，可以对应于原生数据类型。

②嵌入式文档和数组可以减少昂贵的连接操作。

③动态模式支持流畅的多态性。

MongoDB 的特点是高性能、易部署、易使用，存储数据非常方便。主要功能特性有以下几点。

（1）高性能。MongoDB 中提供高性能的数据持久化，尤其表现在以下两点。

①对于嵌入式数据模型支持，减少了数据库系统的 I/O 活动。

②支持索引，用于快速查询。其索引对象可以是嵌入文档或数组的 Key。

（2）丰富的查询语言。MongoDB 支持丰富的查询语言，除读取和写入操作（CRUD）外，还包括数据聚合、文本搜索和地理空间查询。

（3）高可用。MongoDB 的复制设备，称为 replica set，提供了自动故障转移和数据冗余。

replica set 是一组保存相同数据集合的 MongoDB 服务器，提供了数据冗余并提高了数据的可用性。

（4）横向扩展 MongoDB 提供了水平横向扩展作为其核心功能部分，表现在以下两个方面。

①将数据分片到一组计算机集群上。

② tag aware sharding （标签意识分片）允许将数据传送到特定的碎片，如在分片时考虑碎片的地理分布。

15.2.2 MongoDB 核心概念

MongoDB 的核心概念表现在以下几方面。

1. 数据库和集合

MongoDB 存储 BSON 文档（即数据记录）在集合（collection）中，而集合是在数据库（database）中，如图 15-2 所示。

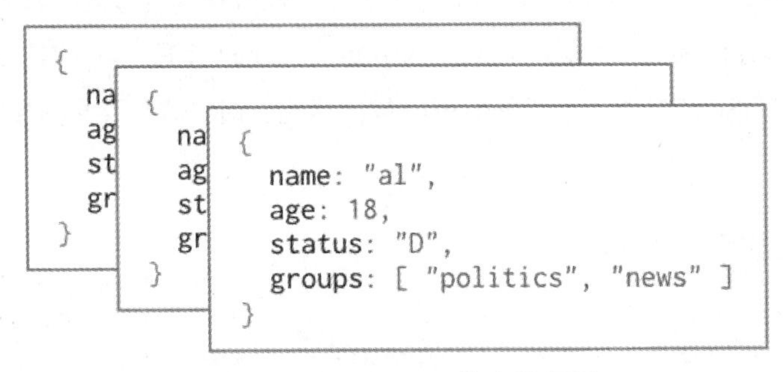

图15-2　MongoDB 集合示意图

在 MongoDB 中，数据库保存文档的集合。

选择要使用的数据库，使用 mongo shell 的 use <db> 语句，示例如下。

```
use myDB
```

2. 限制集合

限制集合（Capped Collection）是固定大小的集合，用于支持基于文档插入顺序的高吞吐率的插入和检索操作。Capped Collection 工作原理在某种程度上类似于 circular buffer（循环缓冲区）：一旦一个文档填满分配给它的空间，它将通过在 Capped Collection 中重写老文档来给新文档让出空间。

查阅 createCollection()（https://docs.mongodb.com/manual/reference/method/db.createCollection/#db.createCollection）或 create（https://docs.mongodb.com/manual/reference/command/create/#dbcmd.create），了解关于创建 Capped Collection 的更多信息。

（1）插入顺序。Capped Collection 能够保留插入顺序。因此，查询是按照文档的插入顺序，而不是使用索引确定插入位置的，这样可以提高增添数据的效率，所以 Capped Collection 可以支持更高的插入吞吐量。

（2）最旧文档的自动删除。为了给新文档腾出空间，在不需要脚本或显式删除操作的前提下，Capped Collection 会自动删除集合中最旧的文档。

（3）_id 索引。Capped Collection 有一个 _id 字段并且默认在 _id 字段上创建索引。

（4）更新。如果要更新 Capped Collection 中的文档，创建一个索引，就可以保证这些更新操作不需要进行集合扫描。

（5）文档大小。在 MongoDB 3.2 版之后，如果一个更新或替换操作改变了文档大小，操作将会失败。

（6）文档删除。不能从一个 Capped Collection 中删除文档，为了从一个集合中删除所有文档，使用 drop() 方法来删除集合，然后重新创建 Capped Collection。

（7）分片（Sharding）。不能对 Capped Collection 进行分片。

（8）查询效率。用自然顺序检索集合中大部分最近插入的元素。这类似于查询日志文件的尾部内容。

（9）聚合 $out。聚合管道操作器 $out 不能将结果写入 Capped Collection。

（10）创建 Capped Collection。必须使用 db.createCollection() 方法显式创建 Capped Collection，在 mongo shell 的 create 命令中可以查看帮助信息。当创建 Capped Collection 时，必须指定以字节为单位的最大集合大小，而 MongoDB 将会预先分配集合。Capped Collection 的大小包括内部消耗的一小部分空间。

```
db.createCollection( "log", { capped: true, size: 100000 } )
```

如果 size 字段小于或等于 4096，该集合将会有 4096 字节。否则，MongoDB 将会在给定大小的基础上增加为 256 的整数倍。

另外，也可以为集合指定最大文档数量，使用 max 字段，用法如下。

```
db.createCollection("log", { capped : true, size : 5242880, max : 5000 } )
```

size 参数始终是必需的，即使你指定文件的 max 数量。如果集合达到最大数量的限制，在达到最大文档计数之前，MongoDB 将删除旧文档。

（11）查询 Capped Collection。如果在 Capped Collection 上执行一个没有指定排序的 find() 方法，MongoDB 将保证结果的顺序和插入顺序相同。

若想实现与插入相反的顺序来检索文档，则需要使用 find() 与 sort() 方法，以及将 $natural 参数设置为 -1，就像下面的例子。

```
db.cappedCollection.find().sort( { $natural: -1 } )
```

3. Document（文档）

MongoDB 是将数据的记录作为 BSON 文档进行存储的。BSON 是 JSON 文档的二进制表示，但拥有比 JSON 更多的数据类型。如果想了解 BSON 规范的相关内容，可以参阅 http://bsonspec.org/ 或 https://docs.mongodb.com/manual/reference/bson-types/。

4. 文档的结构

MongoDB 文档由 field value（字段 / 值）对组成，如下所示。

```
{
   field1: value1,
   field2: value2,
```

```
    field3: value3,
    ...
    fieldN: valueN
}
```

字段的值可以是任意 BSON 数据类型，包括其他文档、数组和文档数组。 例如，下面的文档包含不同类型的值。

```
var mydoc = {
    _id: ObjectId("5099803df3f4948bd2f98391"),
    name: { first: "Alan", last: "Turing" },
    birth: new Date('Jun 23, 1912'),
    death: new Date('Jun 07, 1954'),
    contribs: [ "Turing machine", "Turing test", "Turingery" ],
    views : NumberLong(1250000)
    }
```

上面字段，分别包括了以下数据类型。

- _id 是一个 ObjectId。
- name 是一个嵌入式的文档，包含了字段的 first 和 last。
- birth 和 death 保存的是 Date 类型的值。
- contribs 保存的是 string 的 array。
- views 保存的是 NumberLong 类型的值。

（1）字段名称。字段名称是字符串。文档中对字段名称有如下限制。

- 字段名称 _id 被保留用于作为主键；其值必须是集合中唯一的、不可变的，并且可以是除 array 以外的任何类型。
- 该字段名称不能以美元符号 $ 字符开头。
- 字段名称不能包含点（.）字符。
- 字段名称不能包含空（null）字符。

BSON 文档可能有多个字段可以具有相同名称。大多数的 MongoDB 接口用来代表一个 MongoDB 结构（如 hash table），不支持重复的字段名称。如果需要操纵包含具有相同名称的多个字段的文档，请参阅 MongoDB 驱动程序的相关内容，参见 https://docs.mongodb.com/manual/applications/drivers/。

内部 MongoDB 的进程创建一些文件可能有重复的字段，但是 MongoDB 进程不会不断增加重复字段到现有用户的文档。

（2）字段值限制。索引的集合，其值受到字段值 Maximum Index Key Length 的限制。详情可以参阅 https://docs.mongodb.com/manual/reference/limits/#Index-Key-Limit。

15.3 MongoDB 与 Spring Boot 集成

在 15.2 节中了解了很多 MongoDB 的概念和用法。虽然，看上去 MongoDB 有点高深，但如果结合 Spring Boot 来使用，就会发现使用 MongoDB 也非难事。

本节将讲解 MongoDB 与 Spring Boot 集成。

首先创建一个新项目 mongodb-file-server。

15.3.1 所需环境

本例采用的开发环境如下。

- MongoDB 3.4.6。
- Spring Boot 2.0.0.M2。
- Spring Data Mongodb 2.0.0.M4。
- Thymeleaf 3.0.6.RELEASE。
- Thymeleaf Layout Dialect 2.2.2。
- Embedded MongoDB 2.0.0。

其中，Spring Boot 用于快速构建一个可独立运行的 Java 项目；Thymeleaf 作为前端页面模板，方便展示数据；Embedded MongoDB 则是一款由 Organization Flapdoodle OSS 出品的内嵌 MongoDB，可以在不启动 MongoDB 服务器的前提下，方便进行相关的 MongoDB 接口测试。

15.3.2 build.gradle

本节所演示的项目是采用 Gradle 进行组织及构建的，如果对 Gradle 不熟悉，也可以自行将项目转为 Maven 项目。

build.gradle 文件完整配置内容如下。

```
buildscript { // buildscript代码块中脚本优先执行

    // ext用于定义动态属性
    ext {
        springBootVersion = '2.0.0.M2'
    }

    // 使用了Maven的中央仓库（也可以指定其他仓库）
    repositories {
        //mavenCentral()
        maven { url "https://repo.spring.io/snapshot" }
        maven { url "https://repo.spring.io/milestone" }
        maven { url "http://maven.aliyun.com/nexus/content/groups/
```

```
public/" }
    }

    //依赖关系
    dependencies {

        // classpath声明说明了在执行其余的脚本时，ClassLoader可以使用这些依赖项
        classpath("org.springframework.boot:spring-boot-gradle-plug
in:${springBootVersion}")
    }
}

//使用插件
apply plugin: 'java'
apply plugin: 'eclipse'
apply plugin: 'org.springframework.boot'
apply plugin: 'io.spring.dependency-management'

// 指定了生成的编译文件的版本，默认为jar包
version = '1.0.0'

// 指定编译.java文件的JDK版本
sourceCompatibility = 1.8

// 使用了Maven的中央仓库（也可以指定其他仓库）
repositories {
    //mavenCentral()
    maven { url "https://repo.spring.io/snapshot" }
    maven { url "https://repo.spring.io/milestone" }
    maven { url "http://maven.aliyun.com/nexus/content/groups/public/"
}
}

//依赖关系
dependencies {

    //该依赖用于编译阶段
    compile('org.springframework.boot:spring-boot-starter-web')

    //添加Thymeleaf的依赖
    compile('org.springframework.boot:spring-boot-starter-thymeleaf')

    //添加Spring Data Mongodb的依赖
    compile('org.springframework.boot:spring-boot-starter-data-mongodb')


    //添加Embedded MongoDB的依赖用于测试
    compile('de.flapdoodle.embed:de.flapdoodle.embed.mongo')
```

```
    //该依赖用于测试阶段
    testCompile('org.springframework.boot:spring-boot-starter-test')
}
```

该 build.gradle 文件中的各配置项的注释已经非常详尽了，这里就不再赘述其配置项的含义。

15.4 文件服务器的实现

本节在 mongodb-file-server 项目基础上将实现文件服务器的功能。

15.4.1 领域对象

首先要对文件服务器进行建模。相关的领域模型如下。

1. 文档类File

文档类是类似于 JPA 中的实体的概念。不同的是 JPA 是采用 @Entity 注解，而文档类是采用 @Document 注解。

在 com.waylau.spring.boot.fileserver.domain 包下创建了一个 File 类。

```
import org.bson.types.Binary;
import org.springframework.data.annotation.Id;
import org.springframework.data.mongodb.core.mapping.Document;

...

@Document
public class File {
    @Id  //主键
    private String id;
    private String name; //文件名称
    private String contentType; //文件类型
    private long size;
    private Date uploadDate;
    private String md5;
    private Binary content; //文件内容
    private String path; //文件路径

    // 省略getter/setter方法

    protected File() {
    }

    public File(String name, String contentType, long size,Binary con-
```

```
tent) {
        this.name = name;
        this.contentType = contentType;
        this.size = size;
        this.uploadDate = new Date();
        this.content = content;
    }

    @Override
    public boolean equals(Object object) {
        if (this == object) {
            return true;
        }
        if (object == null || getClass() != object.getClass()) {
            return false;
        }
        File fileInfo = (File) object;
        return java.util.Objects.equals(size, fileInfo.size)
                && java.util.Objects.equals(name, fileInfo.name)
                && java.util.Objects.equals(contentType, fileInfo.con-
tentType)
                && java.util.Objects.equals(uploadDate, fileInfo.uploadD-
ate)
                && java.util.Objects.equals(md5, fileInfo.md5)
                && java.util.Objects.equals(id, fileInfo.id);
    }

    @Override
    public int hashCode() {
        return java.util.Objects.hash(name, contentType, size, upload
Date, md5, id);
    }

    @Override
    public String toString() {
        return "File{"
                + "name='" + name + '\''
                + ", contentType='" + contentType + '\''
                + ", size=" + size
                + ", uploadDate=" + uploadDate
                + ", md5='" + md5 + '\''
                + ", id='" + id + '\''
                + '}';
    }
}
```

需要注意的有以下两点。

（1）文档类主要采用的是 Spring Data MongoDB 中的注解，用于标识这是个 NoSQL 中的文档概念。

（2）文件的内容是用 org.bson.types.Binary 类型来进行存储的。

2. 存储库FileRepository

存储库用于提供与数据库打交道的常用的数据访问接口。其中 FileRepository 接口继承自 org.springframework.data.mongodb.repository.MongoRepository 即可，无须自行实现该接口的功能，Spring Data MongoDB 会自动实现接口中的方法。

```
import org.springframework.data.mongodb.repository.MongoRepository;
import com.waylau.spring.boot.fileserver.domain.File;

public interface FileRepository extends MongoRepository<File, String> {
}
```

3. 服务接口及实现类

FileService 接口定义了对文件的 CURD 操作，其中查询文件接口采用的是分页处理，以有效提高查询性能。

```
public interface FileService {
    /**
     * 保存文件
     * @param File
     * @return
     */
    File saveFile(File file);

    /**
     * 删除文件
     * @param File
     * @return
     */
    void removeFile(String id);

    /**
     * 根据id获取文件
     * @param File
     * @return
     */
    File getFileById(String id);

    /**
     * 分页查询，按上传时间降序
     * @param pageIndex
     * @param pageSize
     * @return
     */
    List<File> listFilesByPage(int pageIndex, int pageSize);
```

```
}
```

FileServiceImpl 实现了 FileService 中所有的接口。

```
@Service
public class FileServiceImpl implements FileService {

    @Autowired
    public FileRepository fileRepository;

    @Override
    public File saveFile(File file) {
        return fileRepository.save(file);
    }

    @Override
    public void removeFile(String id) {
        fileRepository.deleteById(id);
    }

    @Override
    public Optional<File> getFileById(String id) {
        return fileRepository.findById(id);
    }

    @Override
    public List<File> listFilesByPage(int pageIndex, int pageSize) {
        Page<File> page = null;
        List<File> list = null;

        Sort sort = new Sort(Direction.DESC,"uploadDate");
        Pageable pageable = PageRequest.of(pageIndex, pageSize, sort);

        page = fileRepository.findAll(pageable);
        list = page.getContent();
        return list;
    }
}
```

15.4.2 控制层、API 资源层

FileController 控制器作为 API 的提供者，接收用户的请求及响应。API 的定义符合 RESTful 的风格。

```
@CrossOrigin(origins = "*", maxAge = 3600) // 允许所有域名访问
@Controller
public class FileController {
```

```
    @Autowired
    private FileService fileService;

    @Value("${server.address}")
    private String serverAddress;

    @Value("${server.port}")
    private String serverPort;

    @RequestMapping(value = "/")
    public String index(Model model) {
        // 展示最新20条数据
        model.addAttribute("files", fileService.listFilesByPage(0, 20));
        return "index";
    }

    /**
     * 分页查询文件
     *
     * @param pageIndex
     * @param pageSize
     * @return
     */
    @GetMapping("files/{pageIndex}/{pageSize}")
    @ResponseBody
    public List<File> listFilesByPage(@PathVariable int pageIndex, @
PathVariable int pageSize) {
        return fileService.listFilesByPage(pageIndex, pageSize);
    }

    /**
     * 获取文件片信息
     *
     * @param id
     * @return
     */
    @GetMapping("files/{id}")
    @ResponseBody
    public ResponseEntity<Object> serveFile(@PathVariable String id) {

        Optional<File> file = fileService.getFileById(id);

        if (file.isPresent()) {
            return ResponseEntity.ok()
                    .header(HttpHeaders.CONTENT_DISPOSITION, "attach-
ment; fileName=\"" + file.get().getName() + "\"")
                    .header(HttpHeaders.CONTENT_TYPE, "application/
octet-stream")
```

```
                    .header(HttpHeaders.CONTENT_LENGTH, file.get().
getSize() + "").header("Connection", "close")
                    .body(file.get().getContent().getData());
        } else {
            return ResponseEntity.status(HttpStatus.NOT_FOUND).body
("File was not fount");
        }

    }

    /**
     * 在线显示文件
     *
     * @param id
     * @return
     */
    @GetMapping("/view/{id}")
    @ResponseBody
    public ResponseEntity<Object> serveFileOnline(@PathVariable String
id) {

        Optional<File> file = fileService.getFileById(id);

        if (file.isPresent()) {
            return ResponseEntity.ok()
                    .header(HttpHeaders.CONTENT_DISPOSITION, "fileName=\
"" + file.get().getName() + "\"")
                    .header(HttpHeaders.CONTENT_TYPE, file.get().getCon-
tentType())
                    .header(HttpHeaders.CONTENT_LENGTH, file.get().
getSize() + "").header("Connection", "close")
                    .body(file.get().getContent().getData());
        } else {
            return ResponseEntity.status(HttpStatus.NOT_FOUND).body
("File was not fount");
        }

    }

    /**
     * 上传
     *
     * @param file
     * @param redirectAttributes
     * @return
     */
    @PostMapping("/")
    public String handleFileUpload(@RequestParam("file") MultipartFile
file, RedirectAttributes redirectAttributes) {
```

```
        try {
            File f = new File(file.getOriginalFilename(), file.getCon-
tentType(), file.getSize(),
                    new Binary(file.getBytes()));
            f.setMd5(MD5Util.getMD5(file.getInputStream()));
            fileService.saveFile(f);
        } catch (IOException | NoSuchAlgorithmException ex) {
            ex.printStackTrace();
            redirectAttributes.addFlashAttribute("message", "Your " +
file.getOriginalFilename() + " is wrong!");
            return "redirect:/";
        }

        redirectAttributes.addFlashAttribute("message",
                "You successfully uploaded " + file.getOriginalFilename()
+ "!");

        return "redirect:/";
    }

    /**
     * 上传接口
     *
     * @param file
     * @return
     */
    @PostMapping("/upload")
    @ResponseBody
    public ResponseEntity<String> handleFileUpload(@RequestParam("file")
MultipartFile file) {
        File returnFile = null;
        try {
            File f = new File(file.getOriginalFilename(), file.getCon-
tentType(), file.getSize(),
                    new Binary(file.getBytes()));
            f.setMd5(MD5Util.getMD5(file.getInputStream()));
            returnFile = fileService.saveFile(f);
            String path = "//" + serverAddress + ":" + serverPort + "/
view/" + returnFile.getId();
            return ResponseEntity.status(HttpStatus.OK).body(path);

        } catch (IOException | NoSuchAlgorithmException ex) {
            ex.printStackTrace();
            return ResponseEntity.status(HttpStatus.INTERNAL_SERVER_
ERROR).body(ex.getMessage());
        }

    }
```

```
    /**
     * 删除文件
     *
     * @param id
     * @return
     */
    @DeleteMapping("/{id}")
    @ResponseBody
    public ResponseEntity<String> deleteFile(@PathVariable String id) {

        try {
            fileService.removeFile(id);
            return ResponseEntity.status(HttpStatus.OK).body("DELETE 
Success!");
        } catch (Exception e) {
            return ResponseEntity.status(HttpStatus.INTERNAL_SERVER_
ERROR).body(e.getMessage());
        }
    }
}
```

其中 @CrossOrigin(origins ="*",maxAge = 3600) 注解标识了 API 可以被跨域请求。

15.4.3 运行

有多种方式可以运行 Gradle 的 Java 项目。使用 Spring Boot Gradle Plugin 插件运行是较为简便的一种方式，只需要执行以下代码。

```
$ gradlew bootRun
```

项目成功运行后，通过浏览器访问 http://localhost:8081 即可。如图 15-3 所示，首页提供了上传的演示界面，上传后就能看到上传文件的详细信息。

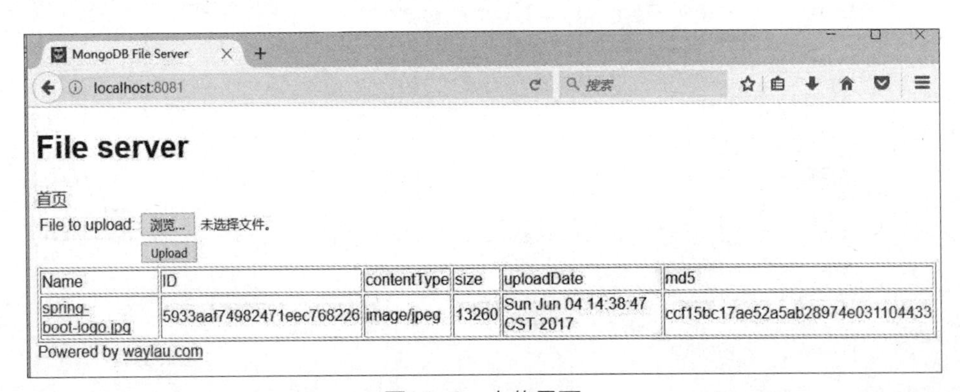

图15-3　上传界面

15.4.4 其他配置项

下面打开 application.properties 配置文件，可以看到如下配置。

```
server.address=localhost
server.port=8081

# Thymeleaf
spring.thymeleaf.encoding=UTF-8
spring.thymeleaf.cache=false
spring.thymeleaf.mode=HTML5

# limit upload file size
spring.http.multipart.max-file-size=1024KB
spring.http.multipart.max-request-size=1024KB

# independent MongoDB server
#spring.data.mongodb.uri=mongodb://localhost:27017/test
```

这些配置的含义如下。

（1）server.address 和 server.port 用来指定文件服务器启动的位置和端口号。

（2）spring.http.multipart.max-file-size 和 spring.http.multipart.max-request-size 是用来限制上传文件的大小，这里设置最大为 1MB。

（3）当 spring.data.mongodb.uri 没有被指定的时候，默认会采用内嵌 MongoDB 服务器。如果要使用独立部署的 MongoDB 服务器，那么设置这个配置，并指定 MongoDB 服务器的地址。同时，将内嵌 MongoDB 的依赖注释掉，操作如下。

```
dependencies {
    //...

    // 注释掉内嵌的 MongoDB
    // compile('de.flapdoodle.embed:de.flapdoodle.embed.mongo')

    //...
}
```

15.4.5 问题解决

在编写本项目的过程中遇到了下面的问题。

```
2017-07-30 01:45:05.988 ERROR 8332 --- [0.1-8081-exec-2] o.
a.c.c.C.[.[.[/].[dispatcherServlet]    : Servlet.service() for servlet
[dispatcherServlet] in context with path [] threw exception [Request
processing failed; nested exception is org.springframework.core.convert.
ConverterNotFoundException: No converter found capable of converting
```

```
from type [org.bson.types.Binary] to type [byte[]]] with root cause

org.springframework.core.convert.ConverterNotFoundException: No con-
verter found capable of converting from type [org.bson.types.Binary] to
type [byte[]]
    at org.springframework.core.convert.support.GenericConversion-
Service.handleConverterNotFound(GenericConversionService.java:319)
~[spring-core-5.0.0.RC2.jar:5.0.0.RC2]

...
```

解决此问题的过程可以参阅 https://jira.spring.io/browse/DATAMONGO-1760。

15.4.6 源码

MongoDB File Server 是一款开源的产品，完整的项目源码参见 https://github.com/waylau/mongodb-file-server。

第16章 博客管理实现

16.1 博客管理的需求回顾

本节将进入博客管理的编码实现阶段。

博客管理可以说是整个博客系统的核心。用户发表自己的博客，当然也可以修改和删除博客，这些都是作为一个博客系统的基本功能。

在 blog-auth 项目基础上构建了一个新的项目 blog-blog，用于演示整个博客管理实现的过程。

16.1.1 所需环境

由于博客正文是采用 Markdown 编辑器来编写的，因此需要 Markdown 格式的解析工具：Markdown parser for the JVM 0.16。

16.1.2 build.gradle

修改 build.gradle 文件，需要添加 Markdown parser 依赖。

```
// 依赖关系
dependencies {
    //...

    // 添加Markdown parser依赖
    compile('es.nitaur.markdown:txtmark:0.16')

    //...
}
```

16.1.3 权限管理的需求

首先来回顾一下博客管理所具有的功能需求。

概括起来，在本节将实现如下功能。

（1）用户主页的实现。用户主页是用于展示博主整体信息的区域，展示的信息包括用户的账号、名称、用户发表的博客文章等。

（2）个人资料设置。用户可以修改个人的资料，包括修改姓名、邮箱、密码等。

（3）个人头像更换。用户能够上传图片来修改个人的头像。

（4）博客模型设计。对博客文章进行建模。

（5）发表博客。用户可以发表博客文章。

（6）编辑博客。用户可以编辑博客文章。

（7）删除博客。用户可以删除博客文章。

（8）查询用户博客。用户可以查询博客文章。

（9）按标题模糊查询。提供模糊查询功能。可以用标题中的关键字来查询。

（10）最新、最热排序查询。提供最新或最热排序的查询。最新是指博客的发表时间是最新的；最热是指博客的热度是最高的。

（11）阅读量统计。对博客的阅读量进行统计。其中阅读量是衡量一篇博客文章热度的重要指标之一。

16.1.4 API 设计

博客管理的整体 API 设计如下。

- user space：用户主页空间。
  - /u/{username}：GET 具体某个用户的主页。
    - username：用户账号。
  - /u/{username}/profile ：GET 获取个人设置页面。
    - username：用户账号。
  - /u/{username}/profile ：POST 保存个人设置。
    - username：用户账号。
    - User：待保存的对象。
  - /u/{username}/avatar：GET 获取个人头像。
    - username：用户账号。
  - /u/{username}/avatar：POST 保存个人头像。
    - username：用户账号。
    - User：待保存的对象。
  - /u/{username}/blogs：GET 查询用户博客，以下 3 个条件任选一个。
    - order: 排序类型，new/hot ， 默认为 new。
    - catalog：博客分类 ID，默认为空。
    - keyword：搜索关键字。博客的标签，即为关键字。
    - async：是否异步请求页面。
    - pageIndex。
    - pageSize。
  - /u/{username}/blogs/edit：GET 获取新增博客的界面。
    - username：用户账号。
  - /u/{username}/blogs/edit：POST 新增、编辑博客。
    - username：用户账号。

  ◦ Blog：待保存的博客对象。
- /u/{username}/blogs/edit/{id}：GET 获取编辑博客的界面。
  ◦ username：用户账号。
  ◦ id：博客的 ID。
- /u/{username}/blogs/edit/{id}：DELETE 删除博客。
  ◦ username：用户账号。
  ◦ id：博客的 ID。

16.2 实现个人设置和头像变更

严格意义上来讲，个人信息设置、头像变更等功能，并非博客管理的领域范畴。之所以把这两个功能放到现阶段来实现，是因为这两个都依赖于前面章节实现的功能。例如，个人信息的设置（密码的修改）依赖于权限管理的功能实现，头像变更（上传头像图片）依赖于文件服务器的功能实现。

16.2.1 个人设置

在 templates/userspace 目录下，创建 profile.html 个人设置页面。

```
<!DOCTYPE html>
<html xmlns="http://www.w3.org/1999/xhtml"
    xmlns:th="http://www.thymeleaf.org"
    xmlns:sec="http://www.thymeleaf.org/thymeleaf-extras-springsecuri-
ty4">
<head th:replace="~{fragments/header :: header}">
</head>
<body>
    <!-- Page Content -->
    <div class="container blog-content-container">
        <div class="row">
            <!-- 左侧栏目 -->
            <div class="col-md-4 col-xl-3">
                <div class="row">
                    <div class="col-md-12">
                        <span> <a class="blog-edit-avatar" data-toggle=
"modal"
                            data-target="#flipFlop" role="button"
                            data-th-attr="userName=${userModel.user.
username}"> <img
                                th:src="${userModel.user.avatar} == null
```

```
                                            ? '/images/avatar-defualt.jpg'
                                            : ${userModel.user.avatar}"
                                    class="blog-avatar blog-avatar-230" />
                            </a>
                            </span>
                        </div>
                    </div>
                </div>
                <!-- 右侧栏目 -->
                <div class="col-md-8 col-xl-9">
                    <!-- 个人设置 -->
                    <div class="card ">
                        <h5 class="card-header">
                            <i class="fa fa-cog" aria-hidden="true"></i> 个
人设置
                        </h5>
                        <div class="card-block">
                            <form th:action="'/u/'+${userModel.user.user
name}+'/profile'"
                                method="post" th:object="${userModel.user}"
id="userForm">
                                <input type="hidden" name="id" id="userId"
th:value="*{id}">
                                <div class="form-group ">
                                    <label for="username" class="col-form-
label">账号</label> <input
                                        type="text" class="form-control"
id="username" name="username"
                                        th:value="*{username}" readonly=
"readonly">

                                </div>
                                <div class="form-group">
                                    <label for="email" class="col-form-label">
邮箱</label> <input
                                        type="email" class="form-control"
id="email" name="email"
                                        th:value="*{email}" maxlength="50"
placeholder="请输入邮箱">
                                </div>
                                <div class="form-group">
                                    <label for="name" class="col-form-label">
姓名</label> <input
                                        type="text" class="form-control"
id="name" name="name"
                                        th:value="*{name}" maxlength="20"
                                        placeholder="请输入姓名，至少2个字符，至
多20个">
                                </div>
```

```
                            <div class="form-group">
                                <label for="password" class="col-form-
label">密码</label> <input
                                    type="password" class="form-control"
id="password"
                                    name="password" th:value="*{pass-
word}" maxlength="30"
                                    placeholder="请输入密码，至少3个字符，至
多30个">
                            </div>
                            <div class="form-group">
                                <button type="submit" class="btn
btn-primary">保存</button>
                            </div>
                        </form>
                    </div>
                </div>
            </div>
        </div>
        <!-- /.row -->
    </div>
    <!-- /.container -->
    <!-- The modal -->
    <div class="modal fade" id="flipFlop" tabindex="-1" role="dialog"
        aria-labelledby="modalLabel" aria-hidden="true">
        <div class="modal-dialog" role="document">
            <div class="modal-content">
                <div class="modal-header">
                    <h4 class="modal-title" id="modalLabel">编辑头像</h4>
                    <button type="button" class="close" data-dismiss=
"modal"
                        aria-label="Close">
                        <span aria-hidden="true">&times;</span>
                    </button>

                </div>
                <div class="modal-body" id="avatarFormContainer"></div>
                <div class="modal-footer">
                    <button class="btn btn-primary" data-dismiss=
"modal"
                        id="submitEditAvatar">提交</button>
                    <button type="button" class="btn btn-secondary"
                        data-dismiss="modal">Close</button>
                </div>
            </div>
        </div>
    </div>

    <div th:replace="~{fragments/footer :: footer}">...</div>
```

```
    <!-- JavaScript -->
    <script th:inline="javascript">
        var fileServerUrl = [[${userModel.fileServerUrl}]];
    </script>
    <script src="../../js/userspace/main.js"
        th:src="@{/js/userspace/main.js}"></script>

</body>
</html>
```

修改 UserspaceController，增加个人设置的方法如下。

```
import org.springframework.beans.factory.annotation.Value;
import org.springframework.security.access.prepost.PreAuthorize;
import org.springframework.security.crypto.bcrypt.BCryptPasswordEncod-
er;
import org.springframework.security.crypto.password.PasswordEncoder;
...

@Autowired
private UserService userService;
@Autowired
private UserDetailsService userDetailsService;
@Value("${file.server.url}")
private String fileServerUrl;
...

/**
    * 获取个人设置页面
    * @param username
    * @param model
    * @return
    */
@GetMapping("/{username}/profile")
@PreAuthorize("authentication.name.equals(#username)")
public ModelAndView profile(@PathVariable("username") String username,
Model model) {
    User  user = (User)userDetailsService.loadUserByUsername(username);
    model.addAttribute("user", user);
    model.addAttribute("fileServerUrl", fileServerUrl);// 文件服务器的地址返
回给客户端
    return new ModelAndView("/userspace/profile", "userModel", model);
}

/**
    * 保存个人设置
    * @param username
    * @param user
    * @return
```

```
    */
@PostMapping("/{username}/profile")
@PreAuthorize("authentication.name.equals(#username)")
public String saveProfile(@PathVariable("username") String username,User
user) {
    User originalUser = userService.getUserById(user.getId()).get();
    originalUser.setEmail(user.getEmail());
    originalUser.setName(user.getName());

    // 判断密码是否做了变更
    String rawPassword = originalUser.getPassword();
    PasswordEncoder  encoder = new BCryptPasswordEncoder();
    String encodePasswd = encoder.encode(user.getPassword());
    boolean isMatch = encoder.matches(rawPassword, encodePasswd);
    if (!isMatch) {
        originalUser.setEncodePassword(user.getPassword());
    }

    userService.saveOrUpateUser(originalUser);
    return "redirect:/u/" + username + "/profile";
}
...
```

这两个方法都增加了 @PreAuthorize("authentication.name.equals(#username)") 注解，这就意味着只有用户自己才有权限修改自己的个人资料。

在保存用户资料前，要先判断用户的密码是否做了修改，如果做了修改，就要在新密码加密之后保存大数据库。

fileServerUrl 是文件服务器的地址，用于客户端文件上传使用（如更改头像）。其值可以配置在 application.properties 文件中。

同时，修改 User 实体，增加加密密码的方法如下。

```
import org.springframework.security.crypto.bcrypt.BCryptPasswordEncoder;
import org.springframework.security.crypto.password.PasswordEncoder;
...
/**
 * 加密密码
 * @param password
 */
public void setEncodePassword(String password) {
    PasswordEncoder  encoder = new BCryptPasswordEncoder();
    String encodePasswd = encoder.encode(password);
    this.password = encodePasswd;
}
...
```

16.2.2 更改头像

更改头像难免会涉及切图，这里引用了 cropbox.js 来实现切图功能。并在 templates/userspace 目录下创建了 avatar.html 个人头像更换页面。

```
<style>
.cropImg {
    position: relative;
    width: 100%;
    height: 300px;
    background-color: #ccc;
    color: #fff;
    line-height: 300px;
    text-align: center;
    border: 1px dashed rgba(0, 0, 0, .4);
}

.cropImg>img {
    position: absolute;
    left: 50%;
    top: 50%;
    -webkit-transform: translate3d(-50%, -50%, 0);
    -moz-transform: translate3d(-50%, -50%, 0);
    -ms-transform: translate3d(-50%, -50%, 0);
    -o-transform: translate3d(-50%, -50%, 0);
    transform: translate3d(-50%, -50%, 0);
}
</style>


<div id="cropImg" class="cropImg">
    <span>图片剪切</span> <img src=""/>
</div>
<div class="crop" id="crop">
    <input type="file" accept="image/*" class="crop-input">
    <div class="crop-mask"></div>
    <div class="crop-wrap">
        <div class="crop-wrap-content">
            <div class="crop-wrap-thum"></div>
            <div class="crop-wrap-spinner">Loading...</div>
        </div>
        <div class="crop-wrap-group">
            <a href="javascript:;" class="croped">剪切</a>
        </div>
        <div class="crop-wrap-group">
            <a href="javascript:;" class="zoomIn">放大</a>
            <a href="javascript:;"
                class="zoomOut">缩小</a>
        </div>
```

```
    </div>
</div>
<form enctype="multipart/form-data" id="avatarformid"></form>

<script>
    var crop = document.querySelector('#cropImg');
    var cropNote = crop.querySelector('span');
    var cropImg = crop.querySelector('img');

    var cropper = new Cropper({
        el : '#crop',
        cp : '#cropImg',
        callback : function(dataURL, dataBlob) {

            cropNote.style.display = 'none';
            cropImg.style.display = 'block';
            cropImg.src = dataURL;
        }
    });
</script>
```

修改 UserspaceController，增加个人头像更换的方法如下。

```
import org.springframework.http.ResponseEntity;
import org.springframework.web.bind.annotation.RequestBody;
import com.waylau.spring.boot.blog.vo.Response;
...

/**
 * 获取编辑头像的界面
 * @param username
 * @param model
 * @return
 */
@GetMapping("/{username}/avatar")
@PreAuthorize("authentication.name.equals(#username)")
public ModelAndView avatar(@PathVariable("username") String username,
Model model) {
    User  user = (User)userDetailsService.loadUserByUsername(username);
    model.addAttribute("user", user);
    return new ModelAndView("/userspace/avatar", "userModel", model);
}

/**
 * 保存头像
 * @param username
 * @param model
 * @return
 */
```

```
@PostMapping("/{username}/avatar")
@PreAuthorize("authentication.name.equals(#username)")
public ResponseEntity<Response> saveAvatar(@PathVariable("username")
String username, @RequestBody User user) {
    String avatarUrl = user.getAvatar();

    User originalUser = userService.getUserById(user.getId());
    originalUser.setAvatar(avatarUrl);
    userService.saveOrUpateUser(originalUser);

    return ResponseEntity.ok().body(new Response(true, "处理成功", ava-
tarUrl));
}
```

在 js 目录下建立了 userspace 目录，用于放置用户空间相关的脚本。新建 main.js。

```
/*!
 * Avatar JS.
 *
 * @since: 1.0.0 2017/4/6
 * @author Way Lau <https://waylau.com>
 */
"use strict";
//# sourceURL=main.js

// DOM加载完再执行
$(function() {
    var avatarApi;

    // 获取编辑用户头像的界面
    $(".blog-content-container").on("click",".blog-edit-avatar", func-
tion () {
        avatarApi = "/u/"+$(this).attr("userName")+"/avatar";
        $.ajax({
             url: avatarApi,
             success: function(data){
                 $("#avatarFormContainer").html(data);
             },
             error : function() {
                 toastr.error("error!");
             }
         });
    });

    /**
     * 将以base64的图片url数据转换为Blob
     * @param urlData
     * 用url方式表示的base64图片数据
     */
    function convertBase64UrlToBlob(urlData){
```

```
        var bytes=window.atob(urlData.split(',')[1]); //去掉url的头，并转
换为byte

        //处理异常,将ASCII码小于0的转换为大于0
        var ab = new ArrayBuffer(bytes.length);
        var ia = new Uint8Array(ab);
        for (var i = 0; i < bytes.length; i++) {
            ia[i] = bytes.charCodeAt(i);
        }

        return new Blob( [ab] , {type : 'image/png'});
    }

    // 提交用户头像的图片数据
    $("#submitEditAvatar").on("click", function () {
        var form = $('#avatarformid')[0];
        var formData = new FormData(form);
        var base64Codes = $(".cropImg > img").attr("src");
        formData.append("file",convertBase64UrlToBlob(base64Codes));

        $.ajax({
            url: fileServerUrl,  // 文件服务器地址
            type: 'POST',
            cache: false,
            data: formData,
            processData: false,
            contentType: false,
            success: function(data){

                var avatarUrl = data;

                // 获取CSRF Token
                var csrfToken = $("meta[name='_csrf']").attr("content");

                var csrfHeader = $("meta[name='_csrf_header']").attr
("content");
                // 保存头像更改到数据库
                $.ajax({
                    url: avatarApi,
                    type: 'POST',
                    contentType: "application/json; charset=utf-8",
                    data: JSON.stringify({"id":Number($("#userId").
val()),
                        "avatar":avatarUrl}),
                    beforeSend: function(request) {
                        request.setRequestHeader(csrfHeader, csrfTo
ken); // 添加  CSRF Token
                    },
```

```
                    success: function(data){
                        if (data.success) {
                           // 成功后，置换头像图片
                            $(".blog-avatar").attr("src", data.ava-
tarUrl);
                        } else {
                            toastr.error("error!"+data.message);
                        }

                    },
                    error : function() {
                        toastr.error("error!");
                    }
                });
            },
            error : function() {
                toastr.error("error!");
            }
        })
    });
});
```

16.2.3 与文件服务器通信

在 16.2.2 节中创建了文件服务器 MongoDB File Server。这里把文件服务器的地址配置在博客系统 application.properties 中，以提供给博客系统来访问。

```
...
# 文件服务器接口的位置
file.server.url=http://localhost:8081/upload
...
```

这样就能实现头像的上传和更改了。

启动MongoDB

在启动项目之前，确保 MongoDB 已经正常启动了。如果是采用内嵌 MongoDB 的方式，请忽略下面的内容。否则在使用博客上传图片前，需要先启动 MongoDB 服务器。

新建 MongoDB 的配置文件 mongod.cfg。

```
systemLog:
    destination: file
    path: d:\mongoData\log\mongod.log
storage:
    dbPath: d:\mongoData\db
```

首先，文件中指明了 MongoDB 存储日志和数据的位置。

其次，在启动服务器时，指明该配置文件的位置。

```
mongod --config mongod.cfg
```

该语句表明了该配置文件和 mongod 程序在同一个目录下。

16.3 博客管理的后台实现

自本节开始将进行博客管理的后台编码实现。

16.3.1 博客模型的设计

在 com.waylau.spring.boot.blog.domain 包下创建 Blog 实体，并建立起与 User 的关系。

```
import com.github.rjeschke.txtmark.Processor;
...

@Entity // 实体
public class Blog implements Serializable {
    private static final long serialVersionUID = 1L;

    @Id // 主键
    @GeneratedValue(strategy = GenerationType.IDENTITY) // 自增长策略
    private Long id; // 用户的唯一标识

    @NotEmpty(message = "标题不能为空")
    @Size(min=2, max=50)
    @Column(nullable = false, length = 50) // 映射为字段，值不能为空
    private String title;

    @NotEmpty(message = "摘要不能为空")
    @Size(min=2, max=300)
    @Column(nullable = false) // 映射为字段，值不能为空
    private String summary;

    @Lob  // 大对象，映射MySQL的Long Text类型
    @Basic(fetch=FetchType.LAZY) // “懒”加载
    @NotEmpty(message = "内容不能为空")
    @Size(min=2)
    @Column(nullable = false) // 映射为字段，值不能为空
    private String content;

    @Lob  // 大对象，映射MySQL的Long Text类型
    @Basic(fetch=FetchType.LAZY) // “懒”加载
    @NotEmpty(message = "内容不能为空")
```

```
    @Size(min=2)
    @Column(nullable = false) // 映射为字段，值不能为空
    private String htmlContent; // 将 md 转为 html

    @OneToOne(cascade = CascadeType.DETACH, fetch = FetchType.LAZY)
    @JoinColumn(name="user_id")
    private User user;

    @Column(nullable = false) // 映射为字段，值不能为空
    @org.hibernate.annotations.CreationTimestamp  // 由数据库自动创建时间
    private Timestamp createTime;

    @Column(name="readSize")
    private Integer readSize = 0; // 访问量、阅读量

    @Column(name="commentSize")
    private Integer commentSize = 0;  // 评论量

    @Column(name="voteSize")
    private Integer voteSize = 0;  // 点赞量

    @Column(name="tags", length = 100)
    private String tags;  // 标签

    protected Blog() {

    }

    public Blog(String title, String summary,String content) {
        this.title = title;
        this.summary = summary;
        this.content = content;
    }

    // 省略其他getter/setter方法

    public Timestamp getCreateTime() {
        return createTime;
    }

    public String getHtmlContent() {
        return htmlContent;
    }

    public String getContent() {
        return content;
    }

    public void setContent(String content) {
```

```
        this.content = content;
        this.htmlContent = Processor.process(content);
    // 将Markdown内容转为HTML格式
    }

}
```

这里特别要注意 setContent 方法。除了设置 content 属性外，同时还要将 content 的内容，经过 Markdown 解析器进行解析，转换为 HTML 格式内容，并赋值给 htmlContent。这样，在访问博客正文的时候，直接显示 htmlContent 的内容，而不是 content 的内容。content 的内容只用于博客编辑时使用。

博客的创建时间 createTime 使用了 Hibernate 的注解 @org.hibernate.annotations.CreationTimestamp，这样，该注解会自动创建一个时间。

16.3.2 仓库接口

在 com.waylau.spring.boot.blog.repository 包下创建 Blog 的仓库接口，用于博客的分页查询。

```
public interface BlogRepository extends JpaRepository<Blog, Long>{

    /**
     * 根据用户名、博客标题分页查询博客列表
     * @param user
     * @param title
     * @param pageable
     * @return
     */
    Page<Blog> findByUserAndTitleLike(User user, String title, Pageable
pageable);

    /**
     * 根据用户名、博客查询博客列表（时间逆序）
     * @param title
     * @param user
     * @param tags
     * @param user2
     * @param pageable
     * @return
     */
    Page<Blog> findByTitleLikeAndUserOrTagsLikeAndUserOrderByCreateTime
Desc(String title,
            User user, String tags, User user2, Pageable pageable);
}
```

16.3.3 服务接口及实现

在 com.waylau.spring.boot.blog.service 包下创建 BlogService。

```
public interface BlogService {
    /**
     * 保存Blog
     * @param EsBlog
     * @return
     */
    Blog saveBlog(Blog blog);

    /**
     * 删除Blog
     * @param id
     * @return
     */
    void removeBlog(Long id);

    /**
     * 根据id获取Blog
     * @param id
     * @return
     */
    Optional<Blog> getBlogById(Long id);

    /**
     * 根据用户进行博客名称分页模糊查询（最新）
     * @param user
     * @return
     */
    Page<Blog> listBlogsByTitleVote(User user, String title, Pageable
pageable);

    /**
     * 根据用户进行博客名称分页模糊查询（最热）
     * @param user
     * @return
     */
    Page<Blog> listBlogsByTitleVoteAndSort(User user, String title,
Pageable pageable);

    /**
     * 阅读量递增
     * @param id
     */
    void readingIncrease(Long id);

}
```

创建 BlogService 的实现类 BlogServiceImpl。

```
@Service
public class BlogServiceImpl implements BlogService {

    @Autowired
    private BlogRepository blogRepository;

    @Transactional
    @Override
    public Blog saveBlog(Blog blog) {
        Blog returnBlog = blogRepository.save(blog);
        return returnBlog;
    }

    @Transactional
    @Override
    public void removeBlog(Long id) {
        blogRepository.deleteById(id);
    }

    @Override
    public Optional<Blog> getBlogById(Long id) {
        return blogRepository.findById(id);
    }

    @Override
    public Page<Blog> listBlogsByTitleVote(User user, String title,
Pageable pageable) {
        // 模糊查询
        title = "%" + title + "%";
        String tags = title;
        Page<Blog> blogs = blogRepository.findByTitleLikeAndUserOrTags-
LikeAndUserOrderByCreateTimeDesc(title,
                user, tags, user, pageable);
        return blogs;
    }

    @Override
    public Page<Blog> listBlogsByTitleVoteAndSort(User user, String
title, Pageable pageable) {
        // 模糊查询
        title = "%" + title + "%";
        Page<Blog> blogs = blogRepository.findByUserAndTitleLike(user,
title, pageable);
        return blogs;
    }

    @Override
    public void readingIncrease(Long id) {
```

```
        Optional<Blog> blog = blogRepository.findById(id);
        Blog blogNew = null;

        if (blog.isPresent()) {
            blogNew = blog.get();
            blogNew.setReadSize(blogNew.getReadSize() + 1);
    // 在原有的阅读量基础上递增1
            this.saveBlog(blogNew);
        }
    }
}
```

16.3.4 控制层

修改 UserspaceController。

```
...
/**
    * 用户的主页
    *
    * @param username
    * @param model
    * @return
    */
@GetMapping("/{username}")
public String userSpace(@PathVariable("username") String username,
Model model) {
    User user = (User) userDetailsService.loadUserByUsername(username);
    model.addAttribute("user", user);
    return "redirect:/u/" + username + "/blogs";
}

/**
    * 获取用户的博客列表
    *
    * @param username
    * @param order
    * @param catalogId
    * @param keyword
    * @param async
    * @param pageIndex
    * @param pageSize
    * @param model
    * @return
    */
@GetMapping("/{username}/blogs")
public String listBlogsByOrder(@PathVariable("username") String user
name,
```

```
        @RequestParam(value = "order", required = false, defaultValue =
"new") String order,
        @RequestParam(value = "catalog", required = false) Long cata
logId,
        @RequestParam(value = "keyword", required = false, defaultValue
= "") String keyword,
        @RequestParam(value = "async", required = false) boolean async,
        @RequestParam(value = "pageIndex", required = false, default
Value = "0") int pageIndex,
        @RequestParam(value = "pageSize", required = false, defaultVal
ue = "10") int pageSize, Model model) {

    User user = (User) userDetailsService.loadUserByUsername(username);

    Page<Blog> page = null;

    if (catalogId != null && catalogId > 0) { // 分类查询
        // TODO
    } else if (order.equals("hot")) { // 最热查询
        Sort sort = new Sort(Direction.DESC, "readSize", "commentSize",
"voteSize");
        Pageable pageable = PageRequest.of(pageIndex, pageSize, sort);
        page = blogService.listBlogsByTitleVoteAndSort(user, keyword,
pageable);
    } else if (order.equals("new")) { // 最新查询
        Pageable pageable = PageRequest.of(pageIndex, pageSize);
        page = blogService.listBlogsByTitleVote(user, keyword, pagea
ble);
    }

    List<Blog> list = page.getContent(); // 当前所在页面数据列表

    model.addAttribute("user", user);
    model.addAttribute("order", order);
    model.addAttribute("catalogId", catalogId);
    model.addAttribute("keyword", keyword);
    model.addAttribute("page", page);
    model.addAttribute("blogList", list);
    return (async == true ? "/userspace/u :: #mainContainerRepleace" :
"/userspace/u");
}

/**
    * 获取博客展示界面
    *
    * @param username
    * @param id
    * @param model
    * @return
```

```
    */
@GetMapping("/{username}/blogs/{id}")
public String getBlogById(@PathVariable("username") String username,
@PathVariable("id") Long id, Model model) {
    User principal = null;
    Optional<Blog> blog = blogService.getBlogById(id);

    // 每次读取，简单地可以认为阅读量增加1次
    blogService.readingIncrease(id);

    // 判断操作用户是否为博客的所有者
    boolean isBlogOwner = false;
    if (SecurityContextHolder.getContext().getAuthentication() != null
            && SecurityContextHolder.getContext().getAuthentication().
isAuthenticated() && !SecurityContextHolder
                    .getContext().getAuthentication().getPrincipal().
toString().equals("anonymousUser")) {
        principal = (User) SecurityContextHolder.getContext().getAu
thentication().getPrincipal();
        if (principal != null && username.equals(principal.getUser
name())) {
            isBlogOwner = true;
        }
    }

    model.addAttribute("isBlogOwner", isBlogOwner);
    model.addAttribute("blogModel", blog.get());

    return "/userspace/blog";
}

/**
    * 获取新增博客的界面
    *
    * @param model
    * @return
    */
@GetMapping("/{username}/blogs/edit")
public ModelAndView createBlog(@PathVariable("username") String user
name, Model model) {
    model.addAttribute("blog", new Blog(null, null, null));
    model.addAttribute("fileServerUrl", fileServerUrl);
    // 文件服务器的地址返回给客户端
    return new ModelAndView("/userspace/blogedit", "blogModel", model);
}

/**
    * 获取编辑博客的界面
    *
```

```
    * @param model
    * @return
    */
@GetMapping("/{username}/blogs/edit/{id}")
public ModelAndView editBlog(@PathVariable("username") String username,
@PathVariable("id") Long id, Model model) {
    model.addAttribute("blog", blogService.getBlogById(id).get());
    model.addAttribute("fileServerUrl", fileServerUrl);
    // 文件服务器的地址返回给客户端
    return new ModelAndView("/userspace/blogedit", "blogModel", model);
}

/**
    * 保存博客
    *
    * @param username
    * @param blog
    * @return
    */
@PostMapping("/{username}/blogs/edit")
@PreAuthorize("authentication.name.equals(#username)")
public ResponseEntity<Response> saveBlog(@PathVariable("username")
String username, @RequestBody Blog blog) {

    try {

        // 判断是修改还是新增
        if (blog.getId() != null) {
            Optional<Blog> optionalBlog = blogService.getBlogById(blog.
getId());
            if (optionalBlog.isPresent()) {
                Blog orignalBlog = optionalBlog.get();
                orignalBlog.setTitle(blog.getTitle());
                orignalBlog.setContent(blog.getContent());
                orignalBlog.setSummary(blog.getSummary());
                blogService.saveBlog(orignalBlog);
            }
        } else {
            User user = (User) userDetailsService.loadUserByUsername(us-
ername);
            blog.setUser(user);
            blogService.saveBlog(blog);
        }

    } catch (ConstraintViolationException e) {
        return ResponseEntity.ok().body(new Response(false, Constraint
ViolationExceptionHandler.getMessage(e)));
    } catch (Exception e) {
        return ResponseEntity.ok().body(new Response(false, e.getMes
```

```
sage()));
    }

    String redirectUrl = "/u/" + username + "/blogs/" + blog.getId();
    return ResponseEntity.ok().body(new Response(true, "处理成功", redi
rectUrl));
}

/**
    * 删除博客
    *
    * @param username
    * @param id
    * @return
    */
@DeleteMapping("/{username}/blogs/{id}")
@PreAuthorize("authentication.name.equals(#username)")
public ResponseEntity<Response> deleteBlog(@PathVariable("username")
String username, @PathVariable("id") Long id) {

    try {
        blogService.removeBlog(id);
    } catch (Exception e) {
        return ResponseEntity.ok().body(new Response(false, e.getMes
sage()));
    }

    String redirectUrl = "/u/" + username + "/blogs";
    return ResponseEntity.ok().body(new Response(true, "处理成功", redi
rectUrl));
}
...
```

其中，获取用户博客列表的方法 listBlogsByOrder 中，对于“分类查询”的判断还没有具体的实现。这部分会在后续章节中补齐。

16.4 博客管理的前台实现

在完成后台接口之后，本节将进行博客管理的前端编码。

16.4.1 用户主页实现

用户主页除了分类管理外，其他的都可以实现了。

用户主页 u.html 左侧栏目修改为:

```
...
<div class="row">
    <div class="col-md-12">
        <span> <a href="/u/waylau"
            th:href="'/u/' + ${user.username}" title="waylau"
            th:title="${user.username}"> <img
                th:src="${user.avatar} == null
                    ? '/images/avatar-defualt.jpg'
                    : ${user.avatar} "
                class="blog-avatar-230">
        </a>
        </span>

    </div>

</div>
<div class="row">

    <div class="col-md-12">
        <h2 class="card-text" th:text="${user.name}">老卫</h2>
        <h4 class="card-text" th:text="${user.username}">waylau</h4>
    </div>
</div>

<div class="row">
    <div class="col-md-12">
        <h5>
            <i class="fa fa-envelope-o" aria-hidden="true"></i> <a
                href="mailto:waylau521@gmail.com"
                th:href="'mailto:'+ ${user.email}"
                th:text="${user.email}">waylau521@gmail.com</a>
        </h5>
    </div>
</div>

<hr>
...
```

右侧工具栏修改为:

```
...
<ul class="nav nav-tabs mr-auto">

    <li class="nav-item"><a class="nav-link "
        data-th-classappend="${order} eq 'new' ? 'active' : ''"
        href="javascript:void(0)"
        th:attr="url='/u/'+${user.username}+'/blogs?order=new'">最新</
a></li>
```

```
    <li class="nav-item"><a class="nav-link"
        data-th-classappend="${order} eq 'hot' ? 'active' : ''"
        href="javascript:void(0)"
        th:attr="url='/u/'+${user.username}+'/blogs?order=hot'">最热</
a></li>

    <form class="form-inline mt-2 mt-md-0">
        <input class="form-control mr-sm-2" type="text" placeholder="搜
索"
            id="keyword" th:value="${keyword}"> <a
            href="javascript:void(0)"
            class="btn btn-outline-secondary my-2 my-sm-0" id="search-
Blogs"><i
            class="fa fa-search" aria-hidden="true"></i></a>
    </form>

</ul>
...
```

博客列表修改为：

```
...
<div id="mainContainer">
    <div id="mainContainerRepleace">
        <div class="card mb-4" th:each="blog : ${blogList}">
            <div class="card-block">
                <h2 class="card-title">
                    <a href="/u/waylau/blogs/1" class="card-link" title=
"waylau"
                        th:href="'/u/' + ${blog.user.username} + '/
blogs/'+ ${blog.id}"
                        th:title="${blog.user.username}" th:text=
"${blog.title}">
                        OAuth 2.0认证的原理与实践 </a>
                </h2>
                <p class="card-text" th:text="${blog.summary}">使用
OAuth 2.0认证的好处是显而易见的。你只需要用同一个账号密码，就能在各个网站进行访问，
而免去了在每个网站都进行注册的烦琐过程。
本文将介绍OAuth 2.0的原理，并基于Spring Security和GitHub账号，来演示OAuth2.0
的认证的过程。</p>
                <div class="card-text">
                    发表于 [[${#dates.format(blog.createTime, 'yyyy-MM-
dd HH:mm')}]] <i
                        class="fa fa-eye" aria-hidden="true">[[${blog.
readSize}]]</i>
                    <i class="fa fa-heart-o" aria-hidden="true">[[${
blog.voteSize}]]</i>
                    <i class="fa fa-comment-o" aria-hidden="true">[[${
blog.commentSize}]]</i>
```

```
                </div>
            </div>
        </div>

        <div th:replace="~{fragments/page :: page}">...</div>
    </div>
</div>
...
```

然后，引入了相关的 js。

```
...
<!-- JavaScript -->
<script th:inline="javascript">
var username = [[${user.username}]];
</script>
<script src="../../js/userspace/u.js" th:src="@{/js/userspace/u.js}"></
script>
...
```

其中，在 userspace 目录下，新增了 u.js 为：

```
$(function() {

    var _pageSize; // 存储用于搜索

    // 根据用户名、页面索引、页面大小获取用户列表
    function getBlogsByName(pageIndex, pageSize) {
         $.ajax({
             url: "/u/"+  username  +"/blogs",
             contentType : 'application/json',
             data:{
                 "async":true,
                 "pageIndex":pageIndex,
                 "pageSize":pageSize,
                 "catalog":null,  //catalogId,
                 "keyword":$("#keyword").val()
             },
             success: function(data){
                 $("#mainContainer").html(data);

                 // 如果是分类查询，则取消最新、最热选中样式
                 if (catalogId) {
                    $(".nav-item .nav-link").removeClass("active");
                 }
             },
             error : function() {
                 toastr.error("error!");
             }
         });
```

```
    }

    // 分页
    $.tbpage("#mainContainer", function (pageIndex, pageSize) {
        getBlogsByName(pageIndex, pageSize);
        _pageSize = pageSize;
    });

    // 关键字搜索
    $("#searchBlogs").click(function() {
        getBlogsByName(0, _pageSize);
    });

    // 最新/最热切换事件
    $(".nav-item .nav-link").click(function() {

        var url = $(this).attr("url");

        // 先移除其他的单击样式，再添加当前的单击样式
        $(".nav-item .nav-link").removeClass("active");
        $(this).addClass("active");

        // 加载其他模块的页面到右侧工作区
         $.ajax({
             url: url+'&async=true',
             success: function(data){
                 $("#mainContainer").html(data);
             },
             error : function() {
                 toastr.error("error!");
             }
         })

         // 清空搜索框内容
         $("#keyword").val('');
    });
});
```

16.4.2 博客展示

修改了 blog.html 博客展示部分。

```
...
<h2 class="card-title">
    <span> <a href="/u/waylau" title="waylau"
        th:href="'/u/' + ${blogModel.user.username}"
        th:title="${blogModel.user.username}"> <img
            src="/images/avatar-defualt.jpg"
```

```
            th:src="${blogModel.user.avatar} == null
                ? '/images/avatar-defualt.jpg'
                : ${blogModel.user.avatar}"
            class="blog-avatar-50">
    </a>
    </span> <a href="/u/waylau/blogs/1" class="card-link" title="way
lau"
        th:href="'/u/' + ${blogModel.user.username} + '/blogs/'+ ${blog-
Model.id}"
        th:title="${blogModel.user.username}"
        th:text="${blogModel.title}"> OAuth 2.0认证的原理与实践 </a>
</h2>
<div class="card-text">
    <a href="/u/waylau" th:href="'/u/' + ${blogModel.user.username}"
        class="card-link" th:text="${blogModel.user.username}">waylau</
a>
    发表于 [[${#dates.format(blogModel.createTime, 'yyyy-MM-dd
    HH:mm')}]] <i class="fa fa-eye" aria-hidden="true">[[${blogModel.
readSize}]]</i>
    <i class="fa fa-heart-o" aria-hidden="true">[[${blogModel.vote
Size}]]</i>
    <i class="fa fa-comment-o" aria-hidden="true">[[${blogModel.com
mentSize}]]</i>
    <a href="/u/waylau" th:if="${isBlogOwner}"
        th:href="'/u/' + ${blogModel.user.username}+ '/blogs/edit/'+
${blogModel.id}"
        class="btn btn-primary float-right">编辑</a> <a
        href="javascript:void(0)" th:if="${isBlogOwner}"
        class="btn btn-primary float-right blog-delete-blog">删除</a>
</div>
<hr>
<article class="post-content" th:utext="${blogModel.htmlContent}">

</article>
<hr>
...
```

引入了脚本，并修正了 blog.js 的位置，将其移动到 /userspace/ 目录下。

```
...
<!-- JavaScript -->
<script th:inline="javascript">
var blogId = [[${blogModel.id}]];
var blogUrl = '/u/' + [[${blogModel.user.username}]] + '/blogs/'+
[[${blogModel.id}]] ;
</script>
<script src="../../js/userspace/blog.js" th:src="@{/js/userspace/blog.
js}"></script>
...
```

其中，blog.js 增加了处理删除博客事件。

```
...
// 处理删除博客事件
$(".blog-content-container").on("click",".blog-delete-blog", function
() {
    // 获取CSRF Token
    var csrfToken = $("meta[name='_csrf']").attr("content");
    var csrfHeader = $("meta[name='_csrf_header']").attr("content");


    $.ajax({
         url: blogUrl,
         type: 'DELETE',
         beforeSend: function(request) {
             request.setRequestHeader(csrfHeader, csrfToken); // 添加
CSRF Token
         },
         success: function(data){
             if (data.success) {
                 // 成功后，重定向
                 window.location = data.body;
             } else {
                 toastr.error(data.message);
             }
         },
         error : function() {
             toastr.error("error!");
         }
     });
});
...
```

16.4.3 博客编辑

将博客编辑界面 blog.js 做了修改，并绑定了 Thymeleaf 的模型。

```
...
<div class="card-block" th:object="${blogModel.blog}">
    <input type="hidden" name="id" th:value="*{id}" id="blogId">
    <input type="text" class="form-control" placeholder="请填写博客标题"
        id="title" name="title" th:value="*{title}" maxlength="50">

    <textarea class="blog-textarea" placeholder="请填写博客摘要" id="sum-
mary"
        name="summary" th:text="*{summary}" maxlength="300"></textarea>

    <hr>
```

```
    <textarea id="md" data-provide="markdown"
        data-hidden-buttons="cmdImage" name="content"
            th:text="*{content}"></textarea>

    <hr>
</div>
...
```

在栏目右侧，增加了图片上传的处理界面。

```
...
<!-- 图片上传 -->
<div class="card ">
    <h5 class="card-header">
        <i class="fa fa-file-image-o" aria-hidden="true"></i> 图片上传
    </h5>
    <div class="card-block">
        <div class="row mt-1">
            <div class="col-lg-12">

                <form enctype="multipart/form-data" id="uploadformid">
                    <input type="file" name="file"
                        accept="image/png,image/gif,image/jpeg" id=
"file">
                    <button class="btn btn-primary float-right"
type="button"
                        id="uploadImage">插入</button>
                </form>

            </div>
        </div>
    </div>

</div>
...
```

发布按钮做了修改，增加了 ID 及 username 属性。

```
...
<button class="btn btn-primary float-right" id="submitBlog"
    th:attr="userName=${#authentication.principal.username}">发布</
button>
...
```

加入了脚本内容，同时将 blogedit.js 的位置做了调整，并移动到 /userspace/ 目录下。

```
...
<!-- JavaScript -->
<script th:inline="javascript">
    var fileServerUrl = [[${blogModel.fileServerUrl}]];
```

```
</script>
<script src="../../js/userspace/blogedit.js"
    th:src="@{/js/userspace/blogedit.js}"></script>
...
```

并在该 js 中增加了如下接口。

```
...
$("#uploadImage").click(function() {
    $.ajax({
        url: fileServerUrl,   // 文件服务器地址
        type: 'POST',
        cache: false,
        data: new FormData($('#uploadformid')[0]),
        processData: false,
        contentType: false,
        success: function(data){
            var mdcontent=$("#md").val();
                $("#md").val(mdcontent + "\n![]("+data +") \n");

            }
    }).done(function(res) {
        $('#file').val('');
    }).fail(function(res) {});
})

// 发布博客
$("#submitBlog").click(function() {

    // 获取CSRF Token
    var csrfToken = $("meta[name='_csrf']").attr("content");
    var csrfHeader = $("meta[name='_csrf_header']").attr("content");

    $.ajax({
        url: '/u/'+ $(this).attr("userName") + '/blogs/edit',
        type: 'POST',
        contentType: "application/json; charset=utf-8",
        data:JSON.stringify({"id":$('#blogId').val(),
            "title": $('#title').val(),
            "summary": $('#summary').val() ,
            "content": $('#md').val(),
            "catalog":{"id":$('#catalogSelect').val()},
            "tags":$('.form-control-tag').val()
            }),
        beforeSend: function(request) {
            request.setRequestHeader(csrfHeader, csrfToken); // 添加
CSRF Token
        },
            success: function(data){
                if (data.success) {
```

```
                // 成功后，重定向
                    window.location = data.body;
                } else {
                    toastr.error("error!"+data.message);
                }

            },
            error : function() {
                toastr.error("error!");
            }
    })
})
```

16.4.4 运行

首先，确保文件服务器是处于启动状态的，然后启动 blog-blog 项目。

通过浏览器访问 http://localhost:8080/，进入博客系统主页，并登录系统。

图 16-1 所示为博客编辑界面的效果。可以在右侧“图片上传”区域进行图片的上传，并将图片的 URL 地址插入正文中。

图16-1　编辑界面

当博客创建成功之后，会重定向到博客的查看界面。图 16-2 所示为博客的查看界面效果。

图16-2　查看界面

在单击“个人设置”按钮后，可以跳转到用户个人设置界面，如图 16-3 所示。

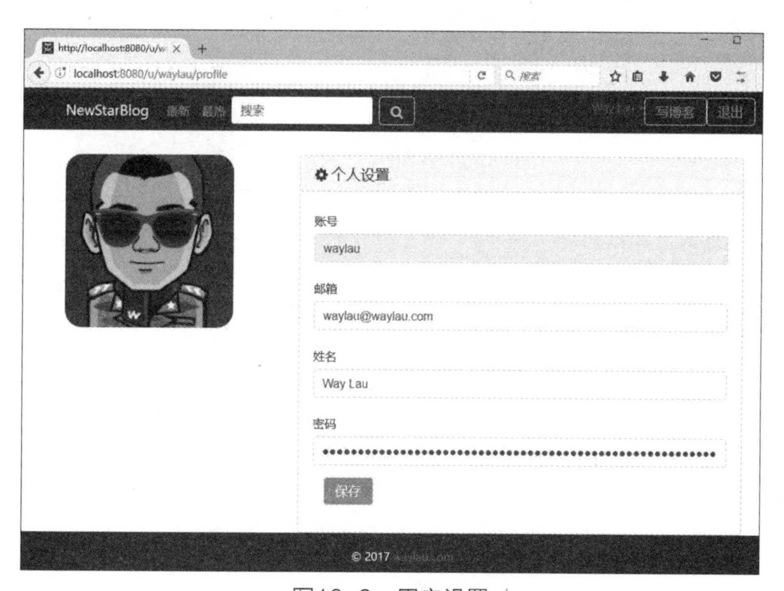

图16-3　用户设置

可以在该“个人设置”界面中对用户的个人信息进行编辑。

单击用户显示的头像，可以进入头像的更换界面，如图 16-4 所示。该头像更换界面具有上传头像、切图的功能。

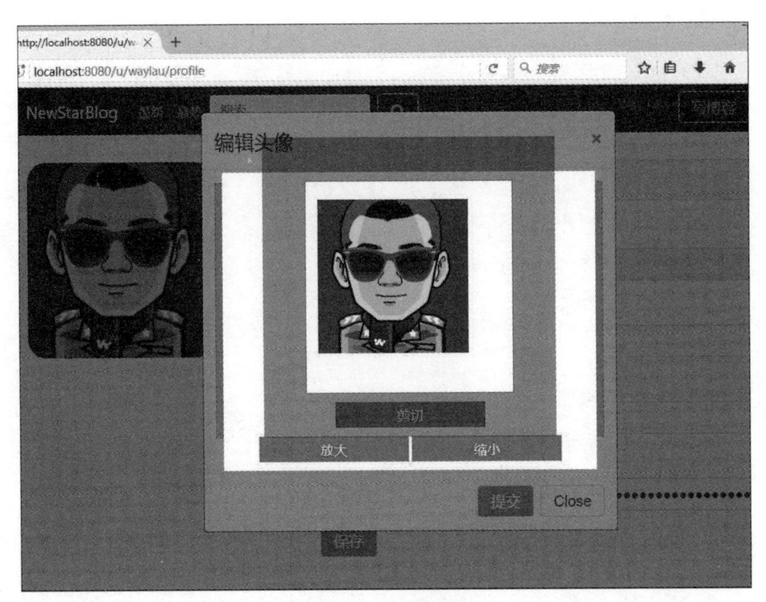

图16-4　头像更换

16.4.5　示例源码

本节示例源码在 blog-blog 目录下。

第17章

评论管理实现

17.1 评论管理的需求回顾

本节将实现博客系统中的评论管理功能。

用户发表博客之后，如果其他用户想要与博主进行交互，那么首要的途径就是评论博客。

在 blog-blog 项目基础上构建了一个新的项目 blog-comment，用于演示整个评论管理的实现过程。

17.1.1 评论管理的需求

首先来回顾一下评论管理所具有的功能需求。

概括起来，在本节将实现如下功能。

（1）评论的模型设计。要设计评论的领域模型。评论不但要跟博客文章关联，还需要跟用户关联。

（2）发表评论。用户在某个博客下面发表评论。一篇博客会对应多个评论内容。

（3）删除评论。用户删除自己发表的评论内容。

（4）统计评论数。实现统计评论数的功能。评论数是衡量一篇博客文章热度的重要指标之一。

17.1.2 API 设计

评论管理的整体 API 设计如下。

- comments：评论管理。
  - /comments：GET 获取评论列表。
    - blogId：博客 ID。
  - /comments：POST 保存评论。
    - blogId：博客 ID。
    - commentContent：评论内容。
  - /comments/{id}：DELETE 删除评论。
    - id：评论 ID。
    - blogId：博客 ID

17.2 评论管理的后台实现

本节将实现评论管理的后台编码实现。

17.2.1 评论的模型设计

评论的模型设计主要有评论实体、用户实体及博客文章实体。

1. 评论的实体Comment与User的关系

在 com.waylau.spring.boot.blog.domain 包下创建评论的实体 Comment，并建立起与 User 的关系。

```
@Entity // 实体
public class Comment implements Serializable {
    private static final long serialVersionUID = 1L;

    @Id // 主键
    @GeneratedValue(strategy = GenerationType.IDENTITY) // 自增长策略
    private Long id; // 用户的唯一标识

    @NotEmpty(message = "评论内容不能为空")
    @Size(min=2, max=500)
    @Column(nullable = false) // 映射为字段，值不能为空
    private String content;

    @OneToOne(cascade = CascadeType.DETACH, fetch = FetchType.LAZY)
    @JoinColumn(name="user_id")
    private User user;

    @Column(nullable = false) // 映射为字段，值不能为空
    @org.hibernate.annotations.CreationTimestamp  // 由数据库自动创建时间
    private Timestamp createTime;

    protected Comment() {
    }

    public Comment(User user, String content) {
        this.content = content;
        this.user = user;
    }

    // 省略其他getter/setter方法

    public Timestamp getCreateTime() {
        return createTime;
    }
}
```

评论的创建时间 createTime 使用了 Hibernate 的注解 @org.hibernate.annotations.CreationTimestamp，这样，该注解会自动创建一个时间。

2. 评论的实体Comment与Blog的关系

同时，修改 Blog 类来建立起 Blog 与 Comment 的关系。

```
@OneToMany(cascade = CascadeType.ALL, fetch = FetchType.EAGER)
@JoinTable(name = "blog_comment", joinColumns = @JoinColumn(name =
"blog_id", referencedColumnName = "id"),
    inverseJoinColumns = @JoinColumn(name = "comment_id", referenced-
ColumnName = "id"))
private List<Comment> comments;

...

public List<Comment> getComments() {
    return comments;
}

public void setComments(List<Comment> comments) {
    this.comments = comments;
    this.commentSize = this.comments.size();
}

/**
 * 添加评论
 * @param comment
 */
public void addComment(Comment comment) {
    this.comments.add(comment);
    this.commentSize = this.comments.size();
}

/**
 * 删除评论
 * @param comment
 */
public void removeComment(Long commentId) {
    for (int index=0; index < this.comments.size(); index ++ ) {
        if (comments.get(index).getId() == commentId) {
            this.comments.remove(index);
            break;
        }
    }

    this.commentSize = this.comments.size();
}
```

这里需要注意的是，每次增加或移除一个评论之后，都要统计评论总数 commentSize。

17.2.2 评论的仓库

在 com.waylau.spring.boot.blog.repository 包下建立继承自 org.springframework.data.jpa.repository.JpaRepository 接口的 CommentRepository。

```
public interface CommentRepository extends JpaRepository<Comment, Long>{

}
```

17.2.3 评论的服务接口及实现

在 com.waylau.spring.boot.blog.service 包下增加评论的服务接口及实现。评论的服务接口 CommentService 较为简单，只定义了获取评论及删除评论的方法。

```
public interface CommentService {

    /**
     * 根据id获取Comment
     * @param id
     * @return
     */
    Optional<Comment> getCommentById(Long id);
    /**
     * 删除评论
     * @param id
     * @return
     */
    void removeComment(Long id);
}
```

以下是服务的实现类 CommentServiceImpl。

```
@Service
public class CommentServiceImpl implements CommentService {

    @Autowired
    private CommentRepository commentRepository;

    @Override
    public Optional<Comment> getCommentById(Long id) {
        return commentRepository.findById(id);
    }

    @Override
    public void removeComment(Long id) {
        commentRepository.deleteById(id);
    }
```

```
}
```

同时，也要在 BlogService 中增加以下方法。

```
...
/**
 * 发表评论
 * @param blogId
 * @param commentContent
 * @return
 */
Blog createComment(Long blogId, String commentContent);

/**
 * 删除评论
 * @param blogId
 * @param commentId
 * @return
 */
void removeComment(Long blogId, Long commentId);
```

在 BlogServiceImpl 中实现上述接口。

```
...
@Override
public Blog createComment(Long blogId, String commentContent) {
    Optional<Blog> optionalBlog = blogRepository.findById(blogId);
    Blog originalBlog = null;
    if(optionalBlog.isPresent()) {
        originalBlog = optionalBlog.get();
        User user = (User)SecurityContextHolder.getContext().getAuthen
tication().getPrincipal();
        Comment comment = new Comment(user, commentContent);
        originalBlog.addComment(comment);
    }

    return this.saveBlog(originalBlog);
}

@Override
public void removeComment(Long blogId, Long commentId) {
    Optional<Blog> optionalBlog = blogRepository.findById(blogId);
    if(optionalBlog.isPresent()) {
        Blog originalBlog = optionalBlog.get();
        originalBlog.removeComment(commentId);
        this.saveBlog(originalBlog);
    }
}
```

17.2.4 控制层

控制层 CommentController 类提供了获取评论列表、发表评论、删除评论的 API。

```
import org.springframework.security.access.prepost.PreAuthorize;
...

@Controller
@RequestMapping("/comments")
public class CommentController {

    @Autowired
    private BlogService blogService;

    @Autowired
    private CommentService commentService;

    /**
     * 获取评论列表
     *
     * @param blogId
     * @param model
     * @return
     */
    @GetMapping
    public String listComments(@RequestParam(value = "blogId", required
= true) Long blogId, Model model) {
        Optional<Blog> optionalblog = blogService.getBlogById(blogId);
        List<Comment> comments = null;

        if (optionalblog.isPresent()) {
            comments = optionalblog.get().getComments();
        }

        // 判断操作用户是否为评论的所有者
        String commentOwner = "";
        if (SecurityContextHolder.getContext().getAuthentication() !=
null
                && SecurityContextHolder.getContext().getAuthentica-
tion().isAuthenticated()
                && !SecurityContextHolder
                        .getContext().getAuthentication().getPrinci-
pal().toString().equals("anonymousUser")) {
            User principal = (User) SecurityContextHolder.getContext().
getAuthentication().getPrincipal();
            if (principal != null) {
                commentOwner = principal.getUsername();
            }
        }
```

```
        model.addAttribute("commentOwner", commentOwner);
        model.addAttribute("comments", comments);
        return "/userspace/blog :: #mainContainerRepleace";
    }

    /**
     * 发表评论
     *
     * @param blogId
     * @param commentContent
     * @return
     */
    @PostMapping
    @PreAuthorize("hasAnyAuthority('ROLE_ADMIN','ROLE_USER')")
    // 指定角色权限才能操作方法
    public ResponseEntity<Response> createComment(Long blogId, String
commentContent) {

        try {
            blogService.createComment(blogId, commentContent);
        } catch (ConstraintViolationException e) {
            return ResponseEntity.ok().body(new Response(false, Con-
straintViolationExceptionHandler.getMessage(e)));
        } catch (Exception e) {
            return ResponseEntity.ok().body(new Response(false, e.get-
Message()));
        }

        return ResponseEntity.ok().body(new Response(true, "处理成功",
null));
    }

    /**
     * 删除评论
     *
     * @return
     */
    @DeleteMapping("/{id}")
    @PreAuthorize("hasAnyAuthority('ROLE_ADMIN','ROLE_USER')")
    // 指定角色权限才能操作方法
    public ResponseEntity<Response> delete(@PathVariable("id") Long id,
Long blogId) {

        boolean isOwner = false;
        Optional<Comment> optionalComment = commentService.getComment-
ById(id);
        User user = null;
```

```
        if (optionalComment.isPresent()) {
            user = optionalComment.get().getUser();
        } else {
            return ResponseEntity.ok().body(new Response(false, "不存在该
评论！"));
        }

        // 判断操作用户是否为评论的所有者
        if (SecurityContextHolder.getContext().getAuthentication() !=
null
                && SecurityContextHolder.getContext().getAuthentica-
tion().isAuthenticated()
                && !SecurityContextHolder
                        .getContext().getAuthentication().getPrinci-
pal().toString().equals("anonymousUser")) {
            User principal = (User) SecurityContextHolder.getContext().
getAuthentication().getPrincipal();
            if (principal != null && user.getUsername().equals(princi-
pal.getUsername())) {
                isOwner = true;
            }
        }

        if (!isOwner) {
            return ResponseEntity.ok().body(new Response(false, "没有操作
权限"));
        }

        try {
            blogService.removeComment(blogId, id);
            commentService.removeComment(id);
        } catch (ConstraintViolationException e) {
            return ResponseEntity.ok().body(new Response(false, Con-
straintViolationExceptionHandler.getMessage(e)));
        } catch (Exception e) {
            return ResponseEntity.ok().body(new Response(false, e.get-
Message()));
        }

        return ResponseEntity.ok().body(new Response(true, "处理成功",
null));
    }
}
```

其中，在创建和删除评论的方法上添加了注解 @PreAuthorize("hasAnyAuthority('ROLE_ADMIN','ROLE_USER')") 用于标识，只有相关角色才能进行操作。

同时，用下面的方式来判断操作用户是否为评论的所有者。

```
if (SecurityContextHolder.getContext().getAuthentication() != null
```

```
                && SecurityContextHolder.getContext().getAuthentica-
tion().isAuthenticated()
                && !SecurityContextHolder
                        .getContext().getAuthentication().getPrinci
pal().toString().equals("anonymousUser"))
```

17.3 评论管理的前台实现

评论的前台页面实现较为简单。

17.3.1 修改 blog.html

修改 blog.html 中的内容，首先是修改评论框的内容。

```
...
<div class="card-block">
    <h5>评论：</h5>
    <div class="row">
        <div class="col-lg-12">
            <textarea class="blog-textarea" placeholder="看帖需留言~"
                id="commentContent"></textarea>
        </div>
    </div>
    <button class="btn btn-primary float-right" id="submitComment">发表评
论</button>
    <button class="btn btn-primary float-right">点赞</button>
</div>
...
```

接着是修改评论展示列表的内容。

```
...
<div class="card-block" id="mainContainer">

    <div class="row" id="mainContainerRepleace"
        th:each="comment,commentStat : ${comments}"
        th:object="${comment}">
        <h2 class="card-title col-lg-1 col-md-2">
            <span> <a href="/u/waylau"
                th:href="'/u/'+ *{user.username}" title="waylau"
                th:title="*{user.username}"> <img
                    src="/images/avatar-defualt.jpg"
                    th:src="*{user.avatar} == null ? '/images/avatar-de-
fualt.jpg' : *{user.avatar}"
```

```
                    class="blog-avatar-50">
                </a>
                </span>
            </h2>
            <div class="card-text col-lg-11 col-md-10">
                <a href="/u/waylau" th:href="'/u/'+ *{user.username}"
                    class="card-link" th:text="*{user.username}">waylau</a>
                [[${commentStat.index} + 1]]楼
                [[${#dates.format(comment.createTime, 'yyyy-MM-dd HH:mm')}]]
<a
                    href="javascript:void(0)" class="blog-delete-comment"
                    th:if="${commentOwner} eq *{user.username}"
                    th:attr="commentId=*{id}"><i class="fa fa-trash-o"
                    aria-hidden="true"></i></a>
                <p th:text="*{content}">不错哦，顶起！</p>
            </div>
        </div>

</div>
...
```

17.3.2 修改 blog.js

修改 blog.js 定义了发起调用获取评论列表、发表评论和删除评论的 API 的事件。

```
...
// 获取评论列表
function getComment(blogId) {

    $.ajax({
         url: '/comments',
         type: 'GET',
         data:{"blogId":blogId},
         success: function(data){
            $("#mainContainer").html(data);

         },
         error : function() {
             toastr.error("error!");
         }
     });
}

// 提交评论
$(".blog-content-container").on("click","#submitComment", function () {
// 获取CSRF Token
    var csrfToken = $("meta[name='_csrf']").attr("content");
    var csrfHeader = $("meta[name='_csrf_header']").attr("content");
```

```
    $.ajax({
         url: '/comments',
         type: 'POST',
         data:{"blogId":blogId, "commentContent":$('#commentContent').
val()},
         beforeSend: function(request) {
             request.setRequestHeader(csrfHeader, csrfToken); // 添加
CSRF Token
         },
         success: function(data){
             if (data.success) {
                 // 清空评论框
                 $('#commentContent').val('');
                 // 获取评论列表
                 getComment(blogId);
             } else {
                 toastr.error(data.message);
             }
         },
         error : function() {
             toastr.error("error!");
         }
     });
});

// 删除评论
$(".blog-content-container").on("click",".blog-delete-comment", func-
tion () {
    // 获取CSRF Token
    var csrfToken = $("meta[name='_csrf']").attr("content");
    var csrfHeader = $("meta[name='_csrf_header']").attr("content");

    $.ajax({
         url: '/comments/'+$(this).attr("commentId")+'?blogId='+blogId,
         type: 'DELETE',
         beforeSend: function(request) {
             request.setRequestHeader(csrfHeader, csrfToken); //添加
CSRF Token
         },
         success: function(data){
             if (data.success) {
                 //获取评论列表
                 getCommnet(blogId);
             } else {
                 toastr.error(data.message);
             }
         },
         error : function() {
             toastr.error("error!");
         }
```

```
        });
});

// 初始化博客评论
getComment(blogId);
```

17.3.3 运行

首先启动 blog-comment 项目。

通过浏览器访问 http://localhost:8080/，进入博客系统主页，并登录系统，撰写博客，进行评论。图 17-1 所示为博客评论界面的效果。

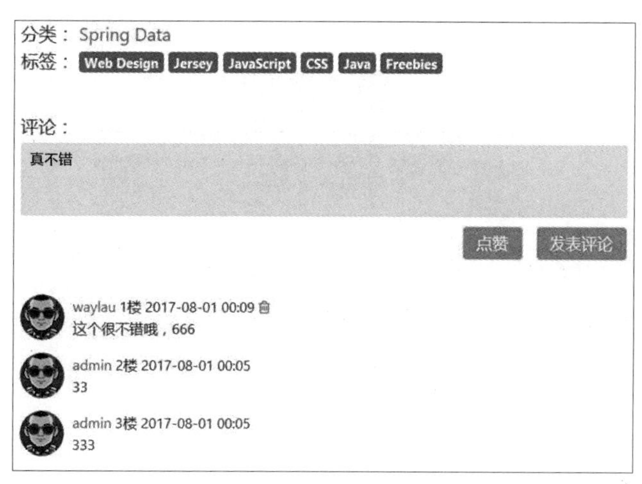

图17-1　评论界面

评论成功后会在评论框下展示评论列表。

17.3.4　示例源码

本节示例源码在 blog-comment 目录下。

第18章 点赞管理实现

18.1 点赞管理的需求回顾

本节将实现博客系统中的点赞管理功能。

点赞是另外一种社会化的交流方式。相对于评论而言，点赞可操作性更强，毕竟点赞只需要按一次，而评论就不得不输入大量的文字。

在 blog-comment 项目基础上构建了一个新的项目 blog-vote，用于演示整个点赞管理的实现过程。

18.8.1 点赞管理的需求

首先来回顾一下点赞管理所具有的功能需求。

概括起来，在本节将实现如下功能。

（1）点赞的模型设计。要设计点赞的领域模型。点赞不但要与博客文章关联，还需要与用户关联。

（2）发表点赞。用户在某个博客下面发表点赞。一篇博客会对应多个点赞。

（3）取消点赞。用户取消自己的点赞操作。

（4）统计点赞量。实现统计点赞量的功能。点赞量是衡量一篇博客文章热度的重要指标之一。

18.8.2 API 设计

点赞管理的整体 API 设计如下。

- votes：点赞管理。
  - /votes：POST 保存点赞。
    - blogId：博客 ID。
  - /votes/{id}：DELETE 删除点赞。
    - id：点赞 ID。
    - blogId：博客 ID。

18.2 点赞管理的后台实现

本节将实现点赞管理的后台编码实现。

18.2.1 点赞的模型设计

在 com.waylau.spring.boot.blog.domain 包下创建点赞的实体 Vote，并建立起与 User 的关系。

```
@Entity // 实体
public class Vote implements Serializable {
    private static final long serialVersionUID = 1L;

    @Id // 主键
    @GeneratedValue(strategy = GenerationType.IDENTITY) // 自增长策略
    private Long id; // 用户的唯一标识

    @OneToOne(cascade = CascadeType.DETACH, fetch = FetchType.LAZY)
    @JoinColumn(name="user_id")
    private User user;

    @Column(nullable = false) // 映射为字段，值不能为空
    @org.hibernate.annotations.CreationTimestamp  // 由数据库自动创建时间
    private Timestamp createTime;

    protected Vote() {
    }

    public Vote(User user) {
        this.user = user;
    }

    // 省略其他getter/setter方法

    public Timestamp getCreateTime() {
        return createTime;
    }
}
```

对于创建时间 createTime，我们使用了 Hibernate 的注解 @org.hibernate.annotations.Creation-Timestamp，这样，该注解会自动创建一个时间。

同时，修改 Blog 类来建立起 Blog 与 Vote 的关系。

```
...
@OneToMany(cascade = CascadeType.ALL, fetch = FetchType.LAZY)
@JoinTable(name = "blog_vote", joinColumns = @JoinColumn(name = "blog_
id", referencedColumnName = "id"),
    inverseJoinColumns = @JoinColumn(name = "vote_id", referencedColumn-
Name = "id"))
private List<Vote> votes;
...

public List<Vote> getVotes() {
    return votes;
```

```
}

public void setVotes(List<Vote> votes) {
    this.votes = votes;
    this.voteSize = this.votes.size();
}

/**
 * 点赞
 * @param vote
 * @return
 */
public boolean addVote(Vote vote) {
    boolean isExist = false;
    // 判断重复
    for (int index=0; index < this.votes.size(); index ++ ) {
        if (this.votes.get(index).getUser().getId() == vote.getUser().
getId()) {
            isExist = true;
            break;
        }
    }

    if (!isExist) {
        this.votes.add(vote);
        this.voteSize = this.votes.size();
    }

    return isExist;
}
/**
 * 取消点赞
 * @param voteId
 */
public void removeVote(Long voteId) {
    for (int index=0; index < this.votes.size(); index ++ ) {
        if (this.votes.get(index).getId() == voteId) {
            this.votes.remove(index);
            break;
        }
    }

    this.voteSize = this.votes.size();
}
```

这里需要注意的是，每次点赞或取消点赞后，都要统计点赞量 voteSize。

18.2.2 点赞的仓库

在 com.waylau.spring.boot.blog.repository 包下，建立继承自 org.springframework.data.jpa.repository.JpaRepository 接口的 VoteRepository。

```
public interface VoteRepository extends JpaRepository<Vote, Long>{

}
```

18.2.3 点赞的服务接口及实现

在 com.waylau.spring.boot.blog.service 包下，增加点赞的服务接口及实现。点赞的服务接口 VoteService 较为简单，只定义了获取点赞及取消点赞的方法。

```
public interface VoteService {
    /**
     * 根据id获取Vote
     * @param id
     * @return
     */
    Optional<Vote> getVoteById(Long id);
    /**
     * 删除Vote
     * @param id
     * @return
     */
    void removeVote(Long id);
}
```

以下是服务的实现类 VoteServiceImpl。

```
@Service
public class VoteServiceImpl implements VoteService {

    @Autowired
    private VoteRepository voteRepository;

    @Override
    public Optional<Vote> getVoteById(Long id) {
        return voteRepository.findById(id);
    }

    @Override
    public void removeVote(Long id) {
        voteRepository.deleteById(id);
    }
```

```
}
```

同时，也要在 BlogService 增加以下方法。

```
...
/**
 * 点赞
 * @param blogId
 * @return
 */
Blog createVote(Long blogId);

/**
 * 取消点赞
 * @param blogId
 * @param voteId
 * @return
 */
void removeVote(Long blogId, Long voteId);
```

在 BlogServiceImpl 中实现上述接口。

```
...
@Override
public Blog createVote(Long blogId) {
    Optional<Blog> optionalBlog = blogRepository.findById(blogId);
    Blog originalBlog = null;

    if (optionalBlog.isPresent()) {
        originalBlog = optionalBlog.get();

        User user = (User)SecurityContextHolder.getContext().getAuthen-
tication().getPrincipal();
        Vote vote = new Vote(user);
        boolean isExist = originalBlog.addVote(vote);
        if (isExist) {
            throw new IllegalArgumentException("该用户已经点过赞了");
        }
    }

    return this.saveBlog(originalBlog);
}

@Override
public void removeVote(Long blogId, Long voteId) {
    Optional<Blog> optionalBlog = blogRepository.findById(blogId);
    Blog originalBlog = null;

    if (optionalBlog.isPresent()) {
```

```
        originalBlog = optionalBlog.get();
        originalBlog.removeVote(voteId);
        this.saveBlog(originalBlog);
    }
}
```

18.2.4 控制层

控制层 VoteController 类提供了发表点赞、取消点赞的 API。

```
@Controller
@RequestMapping("/votes")
public class VoteController {

    @Autowired
    private BlogService blogService;

    @Autowired
    private VoteService voteService;

    /**
     * 发表点赞
     * @param blogId
     * @param VoteContent
     * @return
     */
    @PostMapping
    @PreAuthorize("hasAnyAuthority('ROLE_ADMIN','ROLE_USER')")
    // 指定角色权限才能操作方法
    public ResponseEntity<Response> createVote(Long blogId) {

        try {
            blogService.createVote(blogId);
        } catch (ConstraintViolationException e)  {
            return ResponseEntity.ok().body(new Response(false,
                    ConstraintViolationExceptionHandler.getMessage(e)));

        } catch (Exception e) {
            return ResponseEntity.ok().body(new Response(false, e.get-
Message()));
        }

        return ResponseEntity.ok().body(new Response(true, "点赞成功",
null));
    }

    /**
```

```
     * 删除点赞
     * @return
     */
    @DeleteMapping("/{id}")
    @PreAuthorize("hasAnyAuthority('ROLE_ADMIN','ROLE_USER')")
    // 指定角色权限才能操作方法
    public ResponseEntity<Response> delete(@PathVariable("id") Long id,
Long blogId) {

        boolean isOwner = false;
        Optional<Vote> optionalVote = voteService.getVoteById(id);
        User user = null;
        if (optionalVote.isPresent()) {
            user = optionalVote.get().getUser();
        } else {
            return ResponseEntity.ok().body(new Response(false, "不存在该
点赞！"));
        }

        // 判断操作用户是否为点赞的所有者
        if (SecurityContextHolder.getContext().getAuthentication()
!=null
                && SecurityContextHolder.getContext().getAuthentication
().isAuthenticated()
                && !SecurityContextHolder.getContext().getAuthentica-
tion()
                .getPrincipal().toString().equals("anonymousUser")) {
            User principal = (User)SecurityContextHolder.getContext()
                    .getAuthentication().getPrincipal();
            if (principal !=null && user.getUsername().equals(princi-
pal.getUsername())) {
                isOwner = true;
            }
        }

        if (!isOwner) {
            return ResponseEntity.ok().body(new Response(false, "没有操作
权限"));
        }

        try {
            blogService.removeVote(blogId, id);
            voteService.removeVote(id);
        } catch (ConstraintViolationException e)  {
            return ResponseEntity.ok().body(new Response(false,
                    ConstraintViolationExceptionHandler.getMessage(e)));

        } catch (Exception e) {
```

```
            return ResponseEntity.ok().body(new Response(false, e.get-
Message()));
        }

        return ResponseEntity.ok().body(new Response(true, "取消点赞成功",
null));
    }
}
```

修改 UserspaceController 的方法 getBlogById，增加判断操作用户的点赞情况。

```
@GetMapping("/{username}/blogs/{id}")
public String getBlogById(@PathVariable("username") String username,
@PathVariable("id") Long id, Model model) {
    ...
    // 判断操作用户的点赞情况
    List<Vote> votes = blog.getVotes();
    Vote currentVote = null; // 当前用户的点赞情况

    if (principal !=null) {
        for (Vote vote : votes) {
            if (vote.getUser().getUsername().equals(principal.getUser-
name())) {
                currentVote = vote;
                break;
            }
        }
    }

    model.addAttribute("currentVote",currentVote);
    ...

    return "/userspace/blog";
}
```

18.3 点赞管理的前台实现

点赞的前台页面实现较为简单。

18.3.1 修改 blog.html

修改 blog.html 中的内容。在评论框下方设置点赞按钮及取消点赞的按钮。

```
...
<div class="card-block">
    <h5>评论：</h5>
    <div class="row">
        <div class="col-lg-12">
            <textarea class="blog-textarea" placeholder="看帖需留言~"
                id="commentContent"></textarea>
        </div>
    </div>
    <button class="btn btn-primary float-right" id="submitComment">发表评
论</button>
    <button class="btn btn-primary float-right"
        th:if="${currentVote} == null" id="submitVote">点赞</button>
    <button class="btn btn-primary float-right" th:if="${currentVote}"
        th:attr="voteId=${currentVote.id}" id="cancelVote">取消点赞</
button>
</div>
...
```

18.3.2 修改 blog.js

修改 blog.js 定义了发起调用提交点赞和取消点赞的 API 的事件。

```
...
// 提交点赞
$(".blog-content-container").on("click","#submitVote", function () {
    // 获取CSRF Token
    var csrfToken = $("meta[name='_csrf']").attr("content");
    var csrfHeader = $("meta[name='_csrf_header']").attr("content");

    $.ajax({
         url: '/votes',
         type: 'POST',
         data:{"blogId":blogId},
         beforeSend: function(request) {
             request.setRequestHeader(csrfHeader, csrfToken); // 添加
CSRF Token
         },
         success: function(data){
             if (data.success) {
                 toastr.info(data.message);
                    // 成功后，重定向
                 window.location = blogUrl;
             } else {
                 toastr.error(data.message);
             }
         },
         error : function() {
```

```
                toastr.error("error!");
            }
        });
});

// 取消点赞
$(".blog-content-container").on("click","#cancelVote", function () {
    // 获取CSRF Token
    var csrfToken = $("meta[name='_csrf']").attr("content");
    var csrfHeader = $("meta[name='_csrf_header']").attr("content");

    $.ajax({
        url: '/votes/'+$(this).attr('voteId')+'?blogId='+blogId,
        type: 'DELETE',
        beforeSend: function(request) {
            request.setRequestHeader(csrfHeader, csrfToken);
            // 添加  CSRF Token
        },
        success: function(data){
            if (data.success) {
                toastr.info(data.message);
               // 成功后，重定向
                window.location = blogUrl;
            } else {
                toastr.error(data.message);
            }
        },
        error : function() {
            toastr.error("error!");
        }
    });
});
```

18.3.3 运行

首先，启动 blog-vote 项目。

通过浏览器访问 http://localhost:8080/ 进入博客系统主页，并登录系统，撰写博客，进行点赞。

图 18-1 所示为博客点赞界面的效果。

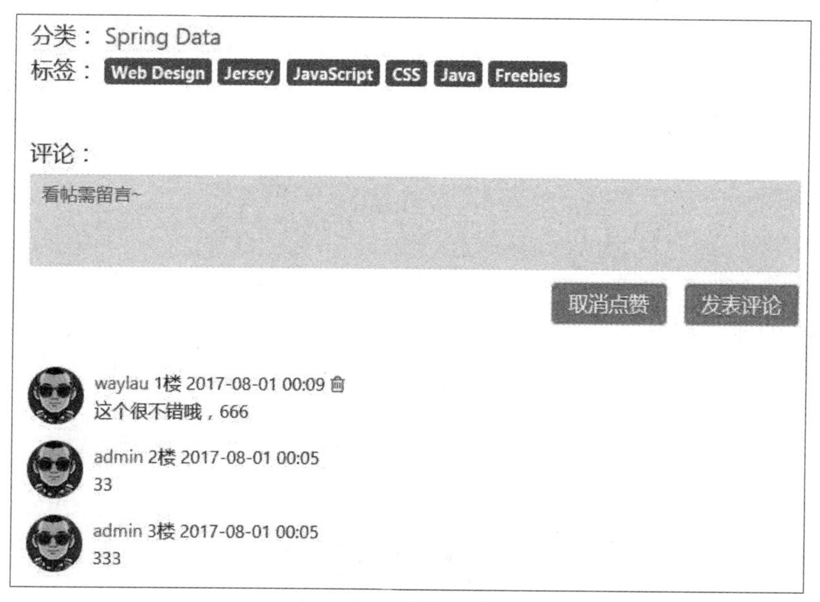

图18-1　点赞界面

点赞成功之后，会在评论框下展示“取消点赞”的按钮。

18.3.4 示例源码

本节示例源码在 blog-vote 目录下。

第19章
分类管理实现

19.1 分类管理的需求回顾

本节将实现博客系统中的分类管理功能。

分类管理可以方便地将自己的个人博客进行分门别类。分类后的博客更加容易管理和查找。

在 blog-vote 项目基础上构建了一个新的项目 blog-catalog，用于演示整个分类管理的实现过程。

19.1.1 分类管理的需求

首先回顾一下分类管理所具有的功能需求。对于分类管理来说，除支持创建、修改、删除分类的功能外，还支持按照分类查询的功能。

概括起来，在本节将实现如下功能。

（1）分类的模型设计。要设计分类的领域模型。分类不但要与博客文章关联，还需要与用户关联。

（2）创建分类。用户可以创建自己的分类，同时，可以将不同的博客归结为某一类。

（3）删除分类。用户可以删除自己创建的分类。

（4）按照分类查询博客。博客文章经过了分类之后，就可以通过分类来方便查找某个分类下的所有博客。

19.1.2 API 设计

分类管理的整体 API 设计如下。

- catalogs：分类管理。
  - /catalogs：GET 获取用户分类列表。
    - username：用户账号。
  - /catalogs：POST 保存用户分类。
    - username：用户账号。
    - CatalogVO：含 username、Catalog。
  - /catalogs/edit：GET 获取编辑分类的界面。
  - /catalogs/edit/{id}：GET 获取某 ID 分类的编辑界面。
  - /catalogs/{id}：DELETE 删除分类。
    - id：分类 id。
    - username：用户账号。

19.2 分类管理的后台实现

本节将实现分类管理的后台编码实现。

19.2.1 分类的模型设计

在 com.waylau.spring.boot.blog.domain 包下创建分类的实体 Catalog，并建立起与 User 的关系。

```
@Entity // 实体
public class Catalog implements Serializable {
    private static final long serialVersionUID = 1L;

    @Id // 主键
    @GeneratedValue(strategy = GenerationType.IDENTITY) // 自增长策略
    private Long id; // 用户的唯一标识

    @NotEmpty(message = "名称不能为空")
    @Size(min=2, max=30)
    @Column(nullable = false) // 映射为字段，值不能为空
    private String name;

    @OneToOne(cascade = CascadeType.DETACH, fetch = FetchType.LAZY)
    @JoinColumn(name="user_id")
    private User user;

    protected Catalog() {
    }

    public Catalog(User user, String name) {
        this.name = name;
        this.user = user;
    }

    // 省略getter/setter方法
}
```

同时，修改 Blog 类来建立起 Blog 与 Catalog 的关系。

```
@OneToOne(cascade = CascadeType.DETACH, fetch = FetchType.LAZY)
@JoinColumn(name="catalog_id")
private Catalog catalog;
...

public Catalog getCatalog() {
    return catalog;
}
public void setCatalog(Catalog catalog) {
    this.catalog = catalog;
```

```
}
```

在 com.waylau.spring.boot.blog.vo 包下创建 CatalogVO，作为前台调用后台的传递用的参数对象。

```
public class CatalogVO {

    private String username;
    private Catalog catalog;

    public CatalogVO() {
    }

    // 省略getter/setter方法
}
```

19.2.2 分类的仓库

建立 CatalogRepository 继承自 org.springframework.data.jpa.repository.JpaRepository 接口即可。

```
public interface CatalogRepository extends JpaRepository<Catalog, Long>{


    /**
     * 根据用户查询
     * @param user
     * @return
     */
    List<Catalog> findByUser(User user);

    /**
     * 根据用户查询
     * @param user
     * @param name
     * @return
     */
    List<Catalog> findByUserAndName(User user,String name);
}
```

同时，修改 BlogRepository，增加了如下方法。

```
...
/**
 * 根据分类查询博客列表
 * @param catalog
 * @param pageable
 * @return
 */
```

```
Page<Blog> findByCatalog(Catalog catalog, Pageable pageable);
```

19.2.3 分类的服务接口及实现

分类的服务接口 CatalogService。

```
public interface CatalogService {
    /**
     * 保存Catalog
     * @param catalog
     * @return
     */
    Catalog saveCatalog(Catalog catalog);

    /**
     * 删除Catalog
     * @param id
     * @return
     */
    void removeCatalog(Long id);

    /**
     * 根据id获取Catalog
     * @param id
     * @return
     */
    Optional<Catalog> getCatalogById(Long id);

    /**
     * 获取Catalog列表
     * @return
     */
    List<Catalog> listCatalogs(User user);
}
```

以下是服务的实现类 CatalogServiceImpl。

```
@Service
public class CatalogServiceImpl implements CatalogService {
    @Autowired
    private CatalogRepository catalogRepository;

    @Override
    public Catalog saveCatalog(Catalog catalog) {
        // 判断重复
        List<Catalog> list = catalogRepository.findByUserAndName(cata
log.getUser(),
                catalog.getName());
```

```
        if(list !=null && list.size() > 0) {
            throw new IllegalArgumentException("该分类已经存在了");
        }
        return catalogRepository.save(catalog);
    }

    @Override
    public void removeCatalog(Long id) {
        catalogRepository.deleteById(id);
    }

    @Override
    public Optional<Catalog> getCatalogById(Long id) {
        return catalogRepository.findById(id);
    }

    @Override
    public List<Catalog> listCatalogs(User user) {
        return catalogRepository.findByUser(user);
    }

}
```

同时，也要在 BlogService 增加以下方法。

```
...
/**
 * 根据分类进行查询
 * @param catalog
 * @param pageable
 * @return
 */
Page<Blog> listBlogsByCatalog(Catalog catalog, Pageable pageable);
```

在 BlogServiceImpl 中实现上述接口。

```
...
@Override
public Page<Blog> listBlogsByCatalog(Catalog catalog, Pageable page
able) {
    Page<Blog> blogs = blogRepository.findByCatalog(catalog, pageable);
    return blogs;
}
```

19.2.4 控制层

在 com.waylau.spring.boot.blog.controller 包下，控制层 CatalogController 类提供了获取分类列表、创建分类、删除分类、获取分类编辑界面等 API。

```
@Controller
@RequestMapping("/catalogs")
public class CatalogController {

    @Autowired
    private CatalogService catalogService;

    @Autowired
    private UserDetailsService userDetailsService;

    /**
     * 获取分类列表
     * @param username
     * @param model
     * @return
     */
    @GetMapping
    public String listComments(@RequestParam(value="username",
required=true) String username, Model model) {
        User user = (User)userDetailsService.loadUserByUsername
(username);
        List<Catalog> catalogs = catalogService.listCatalogs(user);

        // 判断操作用户是否为分类的所有者
        boolean isOwner = false;

        if (SecurityContextHolder.getContext().getAuthentication()
!=null
                && SecurityContextHolder.getContext().getAuthentica-
tion().isAuthenticated()
                 &&  !SecurityContextHolder.getContext().getAuthentica-
tion().getPrincipal().toString()
                 .equals("anonymousUser")) {
            User principal = (User)SecurityContextHolder.getContext().
getAuthentication().getPrincipal();
            if (principal !=null && user.getUsername().equals(princi-
pal.getUsername())) {
                isOwner = true;
            }
        }

        model.addAttribute("isCatalogsOwner", isOwner);
        model.addAttribute("catalogs", catalogs);
        return "/userspace/u :: #catalogRepleace";
    }

    /**
     * 创建分类
     * @param catalogVO
```

```
     * @return
     */
    @PostMapping
    @PreAuthorize("authentication.name.equals(#catalogVO.username)")//
指定用户才能操作方法
    public ResponseEntity<Response> create(@RequestBody CatalogVO cata-
logVO) {

        String username = catalogVO.getUsername();
        Catalog catalog = catalogVO.getCatalog();

        User user = (User)userDetailsService.loadUserByUsername(user
name);

        try {
            catalog.setUser(user);
            catalogService.saveCatalog(catalog);
        } catch (ConstraintViolationException e)  {
            return ResponseEntity.ok().body(new Response(false,
                    ConstraintViolationExceptionHandler.getMessage(e)));

        } catch (Exception e) {
            return ResponseEntity.ok().body(new Response(false, e.get-
Message()));
        }

        return ResponseEntity.ok().body(new Response(true, "处理成功",
null));
    }

    /**
     * 删除分类
     * @param username
     * @param id
     * @return
     */
    @DeleteMapping("/{id}")
    @PreAuthorize("authentication.name.equals(#username)")  // 指定用户才
能操作方法
    public ResponseEntity<Response> delete(String username, @PathVari
able("id") Long id) {
        try {
            catalogService.removeCatalog(id);
        } catch (ConstraintViolationException e)  {
            return ResponseEntity.ok().body(new Response(false,
                    ConstraintViolationExceptionHandler.getMessage(e)));

        } catch (Exception e) {
            return ResponseEntity.ok().body(new Response(false, e.get-
```

```
Message()));
        }

        return ResponseEntity.ok().body(new Response(true, "处理成功",
null));
    }

    /**
     * 获取分类编辑界面
     * @param model
     * @return
     */
    @GetMapping("/edit")
    public String getCatalogEdit(Model model) {
        Catalog catalog = new Catalog(null, null);
        model.addAttribute("catalog",catalog);
        return "/userspace/catalogedit";
    }

    /**
     * 根据Id获取编辑界面
     * @param id
     * @param model
     * @return
     */
    @GetMapping("/edit/{id}")
    public String getCatalogById(@PathVariable("id") Long id, Model
model) {
        Optional<Catalog> optionalCatalog = catalogService.getCatalog-
ById(id);
        Catalog catalog = null;

        if (optionalCatalog.isPresent()) {
            catalog = optionalCatalog.get();
        }

        model.addAttribute("catalog",catalog);
        return "/userspace/catalogedit";
    }

}
```

修改 UserspaceController，查询博客时，增加了对分类查询的判断。

```
@Autowired
private CatalogService catalogService;

...

@GetMapping("/{username}/blogs")
```

```
public String listBlogsByOrder(@PathVariable("username") String user
name,
        @RequestParam(value="order",required=false,defaultValue="new")
String order,
        @RequestParam(value="catalog",required=false ) Long catalogId,
        @RequestParam(value="keyword",required=false,defaultValue="" )
String keyword,
        @RequestParam(value="async",required=false) boolean async,
        @RequestParam(value="pageIndex",required=false,defaultValue="0")
int pageIndex,
        @RequestParam(value="pageSize",required=false,defaultValue="10")
int pageSize,
        Model model) {
    ...

    if (catalogId != null && catalogId > 0) { // 分类查询
        Catalog catalog = catalogService.getCatalogById(catalogId);
        Pageable pageable = new PageRequest(pageIndex, pageSize);
        page = blogService.listBlogsByCatalog(catalog, pageable);
        order = "";
    }
    ...
    model.addAttribute("catalogId", catalogId);
    ...
}
```

以及在获取编辑博客的界面时，将当前用户的分类列表绑定到前台界面。

```
/**
    * 获取新增博客的界面
    *
    * @param model
    * @return
    */
@GetMapping("/{username}/blogs/edit")
public ModelAndView createBlog(@PathVariable("username") String user
name,
        Model model) {
    // 获取用户分类列表
    User user = (User)userDetailsService.loadUserByUsername(username);
    List<Catalog> catalogs = catalogService.listCatalogs(user);

    model.addAttribute("catalogs", catalogs);
    ...
}

/**
    * 获取编辑博客的界面
    *
    * @param model
    * @return
```

```
    */
@GetMapping("/{username}/blogs/edit/{id}")
public ModelAndView editBlog(@PathVariable("username") String username,
        @PathVariable("id") Long id, Model model) {
    // 获取用户分类列表
    User user = (User)userDetailsService.loadUserByUsername(username);
    List<Catalog> catalogs = catalogService.listCatalogs(user);

    model.addAttribute("catalogs", catalogs);
    ...
}
```

同时，在保存博客的时候，对分类进行判空处理，并在修改博客时增加对分类的处理。

```
@PostMapping("/{username}/blogs/edit")
@PreAuthorize("authentication.name.equals(#username)")
public ResponseEntity<Response> saveBlog(@PathVariable("username")
String username,
        @RequestBody Blog blog) {

    // 对Catalog进行空处理
    if (blog.getCatalog().getId() == null) {
        return ResponseEntity.ok().body(new Response(false,"未选择分类
"));
    }
    try {

        // 判断是修改还是新增
        if (blog.getId() != null) {
            Optional<Blog> optionalBlog = blogService.getBlogById(blog.
getId());
            if (optionalBlog.isPresent()) {
                Blog orignalBlog = optionalBlog.get();
                orignalBlog.setTitle(blog.getTitle());
                orignalBlog.setContent(blog.getContent());
                orignalBlog.setSummary(blog.getSummary());
                orignalBlog.setCatalog(blog.getCatalog()); // 增加对分类
的处理
                blogService.saveBlog(orignalBlog);
            }
    ...
}
```

19.3 分类管理的前台实现

本节将会进行分类的前台页面实现的演示。

19.3.1 修改 u.html

用户主页的左侧栏目对分类进行了实现。

```
...
<!-- 分类 -->
<div id="catalogMain">
    <div class="card my-4" id="catalogRepleace">
        <h5 class="card-header">
            <i class="fa fa-bars" aria-hidden="true"></i> 分类 <a
                href="javascript:void(0)" th:if="${isCatalogsOwner}"
                class="blog-add-catalog blog-right" data-toggle="modal"
                data-target="#flipFlop" role="button"><i class="fa
fa-plus"
                aria-hidden="true"></i></a>
        </h5>

        <ul class="list-group" th:each="catalog : ${catalogs}"
            th:object="${catalog}">
            <div class="blog-list-group-item">

                <a href="javascript:void(0)" th:attr="catalogId=*{id}"
                    class="blog-query-by-catalog"> [[*{name}]] </a>
<span
                    class="blog-right"> <a href="javascript:void(0)"
                    th:if="${isCatalogsOwner}" class="blog-edit-cata
log"
                    data-toggle="modal" data-target="#flipFlop"
role="button"
                    th:attr="catalogId=*{id}"><i class="fa fa-pencil-
square-o"
                        aria-hidden="true"></i></a> <a href="javas-
cript:void(0)"
                    th:if="${isCatalogsOwner}" class="blog-delete-cata-
log"
                    th:attr="catalogId=*{id}"><i class="fa fa-times"
                        aria-hidden="true"></i></a>
                </span>

            </div>
        </ul>
    </div>
</div>
...
```

并在该页面的底部增加编辑分类的模式窗口标签代码。

```
...
<!-- 模式窗口 -->
<div class="modal fade" id="flipFlop" tabindex="-1" role="dialog"
```

```
    aria-labelledby="modalLabel" aria-hidden="true">
    <div class="modal-dialog" role="document">
        <div class="modal-content">
            <div class="modal-header">
                <h4 class="modal-title" id="modalLabel">新增/编辑</h4>
                <button type="button" class="close" data-dismiss="mod-
al"
                    aria-label="Close">
                    <span aria-hidden="true">&times;</span>
                </button>

            </div>
            <div class="modal-body" id="catalogFormContainer"></div>
            <div class="modal-footer">
                <button class="btn btn-primary" data-dismiss="modal"
                    id="submitEditCatalog">提交</button>
                <button type="button" class="btn btn-secondary"
                    data-dismiss="modal">Close</button>
            </div>
        </div>
    </div>
</div>
...
```

19.3.2 修改 blogedit.html

修改 blogedit.html，以支持在编辑博客时可以选择博客的分类。

```
...
<!--分类-->
<div class="row mt-1">
    <div class="col-lg-12">
        分类：<select id="catalogSelect"
            class="form-control form-control-chosen"
            data-placeholder="请选择">
            <option th:value="*{id}" th:text="*{name}"
                th:each="catalog : ${blogModel.catalogs}"
                th:object="${catalog}"
                th:selected="${catalog  eq blogModel.blog.catalog
}">Java</option>
        </select>
    </div>
</div>
...
```

19.3.3 修改 blog.html

修改 blog.html，以支持在查看博客时可以看到博客的分类。

```
...
<h5>
    分类：<a
        th:href="'/u/'+${blogModel.user.username} + '/blogs?catalog=
'+${blogModel.catalog.id} "
        th:text="${blogModel.catalog.name}"> Spring Boot </a>
</h5>
...
```

19.3.4 新增 catalogedit.html

在 templates/userspace 目录下新增了 catalogedit.html 页面。

```
<div class="container" th:object="${catalog}">
    <form id="catalogForm">
        <input type="hidden" name="id" th:value="*{id}" id="catalogId">

        <div class="form-group">
            <label for="name" class="col-form-label">名称</label> <input
                type="text" class="form-control" id="catalogName"
name="name"
                th:value="*{name}" maxlength="30" placeholder="请输入姓
名，至少2个字符，至多30个">
        </div>
    </form>
</div>
```

该页面主要用于编辑分类。

19.3.5 修改 u.js

修改 u.js，增加了分类相关的 API 的调用事件。

```
var catalogId;

// 获取分类列表
function getCatalogs(username) {

    // 获取CSRF Token
    $.ajax({
        url: '/catalogs',
        type: 'GET',
```

```
        data:{"username":username},
        success: function(data){
        $("#catalogMain").html(data);
        },
        error : function() {
            toastr.error("error!");
        }
    });
}

// 获取编辑分类的页面
$(".blog-content-container").on("click",".blog-add-catalog", function
() {
    $.ajax({
        url: '/catalogs/edit',
        type: 'GET',
        success: function(data){
            $("#catalogFormContainer").html(data);
        },
        error : function() {
            toastr.error("error!");
        }
    });
});

// 获取编辑某个分类的页面
$(".blog-content-container").on("click",".blog-edit-catalog", function
() {

    $.ajax({
        url: '/catalogs/edit/'+$(this).attr('catalogId'),
        type: 'GET',
        success: function(data){
            $("#catalogFormContainer").html(data);
        },
        error : function() {
            toastr.error("error!");
        }
    });
});


    // 提交分类
$("#submitEditCatalog").click(function() {
    // 获取CSRF Token
    var csrfToken = $("meta[name='_csrf']").attr("content");
    var csrfHeader = $("meta[name='_csrf_header']").attr("content");

    $.ajax({
```

```
        url : '/catalogs',
        type : 'POST',
        contentType : "application/json; charset=utf-8",
        data : JSON.stringify({
            "username" : username,
            "catalog" : {
                "id" : $('#catalogId').val(),
                "name" : $('#catalogName').val()
            }
        }),
        beforeSend : function(request) {
            request.setRequestHeader(csrfHeader, csrfToken);
    // 添加 CSRF Token
        },
        success : function(data) {
            if (data.success) {
                toastr.info(data.message);
                // 成功后，刷新列表
                getCatalogs(username);
            } else {
                toastr.error(data.message);
            }
        },
        error : function() {
            toastr.error("error!");
        }
    });
});

// 删除分类
$(".blog-content-container").on("click",".blog-delete-catalog", func-
tion () {
    // 获取CSRF Token
    var csrfToken = $("meta[name='_csrf']").attr("content");
    var csrfHeader = $("meta[name='_csrf_header']").attr("content");

    $.ajax({
        url: '/catalogs/'+$(this).attr('catalogid')+'?username='+user
name,
        type: 'DELETE',
        beforeSend: function(request) {
            request.setRequestHeader(csrfHeader, csrfToken); // 添加
CSRF Token
        },
        success: function(data){
            if (data.success) {
                toastr.info(data.message);
                // 成功后，刷新列表
                getCatalogs(username);
```

```
            } else {
                toastr.error(data.message);
            }
        },
        error : function() {
            toastr.error("error!");
        }
    });
});

// 根据分类查询
$(".blog-content-container").on("click",".blog-query-by-catalog", func-
tion () {
    catalogId = $(this).attr('catalogId');
    getBlogsByName(0, _pageSize);
});

getCatalogs(username);
```

并修改了按分类查询的方法 getBlogsByName。

```
// 根据用户名、页面索引、页面大小获取用户列表
function getBlogsByName(pageIndex, pageSize) {
     $.ajax({
         url: "/u/"+ username +"/blogs",
         contentType : 'application/json',
         data:{
             "async":true,
             "pageIndex":pageIndex,
             "pageSize":pageSize,
             "catalog": catalogId,
             "keyword":$("#keyword").val()
         },
         success: function(data){
             $("#mainContainer").html(data);

             // 如果是分类查询，则取消最新、最热选中样式
             if (catalogId) {
                $(".nav-item .nav-link").removeClass("active");
             }
         },
         error : function() {
             toastr.error("error!");
         }
     });
}
```

19.3.6 运行

首先，启动 blog-catalog 项目。

通过浏览器访问 http://localhost:8080/，进入博客系统主页，并登录系统，创建分类。

图 19-1 所示为分类的编辑界面效果。

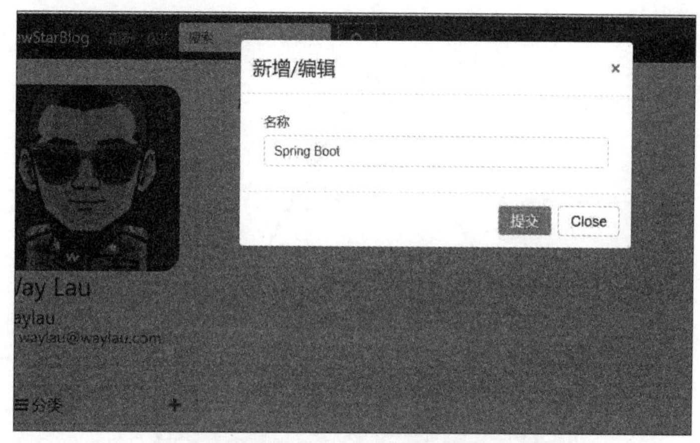

图19-1　分类的编辑界面

分类成功之后，就可以在创建博客时选择该分类。图 19-2 所示为创建博客前选择分类的效果。

图19-2　选择分类

19.3.7 示例源码

本节示例源码在 blog-catalog 目录下。

第20章 标签管理实现

20.1 标签管理的需求回顾

本节将实现博客系统中的标签管理功能。

标签管理可以方便地将自己的个人博客打上标签，这样，一方面，用户可以按照标签来对博客进行分类；另一方面，具有相同标签的博客之间就能建立起关联关系。

对于全文检索来说，标签同样是方便搜索博客的一种重要的关键字。

在 blog-catalog 项目基础上构建了一个新的项目 blog-tag，用于演示整个标签管理的实现过程。

20.1.1 所需环境

本例采用的开发环境为：jQuery Tags Input 1.3.6：http://xoxco.com/projects/code/tagsinput/。

jQuery Tags Input 是一款基于 jQuery 的标签组件。在本博客系统中，主要提供标签功能的界面实现。

20.1.2 标签管理的需求

首先来回顾一下标签管理所具有的功能需求。对于标签管理来说，既支持在博客中添加或修改、删除分类的功能，同时，也支持按照标签来分类查询的功能。

概括起来，在本章将实现如下功能。

（1）标签的模型设计。标签模型主要是与博客发生关联。在博客系统中，标签是博客的一个属性之一。

（2）创建标签。发表博客的时候，就可以来设置标签。

（3）修改、删除标签。编辑博客的时候，既能够修改标签，同时也支持删除标签。

（4）按照标签查询博客。这个功能仍然是基于之前关键字搜索的基础上的扩展，对 tags 字段进行查询。标签本身也是一类关键字。

20.2 标签管理的后台实现

本节将实现标签管理的后台编码实现。

20.2.1 标签的模型设计

标签在之前 Blog 的模型中其实已经是定义了的，就是作为 Blog 中属性之一的 tags。该字段就是通过字符串英文逗号分隔的形式来存储多个标签。这样也就建立起了博客与标签的关系。

```
...
@Column(name="tags", length = 100)
private String tags;  // 标签

public String getTags() {
    return tags;
}
public void setTags(String tags) {
    this.tags = tags;
}
...
```

对于创建、删除、修改标签的操作，本质上只需通对 Blog 对象的属性进行操作就能实现了。

20.2.2 修改 UserspaceController

修改 UserspaceController，在修改博客 saveBlog 方法中增加对标签的处理。

```
public ResponseEntity<Response> saveBlog(@PathVariable("username")
String username,
        @RequestBody Blog blog) {
    ...
    try {

        // 判断是修改还是新增
        if (blog.getId() != null) {
            Optional<Blog> optionalBlog = blogService.getBlogById(blog.
getId());
            if (optionalBlog.isPresent()) {
                Blog orignalBlog = optionalBlog.get();
                orignalBlog.setTitle(blog.getTitle());
                orignalBlog.setContent(blog.getContent());
                orignalBlog.setSummary(blog.getSummary());
                orignalBlog.setCatalog(blog.getCatalog());
                orignalBlog.setTags(blog.getTags());  // 增加对标签的处理
                blogService.saveBlog(orignalBlog);
            }
        }
    ...
}
```

20.3 标签管理的前台实现

标签管理的前台编码工作也比较简单。

20.3.1 修改 blog.html

在 blog.html 页面动态生成标签。单击这些标签，可以触发按照标签查询博客的事件。

```
...
<h5>
    标签： <a
        th:each="tag : ${#strings.arraySplit(blogModel.tags , ',')}"
        th:href="'/u/' + ${blogModel.user.username}  + '/blogs?keyword=
'+ ${tag}">
        <span class="badge badge-default" th:text="${tag}">Web
            Design</span>
    </a>
</h5>
...
```

同样，在 blogedit.html 中绑定了标签模型。

```
...
<!--标签-->
<div class="row mt-1">
    <div class="col-lg-12">
        标签：<input type="text" class="form-control form-control-tag"
            th:value="${blogModel.blog.tags}">
    </div>
</div>
...
```

20.3.2 修改 blogedit.js

修改 blogedit.js，增加了初始化标签的事件。

```
// 初始化标签
$('.form-control-tag').tagsInput({
    'defaultText':'输入标签'
});
```

20.3.3 运行

首先，启动 blog-tag 项目。

通过浏览器访问 http://localhost:8080/，进入博客系统主页，并登录系统，创建分类及博客。同时，可以设置博客的标签，如图 20-1 所示。

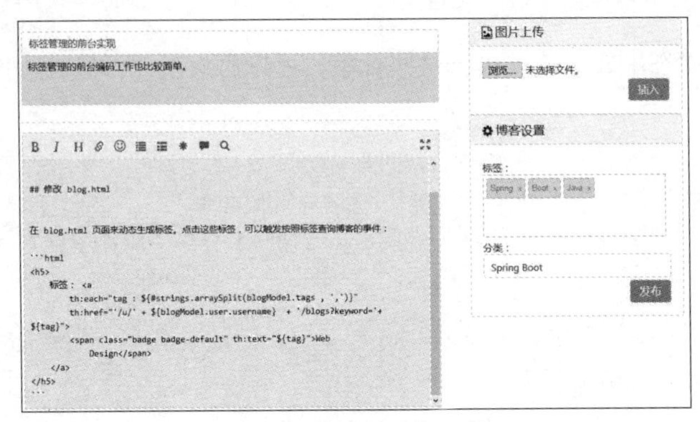

图20-1　设置博客的标签

20.3.4 示例源码

本节示例源码在 blog-tag 目录下。

第21章 首页搜索实现

21.1 首页搜索的需求回顾

对于博客系统来说，首页是整个博客系统的门面。首页所展示的博客内容大多都与搜索分不开。由于博客数量众多，搜索往往会涉及全文搜索的需求，因此需要采用大数据搜索的方式来实现。在本节中是采用 Elasticsearch 来实现首页搜索的功能。同时，本节的功能实现也是整个博客系统的最后一个功能。

有关 Elasticsearch 原理和用法在第 6 章中已经做了详细介绍，这里就不再赘述了。本节将主要关注如何来实现博客系统首页的搜索。

在 blog-tag 项目基础上构建了一个新的项目 blog-search，用于演示整个首页搜索的实现过程。

21.1.1 环境

本节所采用的开发环境与第 6 章中的一致，具体配置如下。

（1）Elasticsearch 5.5.0。

（2）Spring Data Elasticsearch 3.0.0.M4。

（3）IK Analysis for Elasticsearch 5.5.0。

21.1.2 build.gradle

这里需要添加 Spring Data Elasticsearch 的依赖。Spring Boot 已经提供了相关的 Starter 来实现 Spring Data Elasticsearch 开箱即用的功能，所以只需要在 build.gradle 文件中添加 Spring Data Elasticsearch 的 Starter 库即可。

```
// 依赖关系
dependencies {
    //...

    // 添加Spring Data Elasticsearch的依赖
    compile('org.springframework.boot:spring-boot-starter-data-elastic-
search')

    //...
}
```

21.1.3 修改 application.properties

修改 application.properties，增加下面两项配置。

```
# Elasticsearch 服务地址
```

```
spring.data.elasticsearch.cluster-nodes=localhost:9300
# 设置连接超时时间
spring.data.elasticsearch.properties.transport.tcp.connect_timeout=120s
```

21.1.4 实现功能

首先来回顾一下首页搜索所具有的功能需求。对于首页搜索来说，在首页展示各种维度分析的博客信息。这些维度包括最新发布、最热文章、热门标签、热门用户、热门文章、最新发表的文章等。

概括起来，在本节将实现如下功能。

（1）首页的分页模糊查询。在首页的头部（页首）显著位置，提供分页的模糊查询。

（2）最新文章列表。在首页的头部（页首）显著位置，提供按照最新的排序规则来展示文章列表。

（3）最热文章列表。在首页的头部（页首）显著位置，提供按照文章的热度来展示文章列表。

（4）热门标签。在博客系统的右侧展示热门标签。这些标签按照热度来进行排序。

（5）热门用户。在博客系统的右侧展示热门用户。这些用户按照热度来进行排序。

（6）热门文章。在博客系统的右侧展示热门文章。这些文章按照热度来进行排序。

（7）最新发布。在博客系统的右侧展示最新发布的文章。这些文章按照发布时间来进行排序。

21.1.5 API 设计

首页搜索的整体 API 设计如下。

- index : 主页，包含最新、最热文章，最热标签、最热用户等。
  - /blogs：GET。
    - order：排序类型， new/hot，默认为 new。
    - keyword：搜索关键字。博客的标签即为关键字。
    - async：是否异步请求页面。
    - pageIndex。
    - pageSize

21.2 首页搜索的后台实现

本节将完成首页搜索的后台编码实现。

21.2.1 创建文档

创建 com.waylau.spring.boot.blog.domain.es 包，在该包下创建一个注解为 org.springframework.data.elasticsearch.annotations.Document 文档类 EsBlog，专门用于 Elasticsearch 中存储博客的文档。

```
import org.springframework.data.annotation.Id;
import org.springframework.data.elasticsearch.annotations.Document;
import org.springframework.data.elasticsearch.annotations.Field;
import org.springframework.data.elasticsearch.annotations.FieldType;
...
/**
 * EsBlog 文档类
 *
 * @since 1.0.0 2017年3月5日
 * @author <a href="https://waylau.com">Way Lau</a>
 */
@Document(indexName = "blog", type = "blog")
public class EsBlog implements Serializable {

    private static final long serialVersionUID = 1L;

    @Id  // 主键
    private String id;
    @Field(type = FieldType.Long)
    private Long blogId; // Blog实体的id
    @Field(type = FieldType.text)
    private String title;
    @Field(type = FieldType.text)
    private String summary;
    @Field(type = FieldType.text)
    private String content;
    @Field(type = FieldType.keyword, index = false)  // 不做全文检索字段
    private String username;
    @Field(type = FieldType.text,index = false)  // 不做全文检索字段
    private String avatar;
    @Field(type = FieldType.Date,index = false)  // 不做全文检索字段
    private Timestamp createTime;
    @Field(type = FieldType.Integer,index = false)  // 不做全文检索字段
    private Integer readSize = 0; // 访问量、阅读量
    @Field(type = FieldType.Integer,index = false)  // 不做全文检索字段
    private Integer commentSize = 0;  // 评论量
    @Field(type = FieldType.Integer,index = false)  // 不做全文检索字段
    private Integer voteSize = 0;  // 点赞量
    @Field(type = FieldType.text,fielddata = true)
    private String tags;  // 标签

    protected EsBlog() {  // JPA的规范要求无参构造函数；设为protected 防止 直
接使用
    }
```

```
    public EsBlog(Long blogId, String title, String summary, String
content,
            String username, String avatar,Timestamp createTime,
            Integer readSize,Integer commentSize, Integer voteSize ,
String tags) {
        this.blogId = blogId;
        this.title = title;
        this.summary = summary;
        this.content = content;
        this.username = username;
        this.avatar = avatar;
        this.createTime = createTime;
        this.readSize = readSize;
        this.commentSize = commentSize;
        this.voteSize = voteSize;
        this.tags = tags;
    }

    public EsBlog(Blog blog){
        this.blogId = blog.getId();
        this.title = blog.getTitle();
        this.summary = blog.getSummary();
        this.content = blog.getContent();
        this.username = blog.getUser().getUsername();
        this.avatar = blog.getUser().getAvatar();
        this.createTime = blog.getCreateTime();
        this.readSize = blog.getReadSize();
        this.commentSize = blog.getCommentSize();
        this.voteSize = blog.getVoteSize();
        this.tags = blog.getTags();
    }

    public void update(Blog blog){
        this.blogId = blog.getId();
        this.title = blog.getTitle();
        this.summary = blog.getSummary();
        this.content = blog.getContent();
        this.username = blog.getUser().getUsername();
        this.avatar = blog.getUser().getAvatar();
        this.createTime = blog.getCreateTime();
        this.readSize = blog.getReadSize();
        this.commentSize = blog.getCommentSize();
        this.voteSize = blog.getVoteSize();
        this.tags = blog.getTags();
    }

    // 省略getter/setter方法
```

```
    @Override
    public String toString() {
        return String.format(
                "EsBlog[blogId=%d, title='%s', summary='%s']",
                blogId, title, summary);
    }
}
```

在 Elasticsearch 中，主键 id 是用 String 类型的。注解 @Field(index = false) 用来标识该字段，不参与建立索引，也就不参与全文搜索。

21.2.2 创建资源库

在 com.waylau.spring.boot.blog.repository.es 目录下，定义资源库的接口 EsBlogRepository，继承自 org.springframework.data.elasticsearch.repository.ElasticsearchRepository。

```
public interface EsBlogRepository extends ElasticsearchRepository<Es-
Blog, String> {

    /**
     * 模糊查询
     * @param title
     * @param Summary
     * @param content
     * @param tags
     * @param pageable
     * @return
     */
    Page<EsBlog> findByTitleContainingOrSummaryContainingOrContentCon-
tainingOrTagsContaining(
            String title,String summary,String content,String tags,Pageable
pageable);

    /**
     * 根据Blog的id查询EsBlog
     * @param blogId
     * @return
     */
    EsBlog findByBlogId(Long blogId);
}
```

模糊查询方法看上去比较冗长，但好处是 Spring Data 会根据方法签名，自动猜测这个查询语句的意图，从而实现查询数据的目的。从这个方法名称中也能大概读懂，这个方法是用于模糊查询标题，或者是摘要，或者是正文，或者是标签包含有搜索关键字的数据。

21.2.3 创建 Tag 的值对象

在 com.waylau.spring.boot.blog.vo 包下创建 Tag 的值对象 TagVO，用于在前端展示标签信息。

```
public class TagVO {

    private String name;
    private Long count;

    public TagVO(String name, Long count) {
        this.name = name;
        this.count = count;
    }

    // 省略getter/setter方法

}
```

21.2.4 创建服务接口及实现

在 com.waylau.spring.boot.blog.service 包下创建 EsBlog 服务接口。

```
public interface EsBlogService {

    /**
     * 删除EsBlog
     * @param id
     * @return
     */
    void removeEsBlog(String id);

    /**
     * 更新 EsBlog
     * @param EsBlog
     * @return
     */
    EsBlog updateEsBlog(EsBlog esBlog);

    /**
     * 根据Blog的id获取EsBlog
     * @param id
     * @return
     */
    EsBlog getEsBlogByBlogId(Long blogId);

    /**
     * 最新博客列表，分页
     * @param keyword
```

```
     * @param pageable
     * @return
     */
    Page<EsBlog> listNewestEsBlogs(String keyword, Pageable pageable);

    /**
     * 最热博客列表，分页
     * @param keyword
     * @param pageable
     * @return
     */
    Page<EsBlog> listHotestEsBlogs(String keyword, Pageable pageable);

    /**
     * 博客列表，分页
     * @param pageable
     * @return
     */
    Page<EsBlog> listEsBlogs(Pageable pageable);
    /**
     * 最新前5
     * @param keyword
     * @return
     */
    List<EsBlog> listTop5NewestEsBlogs();

    /**
     * 最热前5
     * @param keyword
     * @return
     */
    List<EsBlog> listTop5HotestEsBlogs();

    /**
     * 最热前30标签
     * @return
     */
    List<TagVO> listTop30Tags();

    /**
     * 最热前12用户
     * @return
     */
    List<User> listTop12Users();
}
```

上述接口的服务实现为:

```
@Service
public class EsBlogServiceImpl implements EsBlogService {
```

```
    @Autowired
    private EsBlogRepository esBlogRepository;
    @Autowired
    private ElasticsearchTemplate elasticsearchTemplate;
    @Autowired
    private UserService userService;

    private static final Pageable TOP_5_PAGEABLE = PageRequest.of(0, 5);
    private static final String EMPTY_KEYWORD = "";

    @Override
    public void removeEsBlog(String id) {
        esBlogRepository.deleteById(id);
    }

    @Override
    public EsBlog updateEsBlog(EsBlog esBlog) {
        return esBlogRepository.save(esBlog);
    }

    @Override
    public EsBlog getEsBlogByBlogId(Long blogId) {
        return esBlogRepository.findByBlogId(blogId);
    }

    @Override
    public Page<EsBlog> listNewestEsBlogs(String keyword, Pageable page-
able) throws SearchParseException {
        Page<EsBlog> pages = null;
        Sort sort = new Sort(Direction.DESC, "createTime");
        if (pageable.getSort().isUnsorted()) {
            pageable = PageRequest.of(pageable.getPageNumber(), page
able.getPageSize(), sort);
        }

        pages = esBlogRepository.findByTitleContainingOrSummaryContain-
ingOrContentContainingOrTagsContaining(
                keyword, keyword, keyword, keyword, pageable);

        return pages;
    }

    @Override
    public Page<EsBlog> listHotestEsBlogs(String keyword, Pageable page-
able) throws SearchParseException {

        Sort sort = new Sort(Direction.DESC, "readSize", "commentSize",
"voteSize", "createTime");
        if (pageable.getSort().isUnsorted()) {
```

```
            pageable = PageRequest.of(pageable.getPageNumber(), page
able.getPageSize(), sort);
        }

        return esBlogRepository.findByTitleContainingOrSummaryContainin-
gOrContentContainingOrTagsContaining(
                keyword, keyword, keyword, keyword, pageable);
    }

    @Override
    public Page<EsBlog> listEsBlogs(Pageable pageable) {
        return esBlogRepository.findAll(pageable);
    }

    /**
     * 最新前5
     *
     * @param keyword
     * @return
     */
    @Override
    public List<EsBlog> listTop5NewestEsBlogs() {
        Page<EsBlog> page = this.listHotestEsBlogs(EMPTY_KEYWORD, TOP_5_
PAGEABLE);
        return page.getContent();
    }

    /**
     * 最热前5
     *
     * @param keyword
     * @return
     */
    @Override
    public List<EsBlog> listTop5HotestEsBlogs() {
        Page<EsBlog> page = this.listHotestEsBlogs(EMPTY_KEYWORD, TOP_5_
PAGEABLE);
        return page.getContent();
    }

    @Override
    public List<TagVO> listTop30Tags() {

        List<TagVO> list = new ArrayList<>();

        // 查询条件
        SearchQuery searchQuery = new NativeSearchQueryBuilder().with-
Query(matchAllQuery())
                .withSearchType(SearchType.QUERY_THEN_FETCH).withIndi-
```

```
ces("blog").withTypes("blog")
                .addAggregation(terms("tags").field("tags")
                .order(Terms.Order.count(false)).size(30)).build();

        // 聚合
        Aggregations aggregations = elasticsearchTemplate.query(search-
Query,
                new ResultsExtractor<Aggregations>() {
            @Override
            public Aggregations extract(SearchResponse response) {
                return response.getAggregations();
            }
        });

        StringTerms modelTerms = (StringTerms) aggregations.asMap().
get("tags");

        Iterator<Bucket> modelBucketIt = modelTerms.getBuckets().itera-
tor();
        while (modelBucketIt.hasNext()) {
            Bucket actiontypeBucket = modelBucketIt.next();

            list.add(new TagVO(actiontypeBucket.getKey().toString(),
actiontypeBucket.getDocCount()));
        }
        return list;
    }

    @Override
    public List<User> listTop12Users() {

        List<String> usernamelist = new ArrayList<>();// 存储排序后的用户账号

        // 查询条件
        SearchQuery searchQuery = new NativeSearchQueryBuilder().with-
Query(matchAllQuery())
                .withSearchType(SearchType.QUERY_THEN_FETCH).withIndi-
ces("blog").withTypes("blog")
                .addAggregation(terms("users").field("username")
                .order(Terms.Order.count(false)).size(12)).build();
        // 聚合
        Aggregations aggregations = elasticsearchTemplate.query(search-
Query,
                new ResultsExtractor<Aggregations>() {
            @Override
            public Aggregations extract(SearchResponse response) {
                return response.getAggregations();
            }
        });
```

```
        StringTerms modelTerms = (StringTerms) aggregations.asMap().
get("users");

        Iterator<Bucket> modelBucketIt = modelTerms.getBuckets().itera-
tor();
        while (modelBucketIt.hasNext()) {
            Bucket actiontypeBucket = modelBucketIt.next();
            String username = actiontypeBucket.getKey().toString();
            usernamelist.add(username);
        }

        // 根据用户名查出用户的详细信息
        List<User> list = userService.listUsersByUsernames(username-
list);

        // 按照usernamelist的顺序返回用户对象
        List<User> returnList = new ArrayList<>();

        for (String username : usernamelist) {
            for (User user : list) {
                if (username.equals(user.getUsername())) {
                    returnList.add(user);
                    break;
                }
            }
        }

        return returnList;
    }
}
```

其中，在 UserService 中增加如下接口。

```
...
/**
 * 根据用户名集合，查询用户详细的信息列表
 * @param usernames
 * @return
 */
List<User> listUsersByUsernames(Collection<String> usernames);
```

其实现为：

```
...
@Service
public class UserServiceImpl implements UserService, UserDetailsService
{

    @Autowired
```

```
    private UserRepository userRepository;
    ...

    @Override
    public List<User> listUsersByUsernames(Collection<String> usernames)
{
        return userRepository.findByUsernameIn(usernames);
    }
}
```

其中，UserRepository 也要增加相应的接口。

```
public interface UserRepository extends JpaRepository<User, Long>{
    ...
    /**
     * 根据名称列表查询用户列表
     * @param usernames
     * @return
     */
    List<User> findByUsernameIn(Collection<String> usernames);
}
```

修改 BlogServiceImpl，保存博客及删除博客时要增加处理，修改为：

```
...
@Autowired
private EsBlogService esBlogService;

@Transactional
@Override
public Blog saveBlog(Blog blog) {
    boolean isNew = (blog.getId() == null);
    EsBlog esBlog = null;

    Blog returnBlog = blogRepository.save(blog);

    if (isNew) {
        esBlog = new EsBlog(returnBlog);
    } else {
        esBlog = esBlogService.getEsBlogByBlogId(blog.getId());
        esBlog.update(returnBlog);
    }

    esBlogService.updateEsBlog(esBlog);
    return returnBlog;
}

@Transactional
@Override
public void removeBlog(Long id) {
```

```
    blogRepository.deleteById(id);
    EsBlog esblog = esBlogService.getEsBlogByBlogId(id);
    esBlogService.removeEsBlog(esblog.getId());
}
...
```

21.2.5 修改控制器

修改控制器 BlogController，实现首页查询博客的功能。

```
@Controller
@RequestMapping("/blogs")
public class BlogController {

    @Autowired
    private EsBlogService esBlogService;

    @GetMapping
    public String listEsBlogs(
            @RequestParam(value="order",required=false,defaultValue=
"new") String order,
            @RequestParam(value="keyword",required=false,defaultValue=
"" ) String keyword,
            @RequestParam(value="async",required=false) boolean async,
            @RequestParam(value="pageIndex",required=false,defaultValue=
"0") int pageIndex,
            @RequestParam(value="pageSize",required=false,defaultValue=
"10") int pageSize,
            Model model) {

        Page<EsBlog> page = null;
        List<EsBlog> list = null;
        boolean isEmpty = true; // 系统初始化时，没有博客数据
        try {
            if (order.equals("hot")) { // 最热查询
                Sort sort = new Sort(Direction.DESC,"readSize","com-
mentSize","voteSize","createTime");
                Pageable pageable = PageRequest.of(pageIndex, pageSize,
sort);
                page = esBlogService.listHotestEsBlogs(keyword, page
able);
            } else if (order.equals("new")) { // 最新查询
                Sort sort = new Sort(Direction.DESC,"createTime");
                Pageable pageable = PageRequest.of(pageIndex, pageSize,
sort);
                page = esBlogService.listNewestEsBlogs(keyword, page
able);
            }
```

```
            isEmpty = false;
        } catch (Exception e) {
            Pageable pageable = PageRequest.of(pageIndex, pageSize);
            page = esBlogService.listEsBlogs(pageable);
        }

        list = page.getContent();    // 当前所在页面数据列表


        model.addAttribute("order", order);
        model.addAttribute("keyword", keyword);
        model.addAttribute("page", page);
        model.addAttribute("blogList", list);

        // 首次访问页面才加载
        if (!async && !isEmpty) {
            List<EsBlog> newest = esBlogService.listTop5NewestEsBlogs();
            model.addAttribute("newest", newest);
            List<EsBlog> hotest = esBlogService.listTop5HotestEsBlogs();
            model.addAttribute("hotest", hotest);
            List<TagVO> tags = esBlogService.listTop30Tags();
            model.addAttribute("tags", tags);
            List<User> users = esBlogService.listTop12Users();
            model.addAttribute("users", users);
        }

        return (async==true?"/index :: #mainContainerRepleace":"/in-
dex");
    }
}
```

修改 MainController，将访问 /index 的接口重定向到 /blogs 接口。

```
...
@GetMapping("/index")
public String index() {
    return "redirect:/blogs";
}
...
```

21.3 首页搜索的前台实现

本节将实现首页搜索的前台编码工作。

21.3.1 修改 index.html

修改 index.html，把之前的静态原型页面都修改为 Thymeleaf 模板，并绑定相关的数据。

1. 处理博客列表

```
<!-- Blog Post -->
<div id="mainContainer">
    <div id="mainContainerRepleace">
        <div class="card mb-4" th:each="blog : ${blogList}">
            <div class="card-block">
                <h2 class="card-title">
                    <span> <a href="/u/waylau" title="waylau"
                        th:href="'/u/' + ${blog.username}" th:title="${-
blog.username}">
                            <img src="/images/avatar-defualt.jpg"
                            th:src="${blog.avatar} == null
                            ? '/images/avatar-defualt.jpg'
                            : ${blog.avatar}"
                            class="blog-avatar-50">
                    </a>
                    </span> <a href="/u/waylau/blogs/1" class=
"card-link" title="waylau"
                        th:href="'/u/' + ${blog.username} + '/blogs/'+
${blog.blogId}"
                        th:title="${blog.title}" th:text="${blog.title}">
 OAuth 2.0 认证的原理与实践 </a>
                </h2>
                <p class="card-text" th:text="${blog.summary}">使用 OAuth
 2.0 认证的好处是显而易见的。你只需要用同一个账号密码，就能在各个网站进行访问，而免去
了在每个网站都进行注册的烦琐过程。本文将介绍OAuth2.0的原理，并基于Spring Security
和GitHub账号，来演示 OAuth2.0 的认证过程。</p>
                <div class="card-text">
                    <a href="/u/waylau" th:href="'/u/' + ${blog.user
name}"
                        class="card-link" th:text=${blog.username}>way-
lau</a> 发表于
                    [[${#dates.format(blog.createTime, 'yyyy-MM-dd
HH:mm')}]] <i
                        class="fa fa-eye" aria-hidden="true">[[${blog.
readSize}]]</i>
                    <i class="fa fa-heart-o" aria-hidden="true">[[$
{blog.voteSize}]]</i>
                    <i class="fa fa-comment-o" aria-hidden="true">[[$
{blog.commentSize}]]</i>
                </div>
            </div>
        </div>
        <div th:replace="~{fragments/page :: page}">...</div>
    </div>
```

```
</div>
```

2. 热门标签

热门标签用 th:each 遍历生成。

```
...
<a th:each="tag, tagStat : ${tags}" th:object="${tag}"
    href="/blogs?keyword=1" th:href="'/blogs?keyword=' +  *{name}">
    <span class="badge"
    th:classappend="${tagStat.index} eq 0
        ? 'badge-danger': (${tagStat.index} eq 1
        ? 'badge-warning':(${tagStat.index} eq 2
        ? 'badge-info':'badge-default'))"
    th:text="*{name}">Web Design</span>
</a>
...
```

3. 热门用户

热门用户用 th:each 遍历生成。

```
...
<span th:each="user: ${users}" th:object="${user}">
    <a
    href="/u/waylau" th:href="'/u/' + *{username}" title="waylau"
    th:title="*{username}"> <img
        src="/images/avatar-defualt.jpg"
        th:src="*{avatar} == null ? '/images/avatar-defualt.jpg' : *{av-
atar}"
        class="blog-avatar-50">
    </a>
</span>
...
```

4. 热门文章

热门文章用 th:each 遍历生成。

```
...
<ul class="list-group" th:each="hotblog , hotblogStat : ${hotest}"
    th:object="${hotblog}">
    <a th:href="'/u/'+*{username} +'/blogs/'+ *{blogId}"
        class="list-group-item"> [[*{title}]]<span
        class="badge badge-pill"
        th:classappend="${hotblogStat.index} eq 0
        ? 'badge-danger': (${hotblogStat.index} eq 1
        ? 'badge-warning':(${hotblogStat.index} eq 2
        ? 'badge-info':'badge-default'))">[[*{readSize}]]</span>
    </a>
</ul>
```

```
...
```

5. 最新发布

最新发布的文章用 th:each 遍历生成。

```
...
<ul class="list-group" th:each="newblog : ${newest}"
    th:object="${newblog}">
    <a th:href="'/u/'+*{username} +'/blogs/'+ *{blogId}"
        class="list-group-item" th:text="*{title}"> </a>
</ul>
...
```

6. 引入index.js

在页面底部引入 index.js：

```
...
<!-- JavaScript -->
<script src="../../js/index.js" th:src="@{/js/index.js}"></script>
...
```

21.3.2 修改 header.html

修改 header.html，以支持对最新、最热及搜索功能的支持。

```
...
<ul class="navbar-nav mr-auto">
    <li class="nav-item"><a class="nav-link"
        href="javascript:void(0)" th:attr="url='/blogs?order=new'">最新
            <span class="sr-only">(current)</span>
    </a></li>
    <li class="nav-item"><a class="nav-link"
        href="javascript:void(0)" th:attr="url='/blogs?order=hot'">最热
</a>
    </li>
    <form class="form-inline mt-2 mt-md-0">
        <input class="form-control mr-sm-2" type="text" placeholder="搜
索"
            id="indexkeyword" th:value="${keyword}"> <a
            href="javascript:void(0)"
            class="btn btn-outline-secondary my-2 my-sm-0" id="index-
search"><i
            class="fa fa-search" aria-hidden="true"></i></a>
    </form>

</ul>
...
```

21.3.3 增加 index.js

增加 index.js，用于处理 index.html 的页面事件。

```
$(function() {

    var _pageSize; // 存储用于搜索

    // 根据用户名、页面索引、页面大小获取用户列表
    function getBlogsByName(pageIndex, pageSize) {
         $.ajax({
              url: "/blogs",
              contentType : 'application/json',
              data:{
                  "async":true,
                  "pageIndex":pageIndex,
                  "pageSize":pageSize,
                  "keyword":$("#indexkeyword").val()
              },
              success: function(data){
                  $("#mainContainer").html(data);

                  var keyword = $("#indexkeyword").val();

                  // 如果是分类查询，则取消最新、最热选中样式
                  if (keyword.length > 0) {
                     $(".nav-item .nav-link").removeClass("active");
                  }
              },
              error : function() {
                  toastr.error("error!");
              }
         });
    }

    // 分页
    $.tbpage("#mainContainer", function (pageIndex, pageSize) {
        getBlogsByName(pageIndex, pageSize);
        _pageSize = pageSize;
    });

    // 关键字搜索
    $("#indexsearch").click(function() {
        getBlogsByName(0, _pageSize);
    });

    // 最新/最热切换事件
    $(".nav-item .nav-link").click(function() {
```

```
        var url = $(this).attr("url");

        // 先移除其他的单击样式，再添加当前的单击样式
        $(".nav-item .nav-link").removeClass("active");
        $(this).addClass("active");

        // 加载其他模块的页面到右侧工作区
         $.ajax({
             url: url+'&async=true',
             success: function(data){
                 $("#mainContainer").html(data);
             },
             error : function() {
                 toastr.error("error!");
             }
         })

         // 清空搜索框内容
         $("#indexkeyword").val('');
    });

});
```

21.3.4 运行

首先要确保 Elasticsearch 服务器已经启动。而后再启动 blog-search 项目。

通过浏览器访问 http://localhost:8080/，进入博客系统主页，并登录系统，创建分类及博客。同时，可以设置博客的标签、点赞留言等。图 21-1 所示为博客在首页不同维度的分析方式。

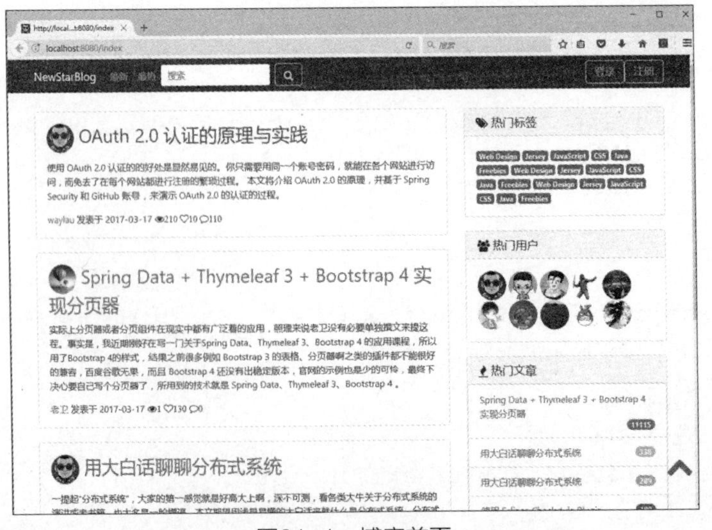

图21-1　博客首页

21.3.5 示例源码

本节示例源码在 blog-search 目录下。

21.4 使用中文分词

至此，所有的博客功能都已经完成了。读者可以愉快地使用博客来发表博客了，但如果他们足够细心，就会发现在输入中文标签的时候，存在一定的问题。

例如，当要把某个博客标记为“大神”时，博客系统却将这个单词简单地分成了如图 21-2 所示的两个词“大”和“神”。显然，这并不符合用户的使用习惯。

图21-2 默认分词的标签效果

这是 Elasticsearch 语言分析器上的限制，它并不能友好地处理所有语言，特别是中文。这种情况下就需要额外的中文分词器来协助了。

本节将演示使用 IK Analysis 插件来帮助实现中文分词。

21.4.1 IK Analysis 插件

IK Analysis 插件（https://github.com/medcl/elasticsearch-analysis-ik/）就是一款专门用于 Elasticsearch 的分词器，可以友好地处理中文。

IK Analysis 插件将 Lucene IK 分析仪（http://code.google.com/p/ik-analyzer/）集成到 Elasticsearch 中，从而支持自定义分词。

21.4.2安装 IK Analysis

安装 IK Analysis 非常简单，主要有以下两种方式。

1. 使用elasticsearch-plugin

如果 Elasticsearch 是 5.5.1 以后的版本，可以使用 elasticsearch-plugin 来安装，安装方式如下。

```
./bin/elasticsearch-plugin install https://github.com/medcl/elastic-
```

```
search-analysis-ik/releases/download/v5.5.1/elasticsearch-analy sis-
ik-5.5.1.zip
```

2. 下载解压到指定目录

另外一种方式是下载安装包，解压到 Elasticsearch 安装目录的 /plugins/ 目录下即可。

图 21-3 所示为 IK 的安装目录。

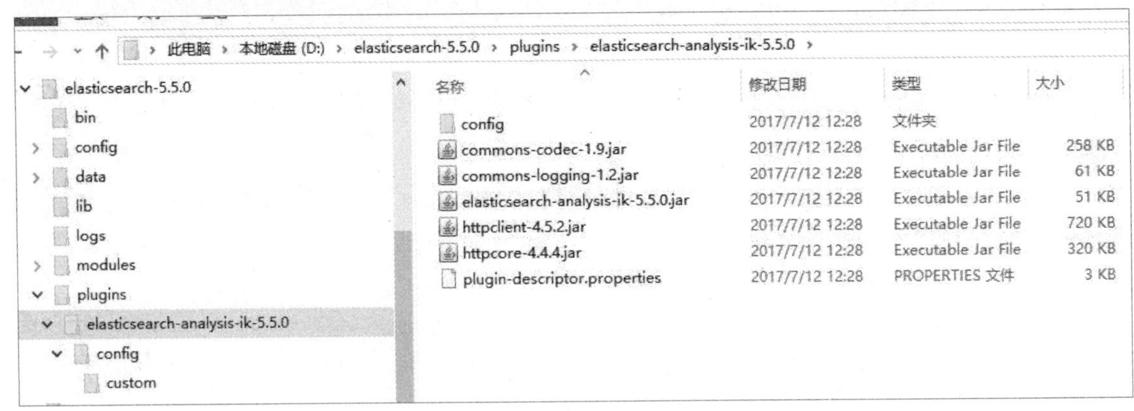

图21-3　IK的安装目录

需要注意的是，不管是哪种安装方式，相应的插件要对应相应的 Elasticsearch 版本，否则可能会安装不成功。下载地址为 https://github.com/medcl/elasticsearch-analysis-ik/releases。本节使用的是 IK Analysis for Elasticsearch 5.5.0 版本。

21.4.3 使用 IK Analysis

要使用 IK Analysis，需要在文档类中指定相应的分词器。在 EsBlog 的 tags 属性上添加 searchAnalyzer ="ik_smart", analyzer = "ik_smart" 的注解内容就可以了。

```
public class EsBlog implements Serializable {

    ...

    @Field(type = FieldType.text,fielddata = true, searchAnalyzer =
"ik_smart", analyzer = "ik_smart")
    private String tags;  // 标签
```

图 21-4 所示为使用了 IK 分词的标签效果。

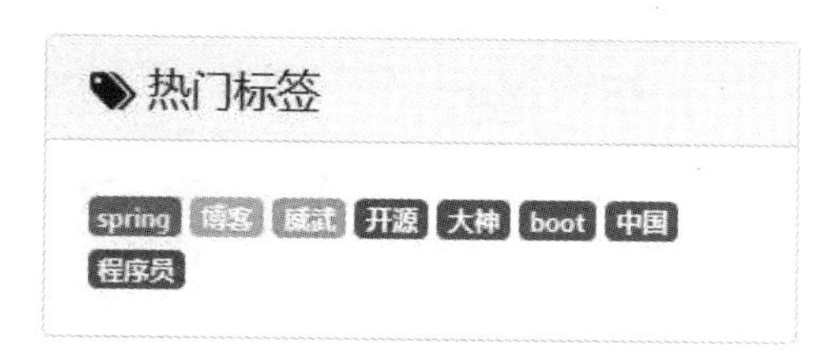

图21-4 使用了IK分词的标签效果

21.4.4 IK Analysis 类型

ik_smart 是 IK Analysis 中的一种分词形式。IK Analysis 主要有两种类型的分词形式，分别是 ik_max_word 和 ik_smart。

（1）ik_max_word: 会将文本做最细粒度的拆分，比如会将“老卫的博客系统”拆分为“老卫”“的”“博客”“系统”“博客系统”“老卫的”“老”“卫”“博”“客”“系”“统”等，会穷尽各种可能的组合。

（2）ik_smart: 会做最粗粒度的拆分，比如会将“老卫的博客系统”拆分为“老卫”“的”“博客系统”。

第22章 总结与展望

22.1 Spring Boot 企业级应用开发的总结

基于 Spring Boot 的企业级应用博客系统，全部功能都已经开发完毕，整个博客系统的全貌也跃然眼前。我们既可以在博客系统中发表个人博客文章，也可以通过评论、点赞等社交化的功能来与广大博客用户发生联系。作为一名技术人员，博客是人们观察世界、了解技术、学习知识非常好的载体。笔者也有写博客的习惯，混迹于广大的技术社区，博客文章也经常能激起网友们的热情讨论。笔者也运营了一个个人博客网站，用于分享个人的一些技术方面的心得体会，内容涵盖了编程语言、系统架构、性能优化等方面的话题。

正是因为笔者如此钟情于博客，所以本书将博客系统作为整个企业级应用开发的案例。本书围绕整合以 Spring Boot 为核心的技术栈，来搭建并实现一个完整的企业级博客系统而展开。该博客系统支持市面上博客系统的常见功能。读者通过学习构建这个博客系统的整个过程，来达到设计和实现一个企业级 Java 应用开发的目的。该博客系统是一个类似于 WordPress 的专注于博客功能的博客平台，支持多用户访问和使用。

那么，读者通过本书能学到哪些内容呢？

22.1.1 了解企业级开发的完整流程

通过本书的学习，可以了解到企业级开发的完整流程。从需求分析到技术选型，再到架构的设计，然后到框架的搭建，最后到功能实现的编写，都是围绕如何来进行企业级开发而贯穿始终的。

当今的企业，对于人才的渴求是非常强烈的，特别是具备有技术的深度及广度的技术人员。很多开发人员，即使是在企业中已经有了很多年的开发经验，但这些经验或者是做一些维护工作，修改 Bug，做一些功能的扩展；或者是做一些简单的增删改查的低级功能，很难有机会接触到一个全新的项目，即便是一个新项目，也很难有机会参与到项目的需求谈论、架构设计、技术选型、编码实现等方面的完整的企业级应用流程。

本书揭示了企业级应用的完整流程。不管是刚刚入行的学生还是初级开发人员，抑或是资深的工程师，都能从本书中获益。

22.1.2 熟悉当前主流、前瞻的技术框架及其应用

本博客系统所涉及的相关技术有 Spring Boot、Spring、Spring MVC、Spring Security 、Spring Data、Hibernate、Gradle、Bootstrap、jQuery、HTML5、JavaScript、CSS、Thymeleaf、MySQL、H2、Elasticsearch、MongoDB 等。本书所涉及的技术面很广，符合当前主流，并富有一定的前瞻性，可以有效提升读者在市场中的核心竞争力。

图 22-1 所示为本书中所涉及的主要的前端和后端技术。

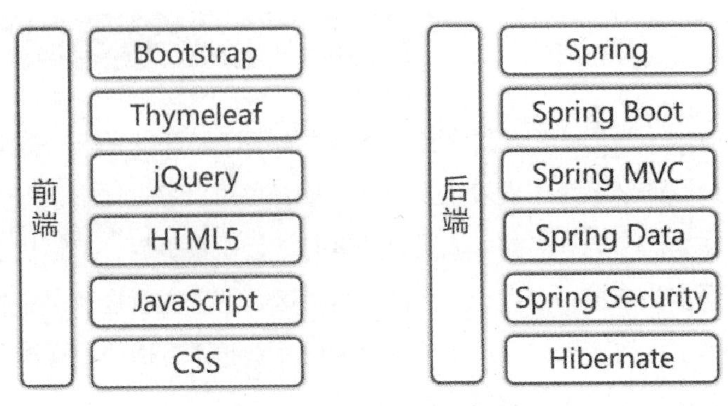

图22-1 前端、后端技术

（1）Bootstrap。对于前端来说，采用了最新的 Bootstrap 4 来作为博客系统的主要样式。Bootstrap 4 相比 Bootstrap 3 来说，两者变动比较大，存在兼容性方面的问题。所以当在选型某些 Bootstrap 插件的时候，也要注意版本之间的兼容性的测试。

（2）Thymeleaf。Thymeleaf 是 Java 开源的一款模板引擎，其“原型即界面”的理念，深受广大开发者的喜爱。同时，Thymeleaf 提供了与 Spring 类似的标准方言，在 Spring 应用中，能有不错的兼容性。

对于 Spring 而言，笔者也编著了一本开源书《Thymeleaf 教程》（https://github.com/waylau/thymeleaf-tutorial）供广大读者免费参阅。

（3）其他前端技术。其他的前端技术，包括 jQuery、HTML5、JavaScript、CSS 可以说是前端开发人员必备的技能，也是相对来说比较成熟和稳定的技术。

（4）Spring 及 SpringBoot。Spring Boot 是本书的核心技术。本质上讲，Spring Boot 并没有提供任何新的功能，它只是在 Spring 的基础上，结合依赖管理，组装成了各种开箱即用的 Starter。所以，学习 Spring Boot，毫无疑问需要自行去学习 Spring 相关的知识。

对于 Spring 而言，笔者也编著了几本开源书《Spring Framework 4.x 参考文档》（https://github.com/waylau/spring-framework-4-reference）及《Spring Boot 教程》（https://github.com/waylau/spring-boot-tutorial）供广大读者免费参阅。

（5）SpringMVC。Spring MVC 是 Java MVC 模式的经典实现，在企业中广泛采用。Spring MVC 支持 RESTful 风格的 API 的编写。但需要注意的是，Spring MVC 的 RESTful API 并非遵循 Java REST 的相关规范。

如果想要了解 Java REST 的规范 JAX-RS （JSR-339），可以参考 Jersey 框架的技术实现。笔者也编著了几本关于 Jersey 的开源书《Jersey 2.x 用户指南》（https://github.com/waylau/Jersey-2.x-User-Guide）及《REST 实战》（https://github.com/waylau/rest-in-action）供读者免费参阅。

（6）SpringData。Spring Data 是一个大家族。本书所介绍的 Spring Data JPA 及 Spring Data Elasticsearch 只是这个大家族中的两个小成员。但它们拥有相同的接口，所以在使用过程中几乎没有学习的成本，学会了其中一个，自然就能掌握另一个。

（7）SpringSecurity。Apache Shiro 和 Spring Security 在 Java 领域中都享有盛誉。一个是以简洁好用而被业界广泛应用，而另一个则以功能强大而著称。

关于这两个框架的用法，读者也可以参考笔者的另外两本开源书《Apache Shiro 1.2.x 参考手册》（https://github.com/waylau/apache-shiro-1.2.x-reference）及《Spring Security 教程》（https://github.com/waylau/spring-security-tutorial）。

下面总结一下数据存储及其他方面的技术。图 22-2 所示为本书中所涉及的主要数据存储技术。

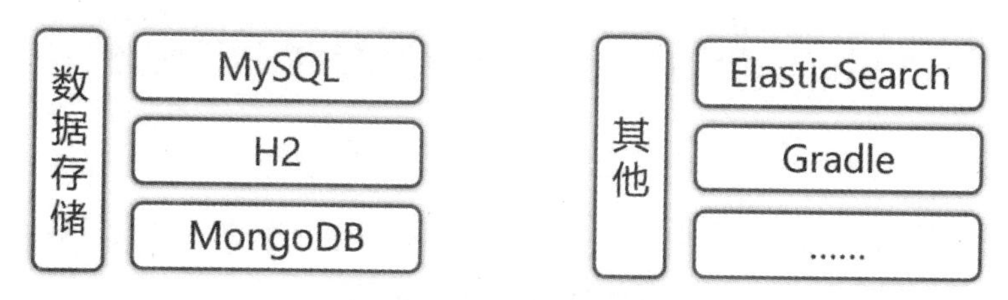

图22-2 数据存储技术

（1）MySQL。在数据存储方面，关系型数据库 MySQL 必不可少。一方面，其开源、免费的特点在企业中广泛采用；另一方面，MySQL 也拥有不错的特性和性能，被开发人员所推崇。

有关 MySQL 的资料市面上也很多，这里就不在一一列举。但如果是涉及本书的一些环境方面的配置，笔者的博客上有很多总结，读者可以参阅 https://waylau.com/categories/#MySQL。

（2）H2。H2 是小型的、开源内存数据库，性能很不错，Spring Boot 对其也有开箱即用的支持。在开发阶段，使用内存数据库可以有效提升开发效率。

关于 H2 的用法，读者也可以参考笔者的另外一本开源书《H2 Database 教程》（https://github.com/waylau/h2-database-doc）。

（3）MongoDB。MongoDB 是一个介于关系数据库和非关系数据库之间的产品，是非关系数据库当中功能最丰富、最像关系数据库的，旨在为 Web 应用提供可扩展的高性能数据存储解决方案。业界盛传，MongoDB 有可能会取代传统的关系数据库。先不管这种传言的准确性，起码能感受到 MongoDB 的前景是无可估量的。

（4）ElasticSearch。ElasticSearch 是全文搜索非常好的技术方案，在目前的互联网公司中广泛采用。

（5）Gradle。Gradle 编写脚本简洁，执行效率高，有可能取代 Maven 成为下一代的构建工具。

22.1.3 理论联系实战，知其然而知其所以然

本书的重点是如何来围绕 Spring Boot 技术栈，通过整合这么多的技术，搭建一个能够用于快速开发的框架，最终实现一个完整的企业级应用。本书的难点是技术点比较多，要完全掌握是有些

难度的。所以本课程的安排是，前半部分先对一些市面上比较前瞻（或者说市面上相关资料介绍比较少的）的技术做一个精讲，在知识点梳理的过程中辅以实战案例；后半部分整合了所有的技术，搭建一个能够用于快速开发的框架，并来逐步构建博客系统的各个功能。理论联系实战，这样读者就能够知其然而知其所以然，相关的案例是“高大上”的，功能是很专业的，技术点是很符合企业级开发应用的要求的。

22.2 博客系统的展望

虽然博客系统的功能已经完全开发完毕了，但博客系统的未来并不止步于此。软件的需求从来都是多变的，用户的喜好也是会不断地演变的，这意味着博客系统也需要紧跟时代的变化，不会因此而停下探索的脚步。毕竟，变化是软件行业唯一不变的定律。

本书对于 Spring Boot 的介绍也要告一段落。在本书的最后，希望读者在编程的道路上一帆风顺。

1. 多看，多思考，多敲代码多实践

（1）多看是指多看书、多查资料。书中给出了很多翔实的案例，供读者参考。但由于篇幅有限，技术点又多，不可能对所有的技术都讲解得很透彻、面面俱到。所以希望广大读者在阅读本书的过程中，对相关的技术难点还要去查其他资料来作为补充。

（2）多思考是指阅读本书过程中，多多思考代码的用意。这段代码可以这么写，那么为什么要这么写，这么写还有没有其他更加高效的方式，等等。

（3）多敲代码多实践，毫无疑问就是要多动手实践，毕竟 Talk is cheap（代码胜于雄辩）。书中提供了非常多的实战案例，都可以提供给大家作为参考。唯有自己真正地敲过代码，用技术真正地去解决实际问题，才能对相关的知识点有更深入的理解。

2. 善于总结，举一反三

学习过程中难免出错，出错不可怕，可怕的是同一个错误反复犯。所以，笔者建议读者在学习过程中要善于去总结，最好是将遇到的错误，以及如何解决的方案记录下来，有利于下次再遇到同样的问题时，可以顺利解决这些问题。

同时，读者也要有举一反三的能力。类似的问题只要解决一个，其他问题自然迎刃而解。

3. 大胆质疑，多交流

（1）大胆质疑是一种学习的态度。任何人都有可能会犯错，笔者也有可能会犯经验性的错误。在这种情况下，读者在阅读本书时，就要抱着大胆质疑的态度。对于书中的某些观点，如果不认同，或者认为有讲得不对的地方，都可以大胆地做出质疑，而后去找相关的资料来验证自己的疑点。这种学习态度是非常有利于读者能力的提高的。一方面，读者在质疑的过程中培养了思考问题的习惯；另一方面，大胆质疑是需要查找相关资料来作为支撑的，这无疑推动了读者的学习积极性，扩大了知识面。

（2）发现疑点，也可以把相关的问题抛出来，让大家一起交流讨论。技术的进步，很大程度上是因为大家不断地交流，不断地做思想上的碰撞，从而激起了创新的火花。

（3）提问是一门艺术。如何将自己想要问的问题，用最精准的语言描述出来，本身就是一个很大的难点。笔者也经常混迹于技术社区，经常在一些技术问答版块做技术的解答，就发现很多网友提出的问题，都存在对问题描述不清的问题。这导致的结果是，问题没有人愿意去回答，即便回答了，也很有可能答非所问。

附录 A

开发环境的搭建

本附录讲述如何来安装和配置开发环境，并将 Spring Boot 项目导入 IDE 中进行开发。本例所选用的 IDE 为 Eclipse，当然也可以自行选择熟悉的 IDE。本例中的源码是与具体 IDE 无关的。

一款好用的 IDE 就如同一件称手的兵器，使用起来自然得心应手。好用的 IDE 有以下优点。

（1）提升编码效率。大部分 IDE 都提供了代码提示、代码自动补全等功能，极大地提升了编码的效率。

（2）纠错。在编码过程中，IDE 也可以对一些运行时、编译时的常见错误做出提示。

（3）养成好的编码规范。IDE 可以对代码格式做校验，这在无形中就帮助用户纠正了错误的编码习惯。

A.1 所需环境

本例采用的开发环境如下：

- JDK 8。
- Gradle 4.0。
- Eclipse 4.7。

A.2 配置 Eclipse

各个版本的 Eclipse 所默认安装的插件不同，例如，Eclipse for Java 中就集成了常用的插件。所以，如果版本中没有相关的插件，请自行安装。本节只介绍插件的基本配置及使用。

1. 安装的Gradle插件（可选）

（1）在 Eclipse 中，单击“Help”→“Install New Software”按钮，在弹出的对话框中添加 Gradle 插件 Buildship 的地址，如附图 A-1 所示。

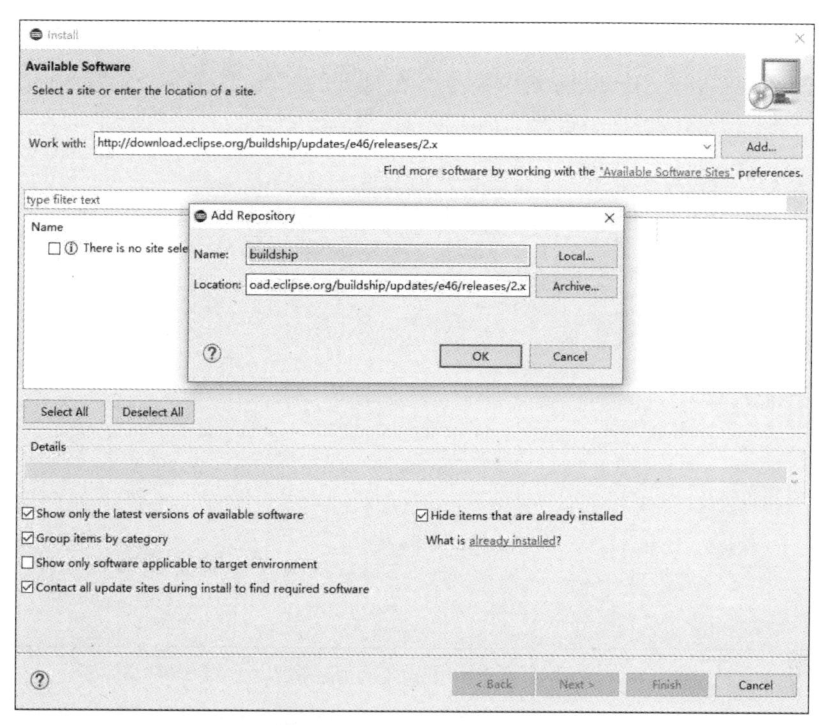

附图A-1　添加 Gradle 插件1

（2）选中插件，进行安装即可，如附图 A-2 所示。

Install
Available Software
Check the items that you wish to install.
Work with: buildship - http://download.eclipse.org/buildship/updates/e46/releases/2.x
Add...
Find more software by working with the 'Available Software Sites' preferences.
type filter text
Name　Version
Buildship: Eclipse Plug-ins for Gradle
Buildship: Eclipse Plug-ins for Gradle　2.0.0.v20170111-1029
Select All　Deselect All　1 item selected
Details
Show only the latest versions of available software
Hide items that are already installed
Group items by category
What is already installed?
Show only software applicable to target environment
Contact all update sites during install to find required software
< Back　Next >　Finish　Cancel

附图A-2　添加 Gradle 插件2

2. 配置用户安装的Gradle（可选）

在选择“Windows”→“Preferences”命令，在打开的“Preferences”对话框中对 Gradle 进行设置，指定一个 Gradle 的用户安装目录，如附图 A-3 所示。

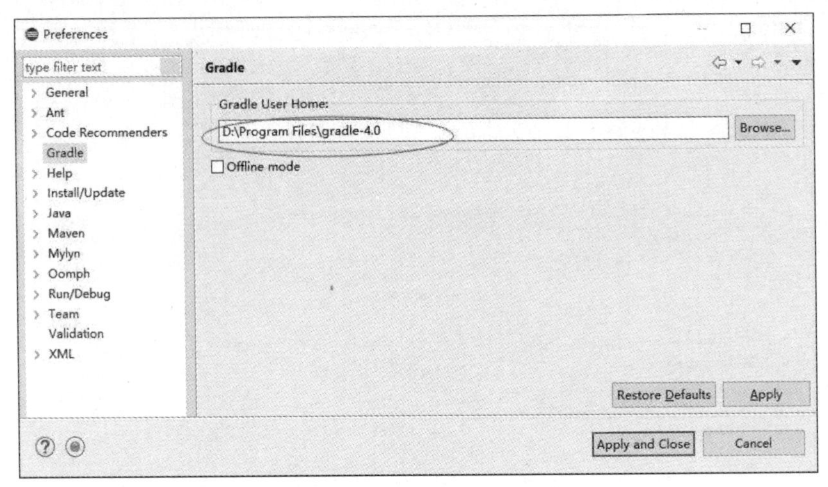

附图A-3　配置 Gradle 插件

注意：如果采用 Gradle Wrapper 形式来导入项目，本步骤也是可选的。

A.3 导入项目到 Eclipse

本节演示导入之前的 hello-world 项目。

（1）在导入类型时，选择“Existing Gradle Project”选项，如附图 A-4 所示。

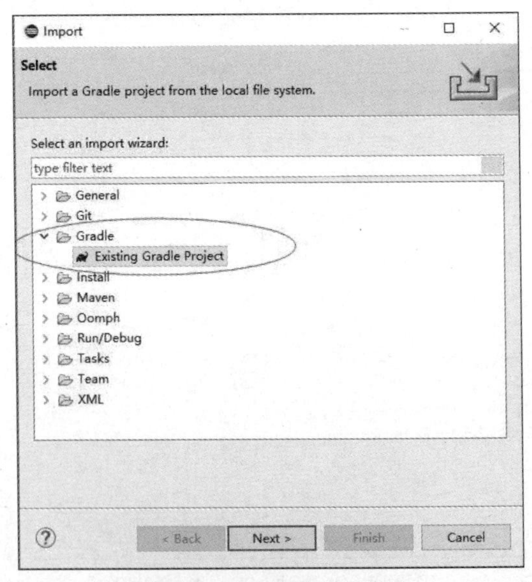

附图A-4　导入项目步骤1

（2）指定要导入项目的路径，如附图 A-5 所示。

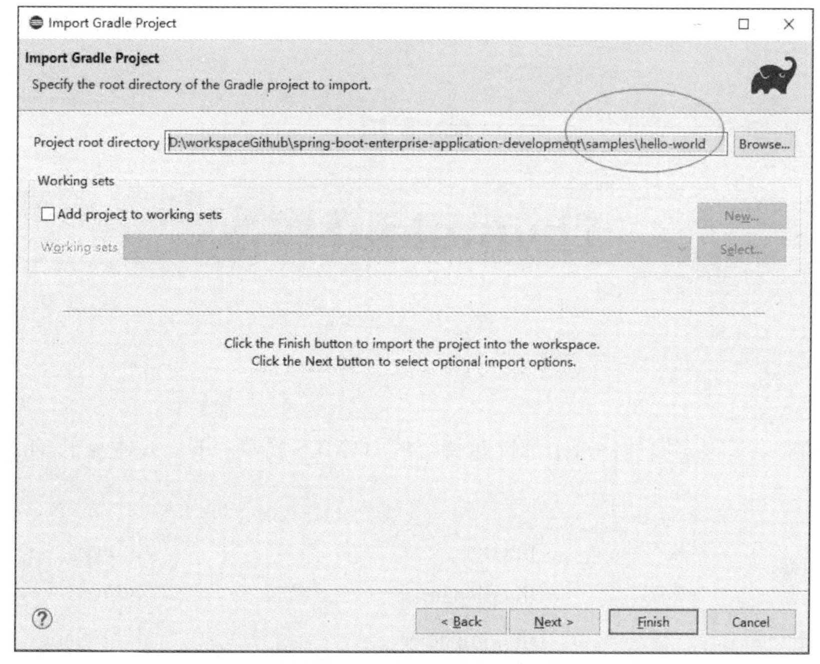

附图A-5 导入项目步骤2

（3）选择 Gradle 分发的类型。支持多种分发形式，本例采用 Gradle Wrapper 形式，如附图 A-6 所示。

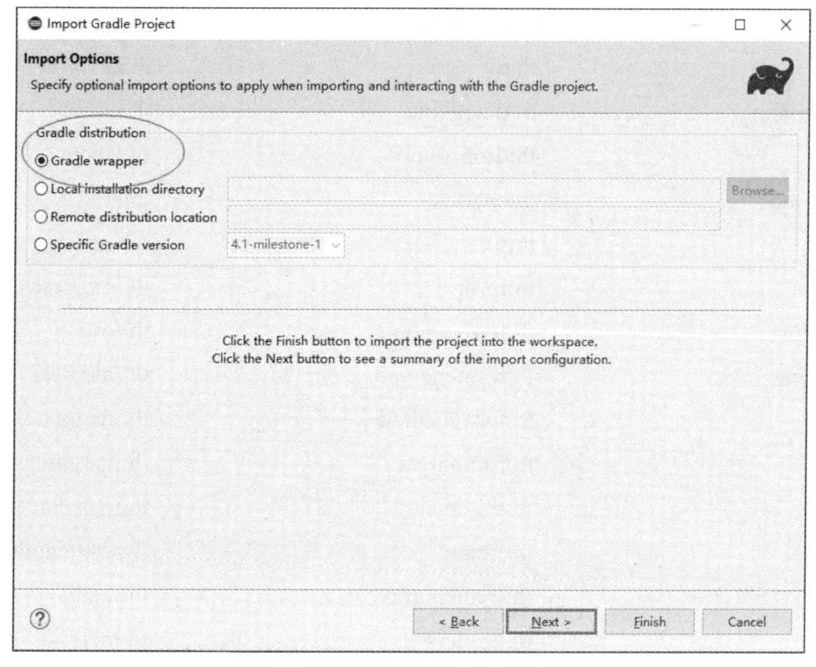

附图A-6 导入项目步骤3

附录 B Thymeleaf 属性

Thymeleaf 提供了很多属性，每个都针对特定的 HTML5 属性。下面是常见的 Thymeleaf 属性。

th:abbr
th:accept
th:accept-charset
th:accesskey
th:action
th:align
th:alt
th:archive
th:audio
th:autocomplete
th:axis
th:background
th:bgcolor
th:border
th:cellpadding
th:cellspacing
th:challenge
th:charset
th:cite
th:class
th:classid
th:codebase
th:codetype
th:cols
th:colspan
th:compact
th:content
th:contenteditable
th:contextmenu
th:data
th:datetime
th:dir
th:draggable
th:dropzone
th:enctype
th:for
th:form
th:formaction
th:formenctype
th:formmethod
th:formtarget
th:fragment
th:frame
th:frameborder
th:headers
th:height
th:high
th:href
th:hreflang
th:hspace
th:http-equiv
th:icon
th:id
th:inline
th:keytype
th:kind
th:label
th:lang
th:list
th:longdesc
th:low
th:manifest
th:marginheight
th:marginwidth
th:max
th:maxlength
th:media
th:method
th:min

th:name
th:onabort
th:onafterprint
th:onbeforeprint
th:onbeforeunload
th:onblur
th:oncanplay
th:oncanplaythrough
th:onchange
th:onclick
th:oncontextmenu
th:ondblclick
th:ondrag
th:ondragend
th:ondragenter
th:ondragleave
th:ondragover
th:ondragstart
th:ondrop
th:ondurationchange
th:onemptied
th:onended
th:onerror
th:onfocus
th:onformchange
th:onforminput
th:onhashchange
th:oninput
th:oninvalid
th:onkeydown
th:onkeypress
th:onkeyup
th:onload
th:onloadeddata
th:onloadedmetadata
th:onloadstart
th:onmessage
th:onmousedown
th:onmousemove
th:onmouseout
th:onmouseover
th:onmouseup
th:onmousewheel
th:onoffline
th:ononline
th:onpause
th:onplay
th:onplaying
th:onpopstate
th:onprogress
th:onratechange
th:onreadystatechange
th:onredo
th:onreset
th:onresize
th:onscroll
th:onseeked
th:onseeking
th:onselect
th:onshow
th:onstalled
th:onstorage
th:onsubmit
th:onsuspend
th:ontimeupdate
th:onundo
th:onunload
th:onvolumechange
th:onwaiting
th:optimum
th:pattern
th:placeholder
th:poster
th:preload
th:radiogroup
th:rel
th:rev
th:rows
th:rowspan
th:rules
th:sandbox
th:scheme
th:scope
th:scrolling
th:size
th:sizes
th:span
th:spellcheck
th:src
th:srclang
th:standby
th:start
th:step
th:style
th:summary
th:tabindex
th:target
th:title
th:type
th:usemap
th:value
th:valuetype
th:vspace
th:width
th:wrap
th:xmlbase
th:xmllang
th:xmlspace

附录 C
Thymeleaf 表达式工具对象

Thymeleaf 表达式工具对象提供非常方便的工具。很多工具对象都与 Java 语言中的工具类用法类似。

1. ExecutionInfo

- #execInfo: 表达式对象提供有关在 Thymeleaf 标准表达式内正在处理模板的有用信息。

```
/*
 * ==========================================================================
* See javadoc API for class org.thymeleaf.expression.ExecutionInfo
 * ==========================================================================
*/

/*
 * Return the name and mode of the 'leaf' template. This means the
template
 * from where the events being processed were parsed. So if this piece of
 * code is not in the root template "A" but on a fragment being inserted
 * into "A" from another template called "B", this will return "B" as a
 * name, and B's mode as template mode.
 */
${#execInfo.templateName}
${#execInfo.templateMode}
/*
 * Return the name and mode of the 'root' template. This means the
template
 * that the template engine was originally asked to process. So if this
 * piece of code is not in the root template "A" but on a fragment being

 * inserted into "A" from another template called "B", this will still
 * return "A" and A's template mode.
 */
${#execInfo.processedTemplateName}
${#execInfo.processedTemplateMode}
```

```
/*
 * Return the stacks (actually, List<String> or List<TemplateMode>) of
 * templates being processed. The first element will be the
 * 'processedTemplate' (the root one), the last one will be the 'leaf'
 * template, and in the middle all the fragments inserted in nested
 * manner to reach the leaf from the root will appear.
 */
${#execInfo.templateNames}
${#execInfo.templateModes}

/*
 * Return the stack of templates being processed similarly (and in the
 * same order) to 'templateNames' and 'templateModes', but returning
 * a List<TemplateData> with the full template metadata.
 */
${#execInfo.templateStack}
```

2. Messages

- #messages: 用于在变量表达式中获取外部消息的实用方法，与使用# {...} 语法获得的方式相同。

```
/*
 * ======================================================================
* See javadoc API for class org.thymeleaf.expression.Messages
 * ======================================================================
*/

/*
 * Obtain externalized messages. Can receive a single key, a key plus
arguments,
 * or an array/list/set of keys (in which case it will return an array/
 * list/set of externalized messages).
 * If a message is not found, a default message (like '??msgKey??') is
returned.
 */
${#messages.msg('msgKey')}
${#messages.msg('msgKey', param1)}
${#messages.msg('msgKey', param1, param2)}
${#messages.msg('msgKey', param1, param2, param3)}
${#messages.msgWithParams('msgKey', new Object[] {param1, param2, param3,
 param4})}
${#messages.arrayMsg(messageKeyArray)}
${#messages.listMsg(messageKeyList)}
${#messages.setMsg(messageKeySet)}

/*
 * Obtain externalized messages or null. Null is returned instead of a
 * default message if a message for the specified key is not found.
 */
${#messages.msgOrNull('msgKey')}
${#messages.msgOrNull('msgKey', param1)}
${#messages.msgOrNull('msgKey', param1, param2)}
```

```
${#messages.msgOrNull('msgKey', param1, param2, param3)}
${#messages.msgOrNullWithParams('msgKey', new Object[] {param1, param2,
param3, param4})}
${#messages.arrayMsgOrNull(messageKeyArray)}
${#messages.listMsgOrNull(messageKeyList)}
${#messages.setMsgOrNull(messageKeySet)}
```

3. URIs/URLs

- #uris：在 Thymeleaf 标准表达式中执行 URI/URL 操作（尤其是转义 / 取消转义）的实用程序对象。

```
/*
 * ======================================================================
* See javadoc API for class org.thymeleaf.expression.Uris
 * ======================================================================
*/

/*
 * Escape/Unescape as a URI/URL path
 */
${#uris.escapePath(uri)}
${#uris.escapePath(uri, encoding)}
${#uris.unescapePath(uri)}
${#uris.unescapePath(uri, encoding)}

/*
 * Escape/Unescape as a URI/URL path segment (between '/' symbols)
 */
${#uris.escapePathSegment(uri)}
${#uris.escapePathSegment(uri, encoding)}
${#uris.unescapePathSegment(uri)}
${#uris.unescapePathSegment(uri, encoding)}

/*
 * Escape/Unescape as a Fragment Identifier (#frag)
 */
${#uris.escapeFragmentId(uri)}
${#uris.escapeFragmentId(uri, encoding)}
${#uris.unescapeFragmentId(uri)}
${#uris.unescapeFragmentId(uri, encoding)}

/*
 * Escape/Unescape as a Query Parameter (?var=value)
 */
${#uris.escapeQueryParam(uri)}
${#uris.escapeQueryParam(uri, encoding)}
${#uris.unescapeQueryParam(uri)}
${#uris.unescapeQueryParam(uri, encoding)}
```

5. Conversions

- #conversions：实用程序对象允许在模板的任何位置执行转换服务。

```
/*
 * ======================================================================
 * See javadoc API for class org.thymeleaf.expression.Conversions
 * ======================================================================
 */

/*
 * Execute the desired conversion of the 'object' value into the
 * specified class.
 */
${#conversions.convert(object, 'java.util.TimeZone')}
${#conversions.convert(object, targetClass)}
```

6. Dates

- #dates：用于 java.util.Date 对象的实用方法。

```
/*
 * ======================================================================
 * See javadoc API for class org.thymeleaf.expression.Dates
 * ======================================================================
 */

/*
 * Format date with the standard locale format
 * Also works with arrays, lists or sets
 */
${#dates.format(date)}
${#dates.arrayFormat(datesArray)}
${#dates.listFormat(datesList)}
${#dates.setFormat(datesSet)}

/*
 * Format date with the ISO8601 format
 * Also works with arrays, lists or sets
 */
${#dates.formatISO(date)}
${#dates.arrayFormatISO(datesArray)}
${#dates.listFormatISO(datesList)}
${#dates.setFormatISO(datesSet)}

/*
 * Format date with the specified pattern
 * Also works with arrays, lists or sets
 */
${#dates.format(date, 'dd/MMM/yyyy HH:mm')}
${#dates.arrayFormat(datesArray, 'dd/MMM/yyyy HH:mm')}
${#dates.listFormat(datesList, 'dd/MMM/yyyy HH:mm')}
${#dates.setFormat(datesSet, 'dd/MMM/yyyy HH:mm')}
```

```
/*
 * Obtain date properties
 * Also works with arrays, lists or sets
 */
${#dates.day(date)}                    // also arrayDay(...), listDay
(...), etc.
${#dates.month(date)}                  // also arrayMonth(...), list-
Month(...), etc.
${#dates.monthName(date)}              // also arrayMonthName(...),
listMonthName(...), etc.
${#dates.monthNameShort(date)}         // also arrayMonthNameShort(...),
listMonthNameShort(...), etc.
${#dates.year(date)}                   // also arrayYear(...), listYear
(...), etc.
${#dates.dayOfWeek(date)}              // also arrayDayOfWeek(...),
listDayOfWeek(...), etc.
${#dates.dayOfWeekName(date)}          // also arrayDayOfWeekName(...),
listDayOfWeekName(...), etc.
${#dates.dayOfWeekNameShort(date)}     // also arrayDayOfWeekNameShort
(...), listDayOfWeekNameShort(...), etc.
${#dates.hour(date)}                   // also arrayHour(...), listHour
(...), etc.
${#dates.minute(date)}                 // also arrayMinute(...), list-
Minute(...), etc.
${#dates.second(date)}                 // also arraySecond(...), list-
Second(...), etc.
${#dates.millisecond(date)}            // also arrayMillisecond(...),
listMillisecond(...), etc.

/*
 * Create date (java.util.Date) objects from its components
 */
${#dates.create(year,month,day)}
${#dates.create(year,month,day,hour,minute)}
${#dates.create(year,month,day,hour,minute,second)}
${#dates.create(year,month,day,hour,minute,second,millisecond)}

/*
 * Create a date (java.util.Date) object for the current date and time
 */
${#dates.createNow()}

${#dates.createNowForTimeZone()}

/*
 * Create a date (java.util.Date) object for the current date (time set
to 00:00)
 */
${#dates.createToday()}
```

```
${#dates.createTodayForTimeZone()}
```

4. Calendars

- #calendars：类似于# dates，对应于 java.util.Calendar 对象。

```
/*
 * ======================================================================
* See javadoc API for class org.thymeleaf.expression.Calendars
 * ======================================================================
*/

/*
 * Format calendar with the standard locale format
 * Also works with arrays, lists or sets
 */
${#calendars.format(cal)}
${#calendars.arrayFormat(calArray)}
${#calendars.listFormat(calList)}
${#calendars.setFormat(calSet)}

/*
 * Format calendar with the ISO8601 format
 * Also works with arrays, lists or sets
 */
${#calendars.formatISO(cal)}
${#calendars.arrayFormatISO(calArray)}
${#calendars.listFormatISO(calList)}
${#calendars.setFormatISO(calSet)}

/*
 * Format calendar with the specified pattern
 * Also works with arrays, lists or sets
 */
${#calendars.format(cal, 'dd/MMM/yyyy HH:mm')}
${#calendars.arrayFormat(calArray, 'dd/MMM/yyyy HH:mm')}
${#calendars.listFormat(calList, 'dd/MMM/yyyy HH:mm')}
${#calendars.setFormat(calSet, 'dd/MMM/yyyy HH:mm')}

/*
 * Obtain calendar properties
 * Also works with arrays, lists or sets
 */
${#calendars.day(date)}                  // also arrayDay(...), listDay
(...), etc.
${#calendars.month(date)}                // also arrayMonth(...), list-
Month(...), etc.
${#calendars.monthName(date)}            // also arrayMonthName(...),
listMonthName(...), etc.
```

```
${#calendars.monthNameShort(date)}      // also arrayMonthNameShort(...),
listMonthNameShort(...), etc.
${#calendars.year(date)}                // also arrayYear(...), listYear
(...), etc.
${#calendars.dayOfWeek(date)}           // also arrayDayOfWeek(...),
listDayOfWeek(...), etc.
${#calendars.dayOfWeekName(date)}       // also arrayDayOfWeekName(...),
listDayOfWeekName(...), etc.
${#calendars.dayOfWeekNameShort(date)} // also arrayDayOfWeekNameShort
(...), listDayOfWeekNameShort(...), etc.
${#calendars.hour(date)}                // also arrayHour(...), listHour
(...), etc.
${#calendars.minute(date)}              // also arrayMinute(...), list-
Minute(...), etc.
${#calendars.second(date)}              // also arraySecond(...), list-
Second(...), etc.
${#calendars.millisecond(date)}         // also arrayMillisecond(...),
listMillisecond(...), etc.

/*
 * Create calendar (java.util.Calendar) objects from its components
 */
${#calendars.create(year,month,day)}
${#calendars.create(year,month,day,hour,minute)}
${#calendars.create(year,month,day,hour,minute,second)}
${#calendars.create(year,month,day,hour,minute,second,millisecond)}

${#calendars.createForTimeZone(year,month,day,timeZone)}
${#calendars.createForTimeZone(year,month,day,hour,minute,timeZone)}
${#calendars.createForTimeZone(year,month,day,hour,minute,second,time-
Zone)}
${#calendars.createForTimeZone(year,month,day,hour,minute,second,milli-
second,timeZone)}

/*
 * Create a calendar (java.util.Calendar) object for the current date
and time
 */
${#calendars.createNow()}

${#calendars.createNowForTimeZone()}

/*
 * Create a calendar (java.util.Calendar) object for the current date
(time set to 00:00)
 */
${#calendars.createToday()}
```

```
${#calendars.createTodayForTimeZone()}
```

7. Numbers

- #numbers：数字对象的实用方法。

```
/*
 * ==========================================================================
* See javadoc API for class org.thymeleaf.expression.Numbers
 * ==========================================================================
*/

/*
 * ==========================
 * Formatting integer numbers
 * ==========================
 */

/*
 * Set minimum integer digits.
 * Also works with arrays, lists or sets
 */
${#numbers.formatInteger(num,3)}
${#numbers.arrayFormatInteger(numArray,3)}
${#numbers.listFormatInteger(numList,3)}
${#numbers.setFormatInteger(numSet,3)}


/*
 * Set minimum integer digits and thousands separator:
 * 'POINT', 'COMMA', 'WHITESPACE', 'NONE' or 'DEFAULT' (by locale).
 * Also works with arrays, lists or sets
 */
${#numbers.formatInteger(num,3,'POINT')}
${#numbers.arrayFormatInteger(numArray,3,'POINT')}
${#numbers.listFormatInteger(numList,3,'POINT')}
${#numbers.setFormatInteger(numSet,3,'POINT')}


/*
 * ==========================
 * Formatting decimal numbers
 * ==========================
 */

/*
 * Set minimum integer digits and (exact) decimal digits.
 * Also works with arrays, lists or sets
 */
```

```
${#numbers.formatDecimal(num,3,2)}
${#numbers.arrayFormatDecimal(numArray,3,2)}
${#numbers.listFormatDecimal(numList,3,2)}
${#numbers.setFormatDecimal(numSet,3,2)}

/*
 * Set minimum integer digits and (exact) decimal digits, and also
decimal separator.
 * Also works with arrays, lists or sets
 */
${#numbers.formatDecimal(num,3,2,'COMMA')}
${#numbers.arrayFormatDecimal(numArray,3,2,'COMMA')}
${#numbers.listFormatDecimal(numList,3,2,'COMMA')}
${#numbers.setFormatDecimal(numSet,3,2,'COMMA')}

/*
 * Set minimum integer digits and (exact) decimal digits, and also thou-
sands and
 * decimal separator.
 * Also works with arrays, lists or sets
 */
${#numbers.formatDecimal(num,3,'POINT',2,'COMMA')}
${#numbers.arrayFormatDecimal(numArray,3,'POINT',2,'COMMA')}
${#numbers.listFormatDecimal(numList,3,'POINT',2,'COMMA')}
${#numbers.setFormatDecimal(numSet,3,'POINT',2,'COMMA')}


/*
 * ======================
 * Formatting currencies
 * ======================
 */

${#numbers.formatCurrency(num)}
${#numbers.arrayFormatCurrency(numArray)}
${#numbers.listFormatCurrency(numList)}
${#numbers.setFormatCurrency(numSet)}


/*
 * ======================
 * Formatting percentages
 * ======================
 */

${#numbers.formatPercent(num)}
${#numbers.arrayFormatPercent(numArray)}
${#numbers.listFormatPercent(numList)}
${#numbers.setFormatPercent(numSet)}
```

```
/*
 * Set minimum integer digits and (exact) decimal digits.
 */
${#numbers.formatPercent(num, 3, 2)}
${#numbers.arrayFormatPercent(numArray, 3, 2)}
${#numbers.listFormatPercent(numList, 3, 2)}
${#numbers.setFormatPercent(numSet, 3, 2)}


/*
 * ==============
 * Utility methods
 * ==============
 */

/*
 * Create a sequence (array) of integer numbers going
 * from x to y
 */
${#numbers.sequence(from,to)}
${#numbers.sequence(from,to,step)}
```

8. Strings

- #strings：String 对象的实用方法。

```
/*
 * ======================================================================
* See javadoc API for class org.thymeleaf.expression.Strings
 * ======================================================================
*/

/*
 * Null-safe toString()
 */
${#strings.toString(obj)}                           // also array*,
list* and set*

/*
 * Check whether a String is empty (or null). Performs a trim() opera-
tion before check
 * Also works with arrays, lists or sets
 */
${#strings.isEmpty(name)}
${#strings.arrayIsEmpty(nameArr)}
${#strings.listIsEmpty(nameList)}
${#strings.setIsEmpty(nameSet)}

/*
```

```
 * Perform an 'isEmpty()' check on a string and return it if false,
defaulting to
 * another specified string if true.
 * Also works with arrays, lists or sets
 */
${#strings.defaultString(text,default)}
${#strings.arrayDefaultString(textArr,default)}
${#strings.listDefaultString(textList,default)}
${#strings.setDefaultString(textSet,default)}

/*
 * Check whether a fragment is contained in a String
 * Also works with arrays, lists or sets
 */
${#strings.contains(name,'ez')}                     // also array*,
list* and set*
${#strings.containsIgnoreCase(name,'ez')}           // also array*,
list* and set*

/*
 * Check whether a String starts or ends with a fragment
 * Also works with arrays, lists or sets
 */
${#strings.startsWith(name,'Don')}                  // also array*,
list* and set*
${#strings.endsWith(name,endingFragment)}           // also array*,
list* and set*

/*
 * Substring-related operations
 * Also works with arrays, lists or sets
 */
${#strings.indexOf(name,frag)}                      // also array*,
list* and set*
${#strings.substring(name,3,5)}                     // also array*,
list* and set*
${#strings.substringAfter(name,prefix)}            // also array*, list*
and set*
${#strings.substringBefore(name,suffix)}           // also array*, list*
and set*
${#strings.replace(name,'las','ler')}               // also array*,
list* and set*

/*
 * Append and prepend
 * Also works with arrays, lists or sets
 */
${#strings.prepend(str,prefix)}                    // also array*, list*
and set*
```

```
${#strings.append(str,suffix)}                          // also array*, list*
and set*

/*
 * Change case
 * Also works with arrays, lists or sets
 */
${#strings.toUpperCase(name)}                           // also array*,
list* and set*
${#strings.toLowerCase(name)}                           // also array*,
list* and set*

/*
 * Split and join
 */
${#strings.arrayJoin(namesArray,',')}
${#strings.listJoin(namesList,',')}
${#strings.setJoin(namesSet,',')}
${#strings.arraySplit(namesStr,',')}                    // returns String[]
${#strings.listSplit(namesStr,',')}                     // returns
List<String>
${#strings.setSplit(namesStr,',')}                      // returns
Set<String>

/*
 * Trim
 * Also works with arrays, lists or sets
 */
${#strings.trim(str)}                                   // also array*,
list* and set*

/*
 * Compute length
 * Also works with arrays, lists or sets
 */
${#strings.length(str)}                                 // also array*,
list* and set*

/*
 * Abbreviate text making it have a maximum size of n. If text is big-
ger, it
 * will be clipped and finished in "..."
 * Also works with arrays, lists or sets
 */
${#strings.abbreviate(str,10)}                          // also array*,
list* and set*

/*
 * Convert the first character to upper-case (and vice-versa)
```

```
 */
${#strings.capitalize(str)}                              // also array*,
list* and set*
${#strings.unCapitalize(str)}                            // also array*,
list* and set*

/*
 * Convert the first character of every word to upper-case
 */
${#strings.capitalizeWords(str)}                         // also array*,
list* and set*
${#strings.capitalizeWords(str,delimiters)}              // also array*,
list* and set*

/*
 * Escape the string
 */
${#strings.escapeXml(str)}                               // also array*,
list* and set*
${#strings.escapeJava(str)}                              // also array*,
list* and set*
${#strings.escapeJavaScript(str)}                        // also array*,
list* and set*
${#strings.unescapeJava(str)}                            // also array*,
list* and set*
${#strings.unescapeJavaScript(str)}                      // also array*,
list* and set*

/*
 * Null-safe comparison and concatenation
 */
${#strings.equals(first, second)}
${#strings.equalsIgnoreCase(first, second)}
${#strings.concat(values...)}
${#strings.concatReplaceNulls(nullValue, values...)}

/*
 * Random
 */
${#strings.randomAlphanumeric(count)}
```

9. Objects

- #objects：一般对象的实用方法。

```
/*
 * ==========================================================================
* See javadoc API for class org.thymeleaf.expression.Objects
 * ==========================================================================
*/
```

```
/*
 * Return obj if it is not null, and default otherwise
 * Also works with arrays, lists or sets
 */
${#objects.nullSafe(obj,default)}
${#objects.arrayNullSafe(objArray,default)}
${#objects.listNullSafe(objList,default)}
${#objects.setNullSafe(objSet,default)}
```

10. Booleans

- #bools：布尔评估的实用方法。

```
/*
 * ======================================================================
* See javadoc API for class org.thymeleaf.expression.Bools
 * ======================================================================
*/

/*
 * Evaluate a condition in the same way that it would be evaluated in a
th:if tag
 * (see conditional evaluation chapter afterwards).
 * Also works with arrays, lists or sets
 */
${#bools.isTrue(obj)}
${#bools.arrayIsTrue(objArray)}
${#bools.listIsTrue(objList)}
${#bools.setIsTrue(objSet)}

/*
 * Evaluate with negation
 * Also works with arrays, lists or sets
 */
${#bools.isFalse(cond)}
${#bools.arrayIsFalse(condArray)}
${#bools.listIsFalse(condList)}
${#bools.setIsFalse(condSet)}

/*
 * Evaluate and apply AND operator
 * Receive an array, a list or a set as parameter
 */
${#bools.arrayAnd(condArray)}
${#bools.listAnd(condList)}
${#bools.setAnd(condSet)}

/*
 * Evaluate and apply OR operator
```

```
 * Receive an array, a list or a set as parameter
 */
${#bools.arrayOr(condArray)}
${#bools.listOr(condList)}
${#bools.setOr(condSet)}
```

11. Arrays

- #arrays：数组的实用方法。

```
/*
 * ======================================================================
* See javadoc API for class org.thymeleaf.expression.Arrays
 * ======================================================================
*/

/*
 * Converts to array, trying to infer array component class.
 * Note that if resulting array is empty, or if the elements
 * of the target object are not all of the same class,
 * this method will return Object[].
 */
${#arrays.toArray(object)}

/*
 * Convert to arrays of the specified component class.
 */
${#arrays.toStringArray(object)}
${#arrays.toIntegerArray(object)}
${#arrays.toLongArray(object)}
${#arrays.toDoubleArray(object)}
${#arrays.toFloatArray(object)}
${#arrays.toBooleanArray(object)}

/*
 * Compute length
 */
${#arrays.length(array)}

/*
 * Check whether array is empty
 */
${#arrays.isEmpty(array)}

/*
 * Check if element or elements are contained in array
 */
${#arrays.contains(array, element)}
${#arrays.containsAll(array, elements)}
```

12. Lists

- #lists：链表的实用方法。

```
/*
 * ==========================================================================
* See javadoc API for class org.thymeleaf.expression.Lists
 * ==========================================================================
*/

/*
 * Converts to list
 */
${#lists.toList(object)}

/*
 * Compute size
 */
${#lists.size(list)}

/*
 * Check whether list is empty
 */
${#lists.isEmpty(list)}

/*
 * Check if element or elements are contained in list
 */
${#lists.contains(list, element)}
${#lists.containsAll(list, elements)}

/*
 * Sort a copy of the given list. The members of the list must imple-
ment
 * comparable or you must define a comparator.
 */
${#lists.sort(list)}
${#lists.sort(list, comparator)}
```

13. Sets

- #sets：集合的实用方法。

```
/*
 * ==========================================================================
* See javadoc API for class org.thymeleaf.expression.Sets
 * ==========================================================================
*/

/*
 * Converts to set
```

```
 */
${#sets.toSet(object)}

/*
 * Compute size
 */
${#sets.size(set)}

/*
 * Check whether set is empty
 */
${#sets.isEmpty(set)}

/*
 * Check if element or elements are contained in set
 */
${#sets.contains(set, element)}
${#sets.containsAll(set, elements)}
```

14. Maps

- #maps：map 的实用方法。

```
/*
 * ======================================================================
* See javadoc API for class org.thymeleaf.expression.Maps
 * ======================================================================
*/

/*
 * Compute size
 */
${#maps.size(map)}

/*
 * Check whether map is empty
 */
${#maps.isEmpty(map)}

/*
 * Check if key/s or value/s are contained in maps
 */
${#maps.containsKey(map, key)}
${#maps.containsAllKeys(map, keys)}
${#maps.containsValue(map, value)}
${#maps.containsAllValues(map, value)}
```

15. Aggregates

- #aggregates：在数组或集合上创建聚合的实用方法。

```
/*
 * ==========================================================================
* See javadoc API for class org.thymeleaf.expression.Aggregates
 * ==========================================================================
*/

/*
 * Compute sum. Returns null if array or collection is empty
 */
${#aggregates.sum(array)}
${#aggregates.sum(collection)}

/*
 * Compute average. Returns null if array or collection is empty
 */
${#aggregates.avg(array)}
${#aggregates.avg(collection)}
```

16. IDs

- #ids : 实用方法用于处理重复的 id 属性（比如作为一个迭代器的结果）。

```
/*
 * ==========================================================================
* See javadoc API for class org.thymeleaf.expression.Ids
 * ==========================================================================
*/

/*
 * Normally used in th:id attributes, for appending a counter to the id
attribute value
 * so that it remains unique even when involved in an iteration process.

 */
${#ids.seq('someId')}

/*
 * Normally used in th:for attributes in <label> tags, so that these
labels can refer to Ids
 * generated by means if the #ids.seq(...) function.
 *
 * Depending on whether the <label> goes before or after the element
with the #ids.seq(...)
 * function, the "next" (label goes before "seq") or the "prev" func-
tion (label goes after
 * "seq") function should be called.
 */
${#ids.next('someId')}
${#ids.prev('someId')}
```

附录 D

Bean Validation 内置约束

下面列出了所有 Bean Validation 内置的约束。

（1）@AssertFalse：用于约束字段或属性值必须是 false。下面是一个用法示例。

```
@AssertFalse
boolean isUnsupported;
```

（2）@AssertTrue：用于约束字段或属性值必须是 true。下面是一个用法示例。

```
@AssertTrue
boolean isActive;
```

（3）@DecimalMax：用于约束字段或属性值必须是一个 decimal 类型的值，且值的大小必须小于或等于注解中的元素值。下面是一个用法示例。

```
@DecimalMax("30.00")
BigDecimal discount;
```

（4）@DecimalMin：用于约束字段或属性值必须是一个 decimal 类型的值，且值的大小必须大于或等于注解中的元素值。下面是一个用法示例。

```
@DecimalMin("5.00")
BigDecimal discount;
```

（5）@Digits：用于约束字段或属性的值必须是指定范围内的数字。整数元素指定数字的最大整数，而小数元素指定数字的最大小数位数。下面是一个用法示例。

```
@Digits(integer=6, fraction=2)
BigDecimal price;
```

（6）@Future：用于约束字段或属性的值必须是将来的日期。下面是一个用法示例。

```
@Future
Date eventDate;
```

（7）@Max：用于约束字段或属性值必须是一个 integer 类型的值，且值的大小必须小于或等于注解中的元素值。下面是一个用法示例。

```
@Max(10)
int quantity;
```

（8）@Min: 用于约束字段或属性值必须是一个 integer 类型的值，且值的大小必须大于或等于注解中的元素值。下面是一个用法示例。

```
@Min(5)
int quantity;
```

（9）@NotNull：用于约束字段或属性值不能为空。下面是一个用法示例。

```
@NotNull
String username;
```

（10）@Null：用于约束字段或属性值必须为空。下面是一个用法示例。

```
@Null
String unusedString;
```

（11）@Past：用于约束字段或属性的值必须是以前的日期。下面是一个用法示例。

```
@Past
Date birthday;
```

（12）@Pattern：用于约束字段或属性的值必须匹配正则表达式的定义。下面是一个用法示例。

```
@Pattern(regexp="\\(\\d{3}\\)\\d{3}-\\d{4}")
String phoneNumber;
```

（13）@Size：对字段或属性的大小进行评估，并且必须与指定的边界匹配。如果字段或属性是 String，则会评估字符串的大小；如果字段或属性是集合，则会对集合的大小进行评估；如果字段或属性是 Map，则会对 Map 的大小进行评估；如果字段或属性是数组，则会对数组的大小进行评估。使用可选的 max 或 min 元素之一来指定边界。

```
@Size(min=2, max=240)
String briefMessage;
```

附录 E

本书所涉及的技术及相关版本

本书所采用的技术及相关版本较新，请读者将相关开发环境设置为与本书所采用的一致，或者不低于本书所列的配置。

- JDK 8
- Gradle 4.0
- Eclipse Oxygen Release (4.7.0)
- Spring Boot 2.0.0.M2
- Spring 5.0.0.RC2
- Thymeleaf 3.0.6.RELEASE
- Thymeleaf Layout Dialect 2.2.2
- MySQL Community Server 5.7.17
- MySQL Workbench 6.3.9
- Spring Data JPA 2.0.0.M4
- Hibernate 5.2.10.Final
- MySQL Connector/J 6.0.5
- H2 Database 1.4.196
- Elasticsearch 5.5.0
- Spring Data Elasticsearch 3.0.0.M4
- Tether 1.4.0
- Bootstrap v4.0.0-alpha.6
- jQuery 3.1.1
- Font Awesome 4.7.0
- NProgress 0.2.0
- Thinker-md
- jQuery Tags Input 1.3.6

- Bootstrap Chosen 1.0.3
- toastr 2.1.1
- Spring Security 5.0.0.M2
- Thymeleaf Spring Security 3.0.2.RELEASE
- Apache Commons Lang 3.6
- Markdown parser for the JVM 0.16
- MongoDB 3.4.6
- Spring Data Mongodb 2.0.0.M4
- Embedded MongoDB 2.0.0
- IK Analysis for Elasticsearch 5.5.0

参考文献

[1] 柳伟卫 . 分布式系统常用技术及案例分析 . 北京：电子工业出版社，2017.
[2] https://spring.io/guides/gs/spring-boot/.
[3] http://docs.spring.io/spring-boot/docs/2.0.0.M2/gradle-plugin//reference/html/.
[4] https://github.com/waylau/thymeleaf-tutorial.
[5] http://www.thymeleaf.org/documentation.html.
[6] http://docs.spring.io/spring-data/commons/docs/2.0.0.M4/api/.
[7] http://hibernate.org/orm/documentation/.
[8] https://www.elastic.co/guide/en/elasticsearch/reference/current/_basic_concepts.html#_cluster.
[9] https://www.elastic.co/guide/en/elasticsearch/reference/current/windows.html#windows-service.
[10] https://github.com/waylau/spring-framework-4-reference.
[11] https://waylau.com/spring-singleton-beans-with-prototype-bean-dependencies/.
[12] http://www.thymeleaf.org/doc/articles/thvsjsp.html.
[13] https://v4-alpha.getbootstrap.com/getting-started/introduction/.
[14] https://waylau.com/new-rbac-resource-based-access-control/.
[15] https://github.com/waylau/apache-shiro-1.2.x-reference.
[16] https://github.com/waylau/spring-security-tutorial.
[17] http://docs.spring.io/spring-security/site/docs/5.0.0.M3/reference/htmlsingle/.
[18] https://github.com/thymeleaf/thymeleaf-extras-springsecurity.
[19] http://stackoverflow.com/questions/28904176/thymeleaf-with-spring-security-how-to-check-if-user-is-logged-in-or-not.
[20] https://docs.mongodb.com/manual/core/gridfs/.
[21] https://github.com/waylau/thymeleaf-tutorial.
[22] https://github.com/waylau/gradle-3-user-guide.
[23] https://github.com/waylau/rest-in-action.
[24] https://waylau.com/mogodb-file-server-with-spring-boot/.
[25] https://github.com/waylau/mongodb-file-server.